Pages listed are first occurrences.

Topography of a bird

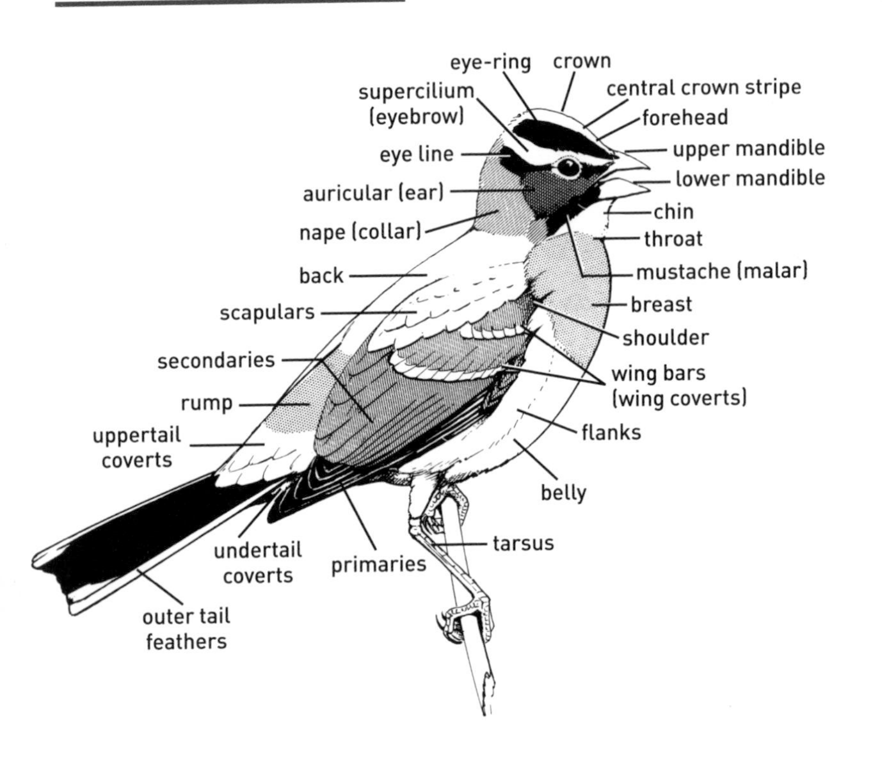

Undersurface of wing

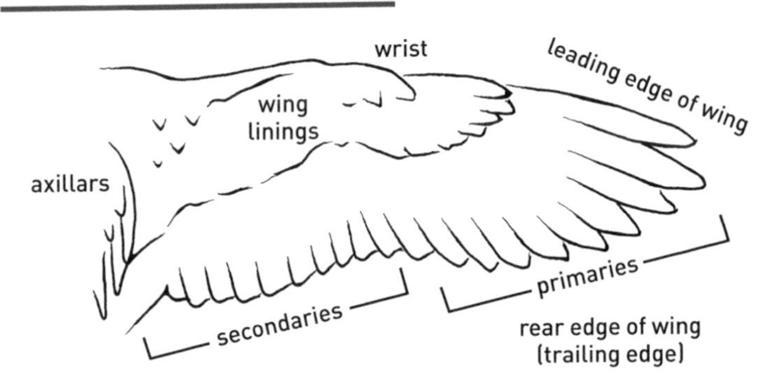

On the upper surface of the secondaries, some waterfowl have a bright-colored patch, called a *speculum.*

PETERSON FIELD GUIDE TO

BIRDS OF EASTERN AND CENTRAL NORTH AMERICA

PETERSON FIELD GUIDE TO

BIRDS OF EASTERN AND CENTRAL NORTH AMERICA

SEVENTH EDITION

Roger Tory Peterson

WITH CONTRIBUTIONS FROM

Michael DiGiorgio
Paul Lehman
Peter Pyle
Larry Rosche

HOUGHTON MIFFLIN HARCOURT
BOSTON NEW YORK

hmhbooks.com

Library of Congress Cataloging-in-Publication Data

Names: Peterson, Roger Tory, 1908–1996, author. | Peterson, Roger Tory,
1908–1996. Field guide to the birds.
Title: Peterson field guide to birds of eastern & central North America /
Roger Tory Peterson ; with contributions from Michael DiGiorgio, Paul
Lehman,Peter Pyle, Larry Rosche.
Other titles: Field guide to birds of eastern and central North America
Description: Seventh edition. | Boston : Houghton Mifflin Harcourt, 2020. |
Series: Peterson field guides | Includes bibliographical references and index.
Identifiers: LCCN 2020004429 (print) | LCCN 2020004430 (ebook) | ISBN
9781328771438 | ISBN 9781328771452 (ebook)
Subjects: LCSH: Birds—North America.
Classification: LCC QL681 .P45 2020 (print) | LCC QL681 (ebook) | DDC
598.097—dc23
LC record available at https://lccn.loc.gov/2020004429
LC ebook record available at https://lccn.loc.gov/2020004430

Book design by Eugenie S. Delaney

Printed in China

SCP 10 9 8 7 6 5 4 3 2
4500812794

ROGER TORY PETERSON INSTITUTE OF NATURAL HISTORY

Continuing the work of Roger Tory Peterson through Art, Education, and Conservation

In 1984, the Roger Tory Peterson Institute of Natural History (RTPI) was founded in Peterson's hometown of Jamestown, New York, as an educational institution charged by Peterson with preserving his lifetime body of work and making it available to the world for educational purposes.

RTPI is the only official institutional steward of Roger Tory Peterson's body of work and his enduring legacy. It is our mission to foster understanding, appreciation, and protection of the natural world. By providing people with opportunities to engage in nature-focused art, education, and conservation projects, we promote the study of natural history and its connections to human health and economic prosperity.

Art—Using Art to Inspire Appreciation of Nature

The RTPI Archives contains the largest collection of Peterson's art in the world—iconic images that continue to inspire an awareness of and appreciation for nature.

Education—Explaining the Importance of Studying Natural History

We need to study, firsthand, the workings of the natural world and its importance to human life. Local surroundings can provide an engaging context for the study of natural history and its relationship to other disciplines such as math, science, and language. Environmental literacy is everybody's responsibility—not just experts and special interests.

Conservation—Sustaining and Restoring the Natural World

RTPI works to inspire people to choose action over inaction, and engages in meaningful conservation research and actions that transcend political and other boundaries. Our goal is to increase awareness and understanding of the natural connections between species, habitats, and people—connections that are critical to effective conservation.

For more information, and to support RTPI, please visit rtpi.org.

CONTENTS

PETERSON FIELD GUIDE TO

BIRDS OF EASTERN AND CENTRAL NORTH AMERICA

How to Identify Birds

Veteran birders will know how to use this book. Beginners, however, should spend some time becoming familiar in a general way with the illustrations. The plates, for the most part, have been grouped into taxonomic families. However, in cases where there is a great similarity of shape and action, similar-appearing families have been grouped outside their current taxonomic order, to aid in field identification.

Birds that could be confused are grouped together when possible and are arranged in identical profile for direct comparison. The arrows point to outstanding field marks, which are explained opposite. The text also gives aids such as voice, behavior, and habitat, not visually portrayable, and under a separate heading discusses species that might be confused. The general range is not described for most species in the text. The three-color maps next to the species accounts provide range information.

In addition to the plates of birds normally found in eastern and central North America, there are also plates depicting accidental vagrants from Eurasia, offshore pelagic areas, and the Tropics, as well as plates of some of the exotic escapees that are sometimes seen.

What Is the Bird's Size?

Acquire the habit of comparing a new bird with some familiar "yardstick"—a House Sparrow, robin, pigeon, etc.—so that you can say to yourself, for example, "Smaller than a robin, a little larger than a House Sparrow." The measurements in this book represent lengths in inches (with centimeters in parentheses) from bill tip to tail tip of specimens on their backs as in museum trays. For species that show considerable size variation, a range of measurements is given. For less variable species, only one measurement is given.

What Is Its Shape?

Is it plump like a starling (left) or slender like a cuckoo (right)?

What Shape Are Its Wings?

Are they rounded like a Northern Bobwhite's (left) or sharply pointed like a Barn Swallow's (right)?

What Shape Is Its Bill?

Is it small and fine like a warbler's (1), stout and short like a seed-cracking sparrow's (2), dagger-shaped like a tern's (3), or hook-tipped like a bird of prey's (4)?

What Shape Is Its Tail?

Is it deeply forked like a Barn Swallow's (1), square-tipped like a Cliff Swallow's (2), notched like a Tree Swallow's (3), rounded like a jay's (4), or pointed like a Mourning Dove's (5)?

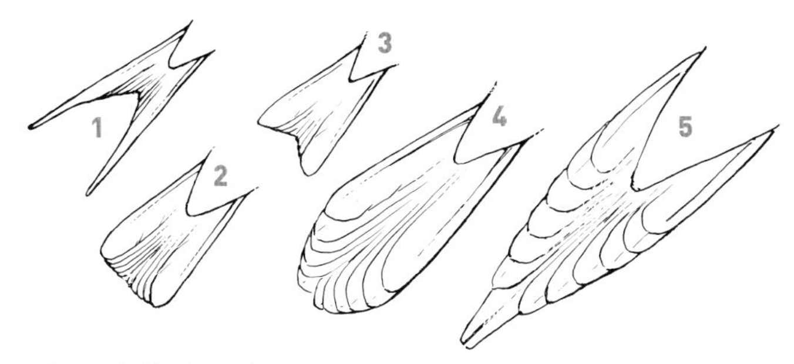

How Does It Behave?

Does it cock its tail like a wren or hold it down like a flycatcher? Does it wag its tail? Does it sit erect on an open perch, dart after an insect, and return as a flycatcher does?

Does It Climb Trees?

If so, does it climb upward in spirals like a creeper (left), in jerks and using its tail as a brace like a woodpecker (center), or go down headfirst like a nuthatch (right)?

How Does It Fly?

Does it undulate (dip up and down) like a flicker (1)? Does it fly straight and fast like a dove (2)? Does it hover like a kingfisher (3)? Does it glide or soar?

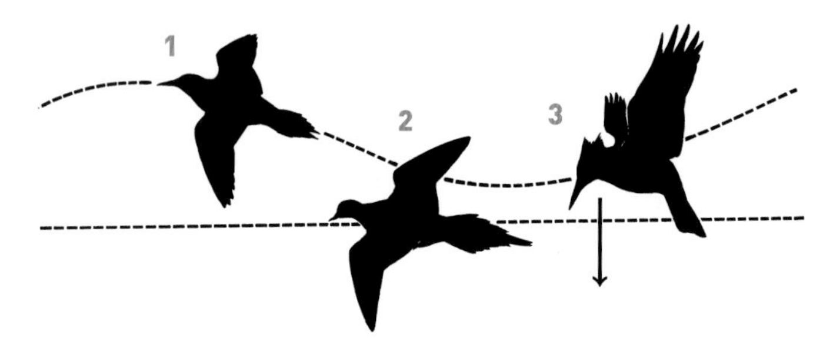

Does It Swim?

Does it sit low in the water like a loon (1) or high like a gallinule (2)? If a duck, does it dive like a scaup or a scoter (3) or dabble and upend like a Mallard (4)?

Does It Wade?

Is it large and long-legged like a heron or small like a sandpiper? If the latter, does it probe the mud or pick at things? Does it teeter or bob?

What Are Its Field Marks?

Some birds can be identified by color alone, but most birds are not that easy. The most important aids are what we call field marks, which are, in effect, the "trademarks of nature." Note whether the breast is spotted as in a thrush (1), streaked as in a thrasher (2), or plain as in a cuckoo (3).

Tail Pattern

Does the tail have a "flash pattern"—a white tip as in the Eastern Kingbird (1), white patches in the outer corners as in the Eastern and Spotted Towhees (2), or white sides as in the juncos (3)?

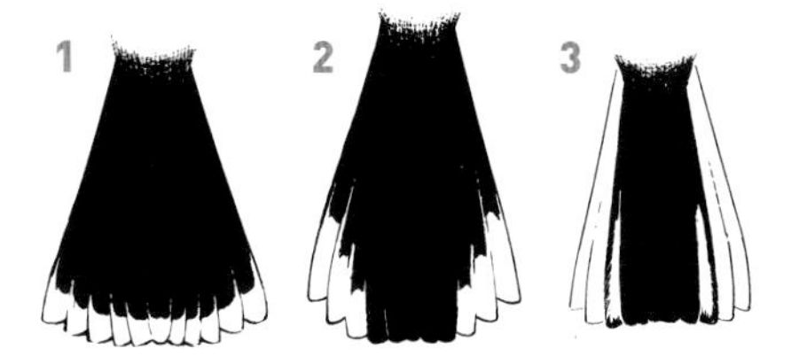

Rump Patch

Does it have a light rump like a Cliff Swallow (1) or flicker (2)? Northern Harrier, Yellow-rumped Warbler, and several shorebirds also have distinctive rump patches.

Eye Stripes and Eye-ring

Does the bird have a stripe above, through, or below the eye, or a combination of these stripes? Does it have a striped crown? A ring around the eye, or "spectacles"? A "mustache" stripe? These details are important in many small songbirds.

Wing Bars

Do the wings have light wing bars or not? Their presence or absence is important in recognizing many warblers, vireos, and flycatchers. Wing bars may be single or double, bold or obscure.

Wing Pattern

The basic wing pattern of ducks (shown below), shorebirds, and other water birds is very important. Notice whether the wings have patches (1) or stripes (2), are solidly colored (3), or have contrasting black tips.

Bird Songs and Calls

Using sounds to identify birds can be just as useful as using visual clues. In fact, in many situations birds are much more readily identified by sound than by sight. Most of the species accounts here include a brief entry on voice, with interpretations of these songs and calls, in an attempt to give birders some handle on the vocalizations they hear. (In the few accounts without a voice entry, it's because the species typically is silent or rarely heard.) Authors of bird books have attempted, with varying success, to fit songs and calls into syllables, words, and phrases. Musical notations, comparative descriptions, and even ingenious systems of symbols have also been employed. To supplement this verbal interpretation, there are recording collections available for nearly every region of the world and for individual groups of birds. The *Peterson Birding by Ear* CDs provide a step-by-step method for learning how to develop your listening and identification skills. Preparation in advance for particular species or groups greatly enhances your ability to identify them. Some birders do a majority of their birding by ear, and there is no substitute for actual sounds—for getting out into the field, tracking down the songster, and committing the song to memory. However, an audio library is a wonderful resource to return home to when attempting to identify a bird heard in the field. Many such collections can now be taken into the field on mobile devices. *Caution:* When using recordings to attract hard-to-see species, limit the number of playbacks, and do not use them on threatened species or in heavily birded areas.

Bird Nests

The more time you spend in the field becoming familiar with bird behavior, the more skilled you'll become at finding bird nests. It is as exciting to keep a bird nest list as it is to keep a life list. Remember, if you happen to find a nest during the breeding season, leave the site as undisturbed as possible. Back away, and do not touch the nest, eggs, or young birds. Often squirrels, raccoons, several other mammals, crows, jays, grackles, and cowbirds are more than happy to have you "point out" a nest and will raid it if you disrupt the site or call attention to it. Many people find juvenile birds that have just left the nest and may appear to be alone. Usually they are not lost but are under the watchful eye of a parent bird and are best left in place rather than scooped up and taken to a foreign environment. In the winter, nest hunting can be great fun and has little impact, as most nests will never be used again. They are easy to see once the foliage is gone, and it can be a challenge to attempt to identify the maker.

The Maps and Ranges of Birds

The ranges of many bird species have changed markedly in recent decades. Some species are expanding because of protection given them, changing habitats, bird feeding, or other factors. Others have declined alarmingly and may have been extirpated from major parts of their range. The primary culprit here has been habitat loss, although other factors, such as increased competition and predation from other species, may sometimes be involved. Many bird species are facing challenges from impending changes in our climate. Species that are in serious decline in North America run the gamut, from Ivory Gull to Lesser Prairie-Chicken and Loggerhead Shrike to Bewick's Wren, Rusty Blackbird, and Red Knot. However, recent studies indicate that most common bird species are also declining, with migratory species and grassland species showing steeper decline. Global climate change will have additional effects on many species, especially rarer and more local ones. Hopeful news is that some managed species, such as waterfowl and some raptors affected by pesticides in the 1960s, are not declining, and in some cases are even increasing despite extensive human-induced changes.

Successful introductions of some species, such as Mute Swan and Eurasian Collared-Dove, have resulted in self-sustaining, growing populations (the latter was either introduced to the Bahamas or flew over from Africa, then dispersed through the U.S. and southwestern Canada on its own). And a good number of additional vagrant species—out-of-range visitors from faraway lands—continue to be found (such as a Red-footed Falcon in Massachusetts, Rufous-necked Wood-Rail in New Mexico, and Chatham Albatross in California). Some species that were formerly thought to occur only exceptionally have, over the past several decades, become much more regular or widespread visitors (such as Cave Swallow and Lesser Black-backed Gull) and sometimes even become local breeders (such as Clay-colored Thrush). Such changes in status can be the result of actual population increases or may reflect better observer coverage and advances in field identification skills.

The maps in this book are approximate, giving the general outlines of the range of each species. Within these broad outlines may be many gaps—areas ecologically unsuitable for the species. A Marsh Wren must have a marsh, a Ruffed Grouse a woodland or a forest. Certain species may be extremely local or sporadic for reasons that may or may not be clear. As noted above, some birds are extending their ranges, a few explosively, while others are declining or even disappearing from large areas where they were formerly found. Winter ranges are often not as well defined as breeding ranges. A species may exist at a very low density near the northern limits of its winter range, surviving through December in mild seasons but often succumbing to or moving south to avoid the bitter conditions of January and February. Varying weather conditions and food supplies from year to year may result in substantial variations in winter bird populations.

The maps are specific only for the area covered by this field guide. The Mallard, for example, is found over a large part of the globe. The map shows only its range in eastern and central North America. The maps are based on data culled from many publications (particularly from monographs detailing the status and distribution of a state or province's avifauna, as well as from breeding bird atlases), from such journals as *North American Birds*, and from communication with many state and provincial experts throughout eastern and central North America.

Range maps don't depict how abundant a particular species is within its range. The following list defines terms of abundance and occurrence used throughout the book. The definitions presume you're in the habitat and season in which a species would occur, but note that this in itself can vary throughout the continent.

COMMON: Always or almost always encountered daily, usually in moderate to large numbers.

FAIRLY COMMON: Usually encountered daily, generally not in large numbers.

UNCOMMON: Occurs in small numbers and may be missed on a substantial number of days.

SCARCE: Present only in small numbers or difficult to find within its normal range.

RARE OR VERY RARE: Annual or probably annual in small numbers but still largely within its normal range.

CASUAL VAGRANT: Beyond its normal range; occurs at somewhat regular intervals but usually less frequently than annually.

ACCIDENTAL VAGRANT: Beyond its normal range; one record or a very few records.

VERY RARE VAGRANT: Beyond its normal range. Annual or probably annual in small numbers.

LOCAL: Limited geographic range within the U.S. and Canada.

ENDEMIC: Found only in the described region.

INTRODUCED OR EXOTIC: Not native; population derived from deliberately released or escaped individuals. These terms can be used for species that are present in limited numbers and may or may not be breeding, or for well-established species such as House Sparrow and European Starling.

UNESTABLISHED EXOTIC: Nonnative releasee or escapee that does not have a naturalized breeding population, though some may be breeding in very localized areas.

Habitats

Gaining a familiarity with a wide range of habitats will greatly enhance your overall knowledge of the birds in a specific region, increase your skills, and add to your enjoyment of birding. It is unlikely you will ever see a meadowlark in an oak woodland or a Wood Thrush in a meadow. Birders know this, and if they want to go out to run up a large day list, they do not remain in one habitat but shift from site to site based on time and species diversity for a given type of habitat.

A few birds do invade habitats other than their own at times, especially on migration. A warbler that spends the summer in the boreal forests of Canada might be seen, on its journey through the southern U.S., in a palm or in coastal scrub. In cities, migrating birds often have to make the best of it, like the American Woodcock found one morning on the window ledge of a New York City office. Strong weather patterns can also alter where a bird happens to appear. Hurricanes, for example, can be a disaster for many species. As these violent storms sweep over the ocean, the eye can often "vacuum" up oceanic species that seek shelter in its calmness. Upon reaching land, these normally offshore species are faced with an entirely strange habitat and account for sightings such as a Yellow-nosed Albatross heading up the Hudson River, a White-tailed Tropicbird in downtown Boston, and numbers of storm-petrels on inland reservoirs in the desert Southwest.

Most species, however, are quite predictable for the major portion of their lives, and for the birder who has learned where to look, the rewards are great.

To start, familiarize yourself with individual habitat types. Become familiar with the dominant plant types that are indicators—for example, oak-beech woods, cactus desert, grass-shrub meadows, salt- or freshwater wetlands—and keep accurate records of what species you find in each. In a short time you will have a working knowledge of the predominant species in each habitat, and this will help you with identification by allowing you to anticipate what might be found there.

The seasonal movements of birds at your sites will provide an overview of migrant species that come through at a given time and will be a reference point for future visits during these migration periods. A forest dotted with migrant warblers in spring may revert to relative quiet accented by the repetitive calls of a Red-eyed Vireo or the drawn-out call of an Eastern Wood-Pewee in midsummer.

Be sure not to overlook cities and towns, where well-adapted species can be found. Peregrine Falcons have shown remarkable adaptability, nesting on strategic ledges in the "walled canyons" of many cities. The fertile grounds for hunting Rock Pigeons and European Starlings seem to suit this raptor well.

Ecotones are edges where two habitat types interface—a forest and a shrub meadow, for example. As this is not a gradual change, ecotones offer habitat for species from both of the adjoining areas and are therefore rich in bird life.

The changes in habitat over the years will also affect your favorite birding areas. Fields turn to shrubby lots and then woodlands. Northern Bobwhite, Savannah Sparrows, and meadowlarks may move on, but Indigo and Lazuli Buntings and Field and Lincoln's Sparrows establish themselves. This dynamic is normal in the natural world. However, humankind's alterations to this process have had a great impact. Forest fragmentation is an example. Land development has affected numerous species. Sudden disruptions have a more drastic effect than slow changes, which allow for adaptation. As we have divided up habitat with roadways, range lands, and agricultural fields, we have created a greater edge effect, and this has allowed Brown-headed Cowbirds to penetrate into forest areas where they would not have ventured in the past. They now parasitize many more species than before, and such parasitization has led to marked declines in total numbers of many species. Forest fragmentation has also affected the success rate of nestling fledging by increasing the access of some predators and by altering prime habitat requirements for obtaining food to raise young.

Some species are obligates to a specific habitat type, and searching these areas greatly improves your chances of finding such birds. These include Golden-crowned Kinglet nesting in coniferous woodlands and Kirtland's Warbler in Michigan, which breeds only in jack pine woodlands of a specific height. Even in migration, many species remain faithful to selected habitats, such as waterthrushes along watercourses. Running or dripping water has proven to be an important attractant for migrating land birds, and in areas where fresh water is scarce, a water drip can be a gold mine for migrant warblers and other passerines.

Subspecies and Geographic Variation

Many species of birds inhabit wide geographic areas. The Song Sparrow (*Melospiza melodia*), for example, breeds throughout North America, from Mexico north into Alaska and from California to Newfoundland. In such a wide-ranging species, there are geographic subsets, or populations, within the species that show distinct plumage patterns and/or song variants. When a population reaches a point when the individuals in it are recognizably different from those in other populations, it may formally be designated as a subspecies by attaching a third, subspecific name to the scientific name of the species. Thus, the pale Song Sparrow of the southwestern deserts of North America is called *Melospiza melodia saltonis,* to distinguish that form from up to 25 other subspecies found throughout North America. The Song Sparrow ranks among the most geographically variable of North American birds.

Often subspecific groups are so distinct that they can be easily recognized in the field by bird watchers. A good example of this is the subspecies of Dark-eyed Junco (*Junco hyemalis*). The roughly 12 subspecies found throughout North America are placed in 5 subspecies groups, 4 of which are found in the East and easily discerned: the "Oregon," "Pink-sided," "White-winged," and "Slate-colored" Juncos (p. 308). For the birder, identification of subspecies can add greater challenges to birding and, when documented, valuable information, especially when subspecies are reclassified to full species status. Such has been the case, for example, with the splitting of Cackling Goose (*Branta hutchinsii*) and its 4 subspecies from Canada Goose (*B. canadensis*) and its 5 subspecies (p. 18) or Bicknell's Thrush (*Catharus bicknelli*) from Gray-cheeked Thrush (*C. minimus*) (p. 236). In the West, field studies of Sage Grouse (*Centrocercus urophasianus*) leading to the separation of Greater Sage-Grouse (*C. urophasianus*) and the rare and threatened Gunnison Sage-Grouse (*C. minimus*) prove how valuable studies of subspecific populations can be. The subtle differences in field marks between Bicknell's and Gray-cheeked Thrushes illustrate why these two had once been considered subspecies. The shifting of this line between subspecies and species is ongoing. Recording data on location and numbers can prove helpful in completing a picture of a species' and subspecies' distribution.

In this edition, distinct subspecies that are easily recognized, such as those of Yellow-rumped Warbler (*Setophaga coronata*) and Dark-eyed Junco (*Junco hyemalis*), are represented. When in the field, challenge yourself to discern the subspecies. It will increase your visual and listening skills and add a new level to your understanding and enjoyment of birds.

Identifying the Age and Sex of Birds

In many species, the ages and sexes of birds can be identified to various degrees, and being able to accurately determine age and sex groups can add fulfillment to your birding experience. It can also be important in assessing the degree to which less common species are reproducing. In this new edition, we have made an effort to point out every identifiable age and sex classification of each species, illustrating many of them. We have also refined and standardized our terminology, replacing such imprecise terms as "immature" with specific age groupings (such as juvenile, adult, first-year, second-winter, etc.), and for plumages we have replaced the labels "breeding" and "nonbreeding" with "spring/summer" and "fall/winter," respectively, to better align with age classifications and because plumage state does not directly equate to breeding state.

Conservation

Birds undeniably contribute to our pleasure and quality of life. But they also are sensitive indicators of the environment, a sort of "ecological litmus paper," and hence more meaningful than just chickadees and cardinals that brighten the suburban garden, grouse and ducks that fill the hunter's bag, or rare warblers and shorebirds that excite the field birder. The observation and recording of bird populations over time lead inevitably to environmental awareness and can signal impending changes. In this edition we have indicated the status of species or populations as threatened or endangered according to the 2019 U.S. Endangered Species List.

To this end, please help the cause of wildlife conservation and education by contributing to or taking part in the work of organizations that are on the forefront of bird conservation. Among the many such groups are American Bird Conservancy (abcbirds.org), American Birding Association (www.aba.org), BirdLife International (www.birdlife.org), Cornell Lab of Ornithology (www.birds.cornell.edu), Defenders of Wildlife (www.defenders.org), Ducks Unlimited (www.ducks.org), National Audubon Society (www.audubon.org), National Wildlife Federation (www.nwf.org), The Nature Conservancy (www.nature.org), Partners in Flight (www.partnersinflight.org), Roger Tory Peterson Institute of Natural History (www.rtpi.org), and World Wildlife Fund (www.wwf.org), as well as your local land trust and natural heritage program and your local Audubon and ornithological societies and bird clubs. These and so many other groups merit your support. Studies suggest that we have lost up to 20 percent of our bird populations in a recent fifty-year period due to habitat destruction, predation by feral and outdoor house cats, and anthropogenic causes for large declines in insect populations, and these and other threats to bird populations may only become more severe with global climate change. It is thus more important now than ever to support these groups.

PLATES

GEESE, SWANS, and DUCKS Family Anatidae

Web-footed waterfowl. **RANGE:** Worldwide.

GEESE

Large, gregarious waterfowl; heavier bodied, longer necked than ducks; bills thick at base. Noisy in flight; some fly in lines or V formations. Sexes similar. Geese are more terrestrial than ducks, often grazing. **FOOD:** Grasses, seeds, waste grain, aquatic plants.

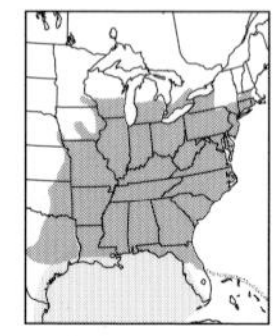

GREATER WHITE-FRONTED GOOSE — Uncommon

Anser albifrons (see also p. 22)

25–31 in. (65–80 cm). Gray-brown with *pink* to orangish-pink bill. *Adult:* Has *white patch on front of face* and sparse to heavy *black bars* on belly. *First-year:* Dusky with dull pinkish bill; gradually acquires white at bill base and black on belly. **VOICE:** A high-pitched tootling, *kah-lah-a-luk,* in chorus. **SIMILAR SPECIES:** Juvenile "Blue" Goose. See Pink-footed Goose. May also be confused with domestic Graylag Goose (p. 48). **HABITAT:** Marshes, prairies, agricultural fields, lakes, bays; in summer, tundra.

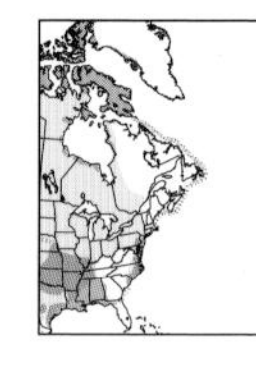

SNOW GOOSE *Anser caerulescens* (see also p. 22) — Locally common

White morph: 25–33 in. (64–84 cm). *White* with *black primaries.* Head sometimes rust-stained from feeding in muddy or iron-rich waters. Bill pink with black "lips." Feet pink. Base of bill curves back slightly toward eye. *Juvenile and first-winter:* Pale gray; dark bill and legs. Dark morph ("Blue" Goose): 25–30 in. (64–76 cm). Has *white throat, dark "lips"; lacks scaly pattern.* Intermediates with white morph observed rarely. *Juvenile and first-winter:* Similar to young Greater White-fronted Goose but blacker, feet and bill *dark.* **VOICE:** A loud, nasal, double-noted *houck-houck,* in chorus. **SIMILAR SPECIES:** Ross's Goose. **HABITAT:** Marshes, grain fields, ponds, bays; in summer, tundra.

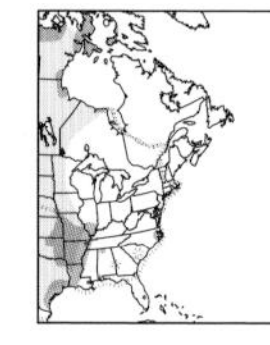

ROSS'S GOOSE *Anser rossii* (see also p. 22) — Uncommon

23 in. (58 cm). Like a small Snow Goose, but neck shorter, head rounder (steeper forehead). Bill has *gray-blue or purple-blue base* and is stubbier (with *vertical border* between base and facial feathering); *lacks distinctive "grinning black lips"*; and warts at bill base can be difficult to see. *Juvenile and first-winter:* Whiter than young Snow Goose. *Rare dark morph* has more extensively dark neck, whiter wing patches and abdomen than "Blue" Snow Goose; hybrids with Snow Goose occur. **VOICE:** Higher than Snow, suggesting Cackling Goose. **SIMILAR SPECIES:** Snow Goose. **RANGE:** Rare vagrant to East Coast. **HABITAT:** Same as Snow Goose.

juvenile/
first-winter
GREATER WHITE-
FRONTED GOOSE
adult
adult
juvenile/
first-fall
adult
adult
SNOW GOOSE
dark morph
dark morph
(rare)
adults
ROSS'S
GOOSE
juvenile/
first-winter
adult
Snow
Ross's
intergrade
between
dark
and white
morphs
SNOW
GOOSE
white
morph
adult
Snow

BRANT *Branta bernicla* (see also p. 22) **Locally common**

24–26 in. (59–66 cm). A small black-necked goose with white vent and undertail, whitish flanks, and band of white on neck. *First-year:* Shows smaller neck patch than adult and thin white wing bars. Travels in large irregular flocks. Eastern subspecies, "Pale-bellied" Brant (*B. b. hrota*), has *light belly, less contrasty flanks,* and smaller neck patches; Pacific Coast subspecies, "Black" Brant (*B. b. nigricans*), a casual vagrant to East Coast, has *dark belly* and more complete white band across foreneck. **VOICE:** A throaty *cr-r-r-ruk* or *krr-onk, krrr-onk.* **SIMILAR SPECIES:** Canada and Cackling Geese not as black, have large white face patches. Brant is more strictly coastal. **RANGE:** Rare migrant or vagrant inland. **HABITAT:** Salt bays, estuaries; in summer, tundra.

BARNACLE GOOSE *Branta leucopsis* **Casual vagrant**

26–27 in. (66–69 cm). Similar in size to Brant. White sides and black chest to waterline, strongly contrasting with white belly. Note white face encircling eye. Back distinctly barred. Ages similar. **VOICE:** Like Snow Goose, but higher-pitched, doglike barks. **SIMILAR SPECIES:** Canada and Cackling Geese browner, have dark faces. Brant has all-dark head. **RANGE:** Casual winter vagrant from Greenland to Atlantic Coast; accidental farther west. Some reports likely represent escapees. **HABITAT:** Ponds, lakes; grazes in fields.

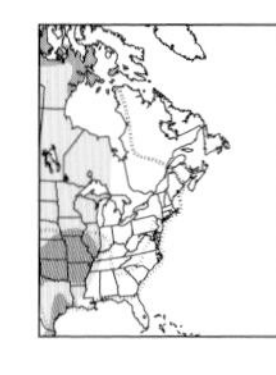

CACKLING GOOSE *Branta hutchinsii* **Locally fairly common**

23–32 in. (58–81 cm). Similar to larger Canada Goose. Ages similar. **VOICE:** A high, cackling *yel-lik.* **SIMILAR SPECIES:** Told from Canada by smaller size, shorter neck, smaller and rounder head, stubbier bill, and higher-pitched voice. Distinctions between larger Cacklings and smaller "Lesser" Canadas can be subtle. **HABITAT:** Lakes, marshes, fields; in summer, tundra. Seen often with flocks of Snow Geese as well as Canadas.

CANADA GOOSE *Branta canadensis* (see also p. 22) **Common**

30–43 in. (76–109 cm). The most widespread goose in N. America. Note black head and neck, or "stocking," that contrasts with pale breast and *white chin strap.* Ages similar. Flocks travel in strings or in Vs, "honking" loudly. Substantial variation in size and neck length exists among populations in the East; e.g., between subspecies *B. c. canadensis* and the semidomestic *B. c. moffitti* ("Greater Canada Goose") and *parvipes* ("Lesser Canada Goose"), not to mention the recently split and even smaller Cackling Goose. **VOICE:** A deep honking, *ka-ronk* or *ka-lunk.* "Lesser" has higher-pitched calls, but not as high as Cackling. **SIMILAR SPECIES:** Cackling Goose. **HABITAT:** Lakes, ponds, bays, marshes, fields. Resident in many areas, frequenting parks, lawns, golf courses.

EGYPTIAN GOOSE *Alopochen aegyptiaca* **Local, exotic**

25–29 in. (63–73 cm). A small stocky goose, largely grayish to tan with variable rufous eye patch and lower back; large white wing patch; blackish green speculum. Bill and legs largely pink. *Juvenile and first-winter:* Duller than adult, head mostly brown with white around bill. **VOICE:** A sharp, repeated *caow-caow-caow.* **RANGE:** Locally common and spreading in se. FL (Martin to Miami-Dade Counties) and around Houston, TX; escapees and incipient populations in other states. **HABITAT:** Ponds, city lakes, coastal wetlands, golf courses.

juvenile/
first-winter
adult
"Pale-bellied"
(Eastern)
adults
Pacific
Pacific
BRANT
"Black"
(Pacific)
BARNACLE
GOOSE
CACKLING
GOOSE
"Lesser"
CANADA
GOOSE
"Greater"
adult
EGYPTIAN
GOOSE

SWANS

Huge, all white to pale gray (juveniles); larger and longer necked than geese. Sexes alike. Swans migrate in lines or Vs. Feed by immersing head and neck or by "tipping up." **FOOD:** Aquatic plants, seeds.

MUTE SWAN *Cygnus olor* — Fairly common, local, introduced

60 in. (152 cm). Introduced from Europe. Swims with an S curve in neck; wings often arched over back with ornamental feathers extended. *Black-knobbed orange bill* tilts downward. *Juvenile and first-winter:* Usually dingy, with dull pinkish bill, lacking knob. **VOICE:** Hissing and wheezing sounds, weak bugling. **SIMILAR SPECIES:** Tundra and Trumpeter Swans have straighter necks, blacker bills; not found in urban parks. **HABITAT:** Ponds, fresh and salt; coastal lagoons, salt bays.

TUNDRA SWAN — Uncommon to locally common

Cygnus columbianus (see also p. 22)

52–53 in. (132–135 cm); wingspan 6–7 ft. (183–213 cm). Our most widespread native swan. Bill *black,* usually with *small yellow basal spot. Juvenile:* Dingy, with pinkish bill variably dark at base and tip; quickly becomes whiter during first winter. **VOICE:** A mellow, high-pitched cooing: *woo-ho, woo-woo, woo-ho.* **SIMILAR SPECIES:** Trumpeter and Mute Swans. **HABITAT:** Lakes, marshes, large rivers, bays, estuaries, grain fields; in summer, tundra.

TRUMPETER SWAN *Cygnus buccinator* — Rare to uncommon, local

58–60 in. (147–152 cm). Larger than Tundra Swan, with longer, heavier, *all-black bill with straight ridge.* Black on lores wider, forming V rather than U shape, *embracing eyes* and lacking yellow spot (some Tundras also lack this spot). *Juvenile and first-year:* Dusky color kept later in spring than in Tundra. **VOICE:** *Deep, nasal calls,* described as bugle-like. **RANGE:** Scarce vagrant well east of normal range, though beware escapees. **HABITAT:** Lakes, ponds; in winter, also marshes, bays, grain fields.

WHISTLING-DUCKS

Somewhat gooselike ducks with long legs and erect necks. Ages and sexes similar. They are named for their high-pitched calls. Gregarious. **FOOD:** Seeds of aquatic plants and grasses.

FULVOUS WHISTLING-DUCK — Uncommon, local

Dendrocygna bicolor (see also p. 42)

20 in. (51 cm). Note *tawny* body, dark back, *pale side stripes.* Flies with neck slightly drooped and feet trailing, showing *black underwings, white band* on rump. **VOICE:** A squealing slurred whistle, *ka-whee-oo.* **SIMILAR SPECIES:** Black-bellied Whistling-Duck, female Northern Pintail. **RANGE:** Casual vagrant to Northeast. **HABITAT:** Freshwater marshes, ponds, irrigated land, rice fields. Active at dusk and night. Seldom perches in trees.

BLACK-BELLIED WHISTLING-DUCK — Locally common

Dendrocygna autumnalis

21 in. (53 cm). Rusty with *black belly,* gray face, bright *coral red* bill; long pink legs. Broad *white patch* along forewing, visible in flight. Frequently perches in trees. **VOICE:** A high-pitched squealing whistle. **RANGE:** Casual vagrant to Northeast. **HABITAT:** Ponds, freshwater marshes.

SWANS AND WHISTLING-DUCKS

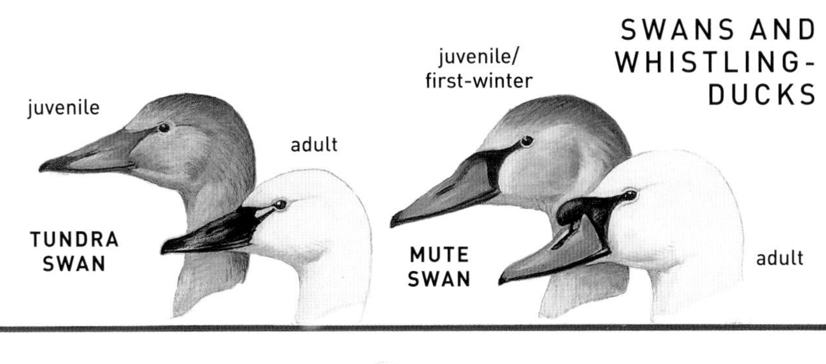

GEESE and SWANS in FLIGHT

CANADA GOOSE *Branta canadensis* p. 18

Large, slow wingbeats. Cackling Goose (p. 18) smaller, more Brantlike, faster wingbeats.

BRANT *Branta bernicla* p. 18

Small; black head and neck, white stern.

GREATER WHITE-FRONTED GOOSE *Anser albifrons* p. 16

Adult: Gray-brown neck, black bars or splotches on belly.
Juvenile and first-winter: Dusky, with light bill and feet.

TUNDRA SWAN *Cygnus columbianus* p. 20

Very long neck. *Adult:* Plumage entirely white. Trumpeter Swan (p. 20) similar, larger.

SNOW GOOSE (WHITE MORPH) *Anser caerulescens* p. 16

Adult: White with black primaries.
Juvenile and first-winter: Grayer than adult.

SNOW GOOSE (DARK MORPH, "BLUE" GOOSE) p. 16
Anser caerulescens

Adult: Dark body, white head.
Juvenile and first-winter: Dusky, with dark bill and feet.

ROSS'S GOOSE *Anser rossii* p. 16

Smaller, slightly shorter necked and shorter billed than Snow Goose.
Juvenile and first-winter: Grayer than adult.
Rare dark morph: Blacker than larger "Blue" Goose.

Many geese and swans fly in line or V formation.

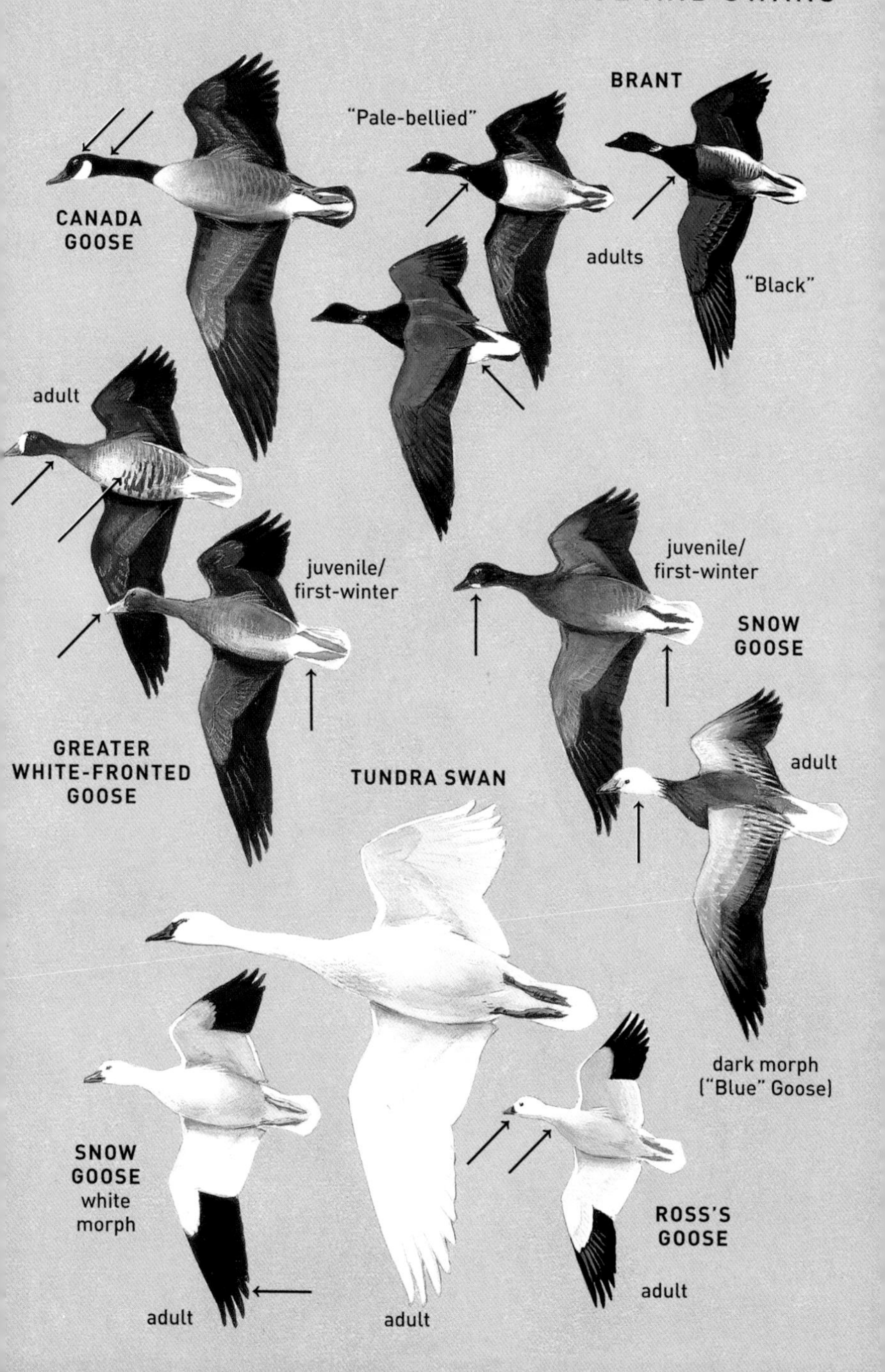
BRANT
"Pale-bellied"
CANADA GOOSE
adults
"Black"
adult
juvenile/ first-winter
juvenile/ first-winter
SNOW GOOSE
adult
GREATER WHITE-FRONTED GOOSE
TUNDRA SWAN
dark morph ("Blue" Goose)
SNOW GOOSE white morph
ROSS'S GOOSE
adult
adult
adult

DABBLING DUCKS

Feed by dabbling and upending; sometimes feed on land. Take flight directly into air. Most species have an iridescent "speculum" on secondaries from above. In most species, adult males brighter than females; in midsummer, males molt into drab "eclipse" (alternate) plumage, usually resembling females. Juvenile males also resemble females but gain colorful plumage in first fall. **FOOD:** Aquatic plants, seeds, grass, waste grain, small aquatic life, insects.

MUSCOVY DUCK *Cairina moschata* **Scarce, local**

Male 32 in. (81 cm); female 28 in. (71 cm). A black, gooselike duck with large white wing patch and underwing coverts. *Male:* Bare, knobby, red face. *Female:* Duller, has reduced facial knobs. *Juvenile and first-winter:* Head and neck brown, bill marked dark, facial knobs absent or reduced. Flight slow, heavy. **VOICE:** Usually silent. **RANGE:** Recent colonizer from Mex. of lower Rio Grande Valley, TX. Feral populations established in FL and near Brownsville, TX. **HABITAT:** Freshwater ponds and backwaters; wooded river corridors.

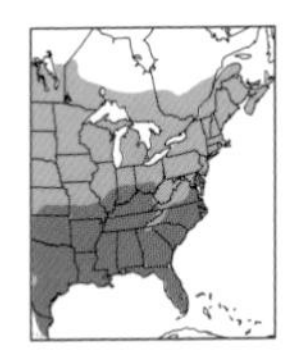

WOOD DUCK *Aix sponsa* (see also p. 40) **Fairly common**

18–19 in. (45–49 cm). Highly colored; often perches in trees. Speculum steely blue and purple. In flight, white belly contrasts with dark breast and wings; tail dark, long, almost square; neck short; looks squat and large-headed. *Male:* Striking face pattern, sweptback crest, red coloration on bill, and rainbow iridescence are unique. Juvenile and eclipse male more like female but have reddish bill and muted head pattern. *Female:* Dull colored; note dark crested head, gray bill, and *white eye patch;* similar female Mandarin Duck (p. 48) has smaller bill, lacks blue speculum, has whiter, less-patterned underwing. **VOICE:** Male, a hissing *jeeeeeeb,* with rising inflection. Female, a loud, rising squeal, *oo-eek,* and sharp *crrek, crrek.* **HABITAT:** Wooded swamps, rivers, ponds, marshes.

EURASIAN WIGEON *Mareca penelope* **Rare**

19–20 in. (48–51 cm). *Male:* Head *red-brown* with *buff* crown; *gray-sided;* rufous-pinkish breast. *Female:* Similar to female American Wigeon, but head less grayish, *brown or reddish brown.* In flight, axillars grayish (not white). **RANGE:** Rare winterer on coast; casual vagrant inland. **HABITAT:** Same as American Wigeon, with which it is usually found.

AMERICAN WIGEON **Fairly common**

Mareca americana (see also p. 40)

19–20 in. (48–51 cm). Speculum green. In flight, note *large white patch on forewing. Male:* Warm brownish; head pale gray with green eye patch. Note *white crown* (nicknamed "Baldpate"). *Female:* Brown; gray head and neck; whitish belly and forewing. **VOICE:** Male, a two-part whistled *whee whew.* Female, *qua-ack.* **SIMILAR SPECIES:** See Eurasian Wigeon. Squarish head, whitish patch on forewing, and small bluish bill separate wigeons from other ducks. **HABITAT:** Marshes, lakes, bays, fields, grass.

SILHOUETTES OF DUCKS ON LAND

dabbling ducks (dabblers)

sea and bay ducks (divers)

mergansers (divers)

Ruddy Duck (diver)

whistling-ducks (dabblers)

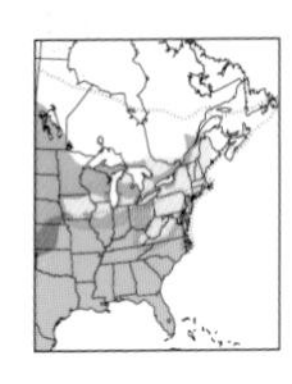

GADWALL *Mareca strepera* (see also p. 40) **Fairly common**

19–20 in. (48–51 cm). *Male: Gray* body with brown head and *black rump, white inner speculum* on rear edge of wing; when swimming, often shows as a white square patch near flank. Belly white, feet yellow, bill dark. *Female:* Brown, mottled, with *white inner speculum,* yellow feet, orange sides on gray bill. **VOICE:** Male, a low, reedy *bek;* a whistling call. Female, a nasal quack. **SIMILAR SPECIES:** Female Mallard has sloping forehead, different wing pattern, more quacklike call. **HABITAT:** Lakes, ponds, marshes.

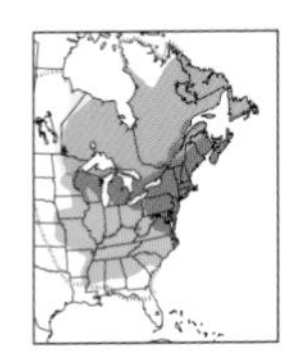

AMERICAN BLACK DUCK **Fairly common**

Anas rubripes (see also p. 42)

22–23 in. (55–58 cm). Darker than female Mallard. In flight, shows flashing *white underwing linings.* Sooty brown with paler head, violet speculum with thin white trailing edge; feet red or brown. Sexes similar except for bill (yellow in male, dull green in female). Hybridizes extensively with Mallard. **VOICE:** Male, a low croak. Female quacks like female Mallard. **SIMILAR SPECIES:** Mallard, Mottled Duck. **HABITAT:** Marshes, bays, estuaries, ponds, rivers, lakes.

MOTTLED DUCK *Anas fulvigula* **Fairly common**

22–23 in. (55–58 cm). Speculum bluish green with narrow white tips. Note tan head, unstreaked buffy throat, and unmarked yellow bill with *dark spot at base of bill at gape.* Sexes similar, although female has duller bill and speculum and less black at base of bill. Female Mallard paler with more black on bill, broader white border to speculum. **VOICE:** Very similar to Mallard. **SIMILAR SPECIES:** American Black Duck, Mallard. **RANGE:** Casual vagrant north of breeding range. **HABITAT:** Marshes, ponds.

MALLARD *Anas platyrhynchos* (see also p. 42) **Common**

22–23 in. (55–59 cm). Speculum greenish blue to blue with broad white tips. *Male: Glossy green head* and *white neck ring,* grayish body, chestnut chest, white tail, yellowish bill, orange feet. *Female:* Mottled brown with *whitish tail.* Dark bill with orange patches; orange feet. In flight, shows white bars *on both sides* of blue speculum. "Mexican" Mallard of nw. Mexico and sw. U.S. may be a separate species and is a casual visitor to the lower Rio Grande Valley. **VOICE:** Male, *yeeb;* a low *kwek.* Female, boisterous quacking. **SIMILAR SPECIES:** Female Gadwall, American Black Duck. **HABITAT:** Marshes, wooded swamps, grain fields, ponds, rivers, lakes, bays, city parks.

NORTHERN PINTAIL *Anas acuta* (see also p. 40) **Fairly common**

Male 25–26 in. (63–66 cm); female 20–21 in. (51–54 cm). Speculum brownish with pale tips. *Male:* Slender, slim-necked, white-breasted, with long, *needle-pointed tail.* A conspicuous *white point* runs onto side of dark head. *Female:* Variably mottled grayish brown to cinnamon brown; note rather pointed tail, slender neck, *gray bill.* In flight both sexes have a *single light border* on rear edge of speculum. **VOICE:** Male, a double-toned whistle: *prrip, prrip;* wheezy notes. Female, a low *quack.* **SIMILAR SPECIES:** Female's neck and bill thinner and longer than in other dabbling ducks. **HABITAT:** Marshes, prairies, ponds, lakes, salt bays.

DABBLING DUCKS
dabbling ducks tip up
female
male
GADWALL
dabbling ducks spring directly from the water
MOTTLED DUCK
male
male
(females similar except bill orange mottled dusky)
AMERICAN BLACK DUCK
"Mexican" Mallard male
female
male
MALLARD
NORTHERN PINTAIL
female
male

BLUE-WINGED TEAL *Spatula discors* (see also p. 40) **Fairly common**

15–16 in. (38–41 cm). A medium-small dabbling duck; speculum green. *Male:* Note *white facial crescent* and large *chalky blue* patch on *forewing.* Molting males hold eclipse plumage later in year than other dabbling ducks and resemble females. *Female, juvenile, and first-winter male:* Mottled brown; dark eye line; partial eye-ring; pale loral spot; blue on forewing duller. **VOICE:** Male, quiet whistled peeping notes. Female, a high *quack.* **SIMILAR SPECIES:** Cinnamon and Green-winged Teal. **HABITAT:** Ponds, marshes, mudflats, flooded fields.

CINNAMON TEAL *Spatula cyanoptera* **Scarce**

16–17 in. (41–43 cm). *Male:* A small, *dark chestnut* duck with large chalky blue patch on forewing. Adult has *red eye,* which it retains in eclipse plumage. *Female, juvenile, and first-winter male:* Very similar to female Blue-winged Teal but tawnier; bill slightly larger (more shoveler-like), face pattern duller. In flight suggest Blue-winged. **VOICE:** Like Blue-winged. **RANGE:** Very rare vagrant to most of East, casually to coast. **HABITAT:** Marshes, freshwater ponds, flooded fields.

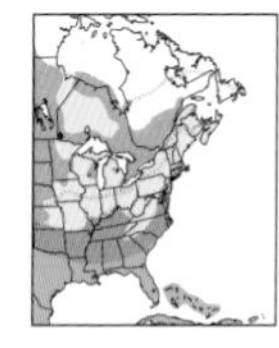

NORTHERN SHOVELER **Fairly common**

Spatula clypeata (see also p. 40)

18–19 in. (46–49 cm). The long *spoon-shaped bill* gives a front-heavy look distinctive among dabbling ducks. When swimming, sits low, with bill angled toward or in water; often strains water. Speculum green. *Male: Rufous* belly and sides; *white breast;* pale blue patch on forewing; orange feet; dark bill. *Juvenile and female:* Brown. Note large, spatulate, dusky orange bill, blue-gray forewing patch, white tail, orange feet. *First-winter male:* Variable between female and male; can have dark head with white crescent in front of the eye. **VOICE:** Male, a soft *thup-thup.* Female, short *quacks.* **SIMILAR SPECIES:** Cinnamon Teal. **HABITAT:** Marshes, ponds, sloughs; in winter, also salt bays.

GREEN-WINGED TEAL *Anas crecca* (see also p. 40) **Common**

14–15 in. (36–39 cm). Our smallest dabbling duck; flies in tight flocks. Green-wingeds lack light wing patches (speculum *deep green*). *Male:* Small, compact, gray with brown head (a green head patch shows in sunlight). On swimming birds, note butter-colored streak near tail and, on common N. American subspecies (*A. c. carolinensis*), *vertical white mark* near shoulder. Rare Eurasian subspecies, known as "Common" Teal (*A. c. crecca*), shows *longitudinal* (not vertical) white stripe above wing, bolder buffy borders to eye patch. *Female:* A nondescript, small speckled duck with *green* speculum, pale undertail coverts; subspecies not distinguishable. **VOICE:** Male, a high, froglike *dreep.* Female, a sharp *quack.* **SIMILAR SPECIES:** Female Blue-winged and Cinnamon Teal slightly larger and larger-billed, have light blue wing patches; in flight, males have dark belly, whereas Green-winged has white belly, broader dark border to underwing. **HABITAT:** Marshes, rivers, bays, mudflats, flooded fields.

TEAL

female
male
BLUE-WINGED TEAL

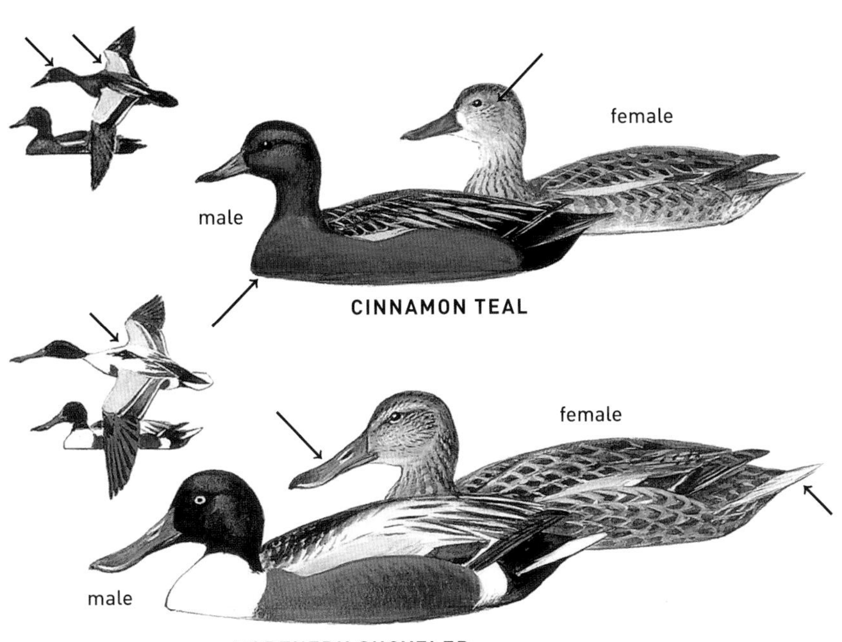
female
male
CINNAMON TEAL
female
male
NORTHERN SHOVELER

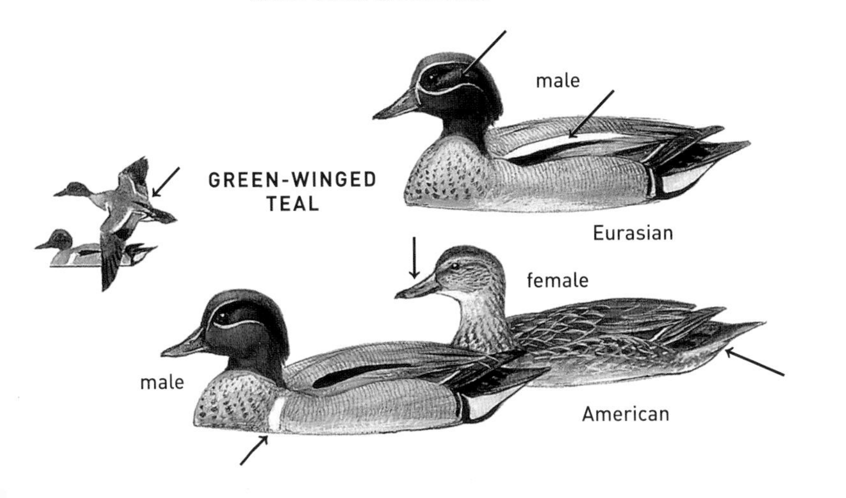
male
GREEN-WINGED TEAL
Eurasian
female
male
American

DIVING DUCKS

Often grouped as "bay ducks" or "sea ducks." All species dive. Adult sexes differ, but juvenile and first-winter males are female-like as are males in "eclipse" plumage in late summer. **FOOD:** Small aquatic animals and plants. Seagoing species eat mollusks and crustaceans. All "bay duck" species on this plate inhabit lakes, salt bays, and estuaries in winter and freshwater marshes and lakes in summer.

LABRADOR DUCK *Camptorhynchus labradorius*

Formerly bred in ne. Canada, wintered to NJ; became extinct around 1878.

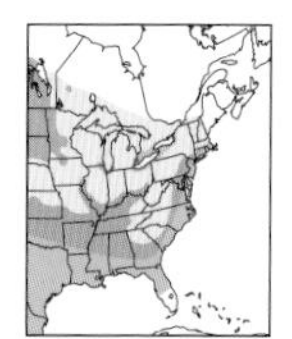

CANVASBACK *Aythya valisineria* (see also p. 44) **Uncommon**

21–22 in. (53–56 cm). Large with *long, sloping head profile. Adult male:* White with *chestnut red* head, red eye, black chest. *Female, juvenile, and first-year male:* Pale grayish brown with rust head and neck. **VOICE:** Cooing notes, a raspy *krrrr,* etc. **SIMILAR SPECIES:** Redhead lacks sloping forehead and bill.

REDHEAD *Aythya americana* (see also p. 44) **Uncommon**

19–20 in. (48–51 cm). *Adult male:* Gray; black chest and *round rufous head;* bluish bill with black tip. *Female, juvenile, and first-year male:* Brown overall; *diffuse light patch* near bill. Both sexes have indistinct *gray* wing stripe. **VOICE:** A harsh *meow,* soft *krrr* notes. **SIMILAR SPECIES:** Female Ring-necked Duck, scaup.

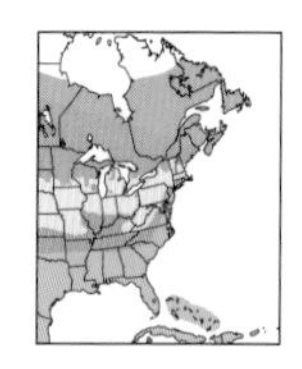

RING-NECKED DUCK *Aythya collaris* (see also p. 44) **Fairly common**

17–17½ in. (43–46 cm). *Adult male:* Like scaup but with *black back* and *indistinct gray* wing stripe in flight. Bill with white ring. *Female, juvenile, and first-year male:* Similar to female scaup but with *indistinct* light face patch, darker eye, *white eye-ring, grayer* wing stripe, and *pale ring on bill.* **VOICE:** A low-pitched whistle, quacking growl. **SIMILAR SPECIES:** Redhead has rounder head, paler crown, browner (less gray) face.

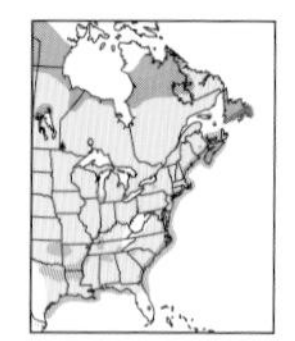

GREATER SCAUP *Aythya marila* (see also p. 44) **Common**

18–18½ in. (46–48 cm). Slightly larger than Lesser Scaup, head more rounded, bill wider with *larger black tip* (nail), *white wing stripe extends onto primaries. Adult male:* Dark at both ends, whitish in middle; head often glossed dull green. *Female, juvenile, and first-year male* (not shown): Like Lesser Scaup, but white facial patch often larger; sometimes shows pale ear crescent. **VOICE:** Wheezy whistles, a raspy *scaup-scaup.* **SIMILAR SPECIES:** Lesser Scaup, Ring-necked Duck, Redhead, Tufted Duck (p. 48).

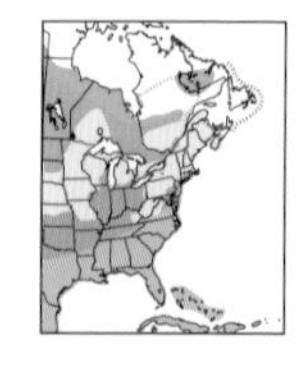

LESSER SCAUP *Aythya affinis* (see also p. 44) **Common**

16½–17 in. (42–44 cm). Lesser Scaup has narrower neck, more "peaked" head, and smaller bill than Greater Scaup. *Adult male:* Head often glossed dull purple rather than dark green, but use this with caution. *Female, juvenile, and first-year male:* Dark brown, usually with white patch near bill. **VOICE:** A soft whistle; a loud *scaup;* also purring notes. **SIMILAR SPECIES:** Greater Scaup, Ring-necked Duck, Redhead, Tufted Duck (p. 48). Tends to inhabit fresher-water, less-marine habitats in winter than Greater Scaup.

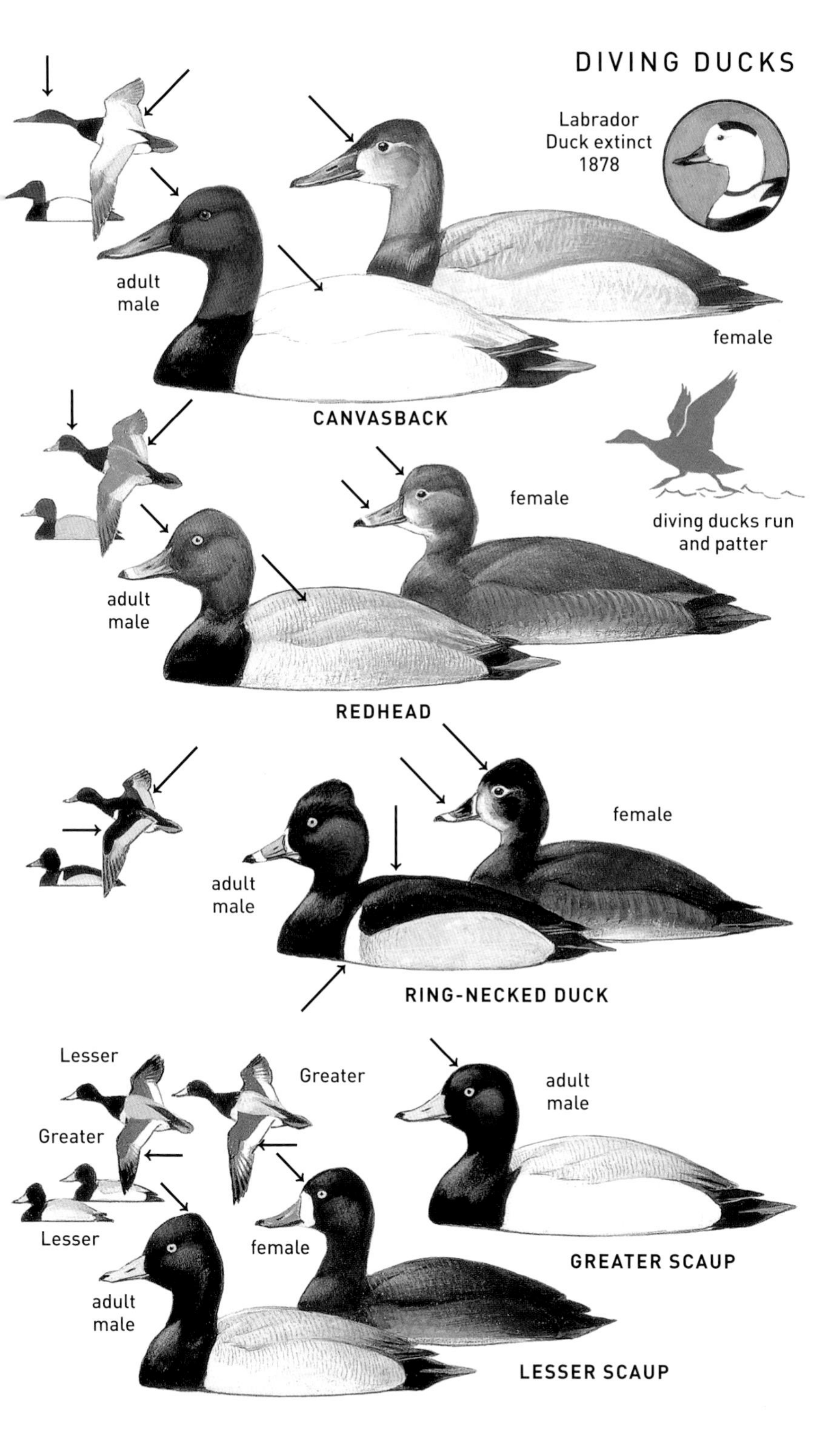
DIVING DUCKS
Labrador
Duck extinct
1878
adult
male
female
CANVASBACK
female
diving ducks run
and patter
adult
male
REDHEAD
female
adult
male
RING-NECKED DUCK
Lesser
Greater
Greater
Lesser
adult
male
female
GREATER SCAUP
adult
male
LESSER SCAUP

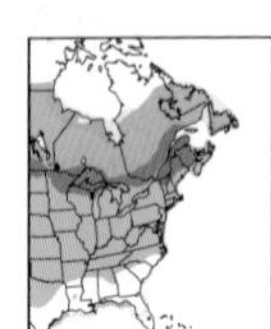

COMMON GOLDENEYE Fairly common
Bucephala clangula (see also p. 44)
18½–19 in. (47–49 cm). *Adult male:* Note large, *round white spot* before eye. White with black back and blocky, green-glossed head. In flight, short-necked; wings whistle or "sing," show large white patches. *Female, juvenile, and first-year male:* Gray, with white collar and dark brown head; large square white patches may show on closed wing. Bill black in males, with some yellow in winter/spring females. **VOICE:** Courting male, a harsh, nasal, double note. Female, a harsh *gaak.* **SIMILAR SPECIES:** Barrow's Goldeneye. **HABITAT:** Forested lakes, rivers; in winter, also open lakes, salt bays, seacoasts.

BARROW'S GOLDENEYE *Bucephala islandica* Scarce
18 in. (46 cm). Similar to Common Goldeneye, but *bill smaller* giving a "cuter" look to the head. *Adult male:* Note *white facial crescent.* Blacker (less white) above than male Common; head often glossed with *purple* (not green); nape blockier; shows *dark "spur"* on shoulder toward waterline, less white in back and wings. *Female and first-year male:* Similar to female Common but head slightly darker, with steeper forehead and blockier nape, less white in wing. Besides being smaller, bill can become entirely *yellow,* subject to seasonal change. Female Common's bill rarely all yellow. **VOICE:** Usually silent. Courting male, a grunting *kuk, kuk.* Female near nest, a soft *coo-coo-coo.* Wings of both species whistle in flight. **SIMILAR SPECIES:** Common Goldeneye, Bufflehead. **RANGE:** Rare winter visitor or vagrant to most interior West areas; accidental to TX. **HABITAT:** Wooded lakes, ponds; in winter, lakes and rivers, protected coastal waters.

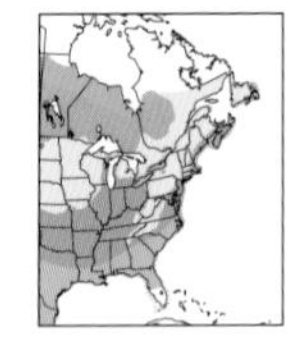

BUFFLEHEAD *Bucephala albeola* (see also p. 44) Common
13½–14 in. (34–36 cm). Small. *Adult male:* Mostly white with black back; blocky head with *large, bonnetlike white patch.* In flight, shows large white wing patches. *Female and first-year male:* Dark and compact, with *white cheek spot,* small bill, smaller wing patch. **VOICE:** Male, in display, a hoarse rolling note. Female, a harsh *ec-ec-ec.* **SIMILAR SPECIES:** Male Hooded Merganser has spikelike bill, brown sides. **HABITAT:** Lakes, ponds, rivers; in winter, also salt bays.

STIFF-TAILED DUCKS

Small, chunky divers, nearly helpless on land. Spiky tail. Adult sexes not alike. **FOOD:** Aquatic life, insects, water plants.

RUDDY DUCK *Oxyura jamaicensis* (see also p. 44) Fairly common
15 in. (38 cm). Small, chubby; note *white cheek* and dark cap. Often cocks tail upward. Flight "buzzy." *Spring/summer male:* Vivid rusty red with white cheek, black cap, large and strikingly *blue* bill. *Fall/winter male:* Gray with *white cheek,* dull blue or gray bill. *Female and juvenile male:* Similar to fall/winter male, but duskier cheek crossed by dark line. **VOICE:** Courting male, a sputtering *chick-ik-ik-ik-k-k-k-kurrrr,* accompanied by head bobbing. **SIMILAR SPECIES:** Female Bufflehead, Black Scoter; see female Masked Duck (rare, p. 48). **HABITAT:** Freshwater marshes, ponds, lakes; in winter, also salt bays, harbors.

DIVING DUCKS

COMMON GOLDENEYE

BARROW'S GOLDENEYE

BUFFLEHEAD

RUDDY DUCK

KING EIDER *Somateria spectabilis* (see also p. 46) Rare to uncommon

22 in. (56 cm). *Adult male:* Foreparts appear white, rear parts black; crown and nape powder blue. Note protruding *orange bill-shield. Female, juvenile, and first-winter male:* Warm brown, with weak pale eye-ring and facial stripe behind eye, flanks barred with crescent-shaped marks. Note facial profile. *First-spring and second-year males:* Dusky brown with light breast; bill becomes orange; gradually gain adult plumage. **VOICE:** Courting male, a low crooning phrase. Female, grunting croaks. **SIMILAR SPECIES:** Common Eider larger, with flatter head profile, longer bill-lobe before eye; adult male has *white* back, female evenly barred flanks. First-spring male King has darker head and lacks white shoulder stripe. In flight, note upperwing pattern. **RANGE:** Casual vagrant well south of winter range; accidental inland. **HABITAT:** Rocky coasts, ocean. Nests on tundra.

COMMON EIDER Fairly common

Somateria mollissima (see also p. 46)

24–25 in. (61–64 cm). *Adult male: Black belly and white back.* Forewing and back white; head white with black crown, greenish nape. *Female, juvenile, and first-winter male:* Large, brown, *closely barred, with pale eyebrow;* long, flat profile. *First-spring and second-year males:* Dusky or chocolate with white breast and collar, becoming irregularly adultlike; bill slowly becomes brighter yellow. **VOICE:** Male, a moaning *ow-ooo-urr.* Female, a grating *kor-r-r.* **SIMILAR SPECIES:** See King Eider. **RANGE:** Casual vagrant well south of winter range; accidental inland. **HABITAT:** Rocky coasts, shoals; in summer, also islands, tundra.

HARLEQUIN DUCK Uncommon

Histrionicus histrionicus (see also p. 46)

16–17 in. (41–44 cm). A smallish dark duck. *Adult male:* Spectacularly patterned, slaty with chestnut sides and elaborate white patches and spots. *Female:* A small dusky duck with three round white spots on each side of head; no wing patch. *First-year and eclipse males:* Intermediate between male and female. **VOICE:** Usually silent. Male, a squeak; also *gwa gwa gwa.* Female, *ek-ek-ek-ek.* **SIMILAR SPECIES:** Female Bufflehead has white wing patch and only one white facial patch. Female scoters larger, with larger bills. **RANGE:** Casual to accidental vagrant inland and well south of range. **HABITAT:** Turbulent mountain streams in summer; rocky coastal waters in winter.

LONG-TAILED DUCK Fairly common

Clangula hyemalis (see also p. 46)

Male 21–22 in. (53–56 cm); female 16 in. (41 cm). A small duck except for long tail in adult male. The only sea duck combining much *white on body and unpatterned dark wings.* Flies in bunched, irregular flocks, individuals rocking side to side as they fly. *Adult male:* Note needlelike tail, pied pattern, dark cheek in spring/summer, dark with white flanks and belly in fall/winter. Note white eye patch, pink on bill. *Juvenile, females, and first-winter male:* Dark unpatterned wings, white face with dark cheek spot, lack long tail feathers. Much individual variation in plumages. **VOICE:** Talkative; a musical *ow-owdle-ow* or *owl-omelet.* **SIMILAR SPECIES:** Bufflehead. **RANGE:** Widespread but rare winter visitor inland across East. **HABITAT:** Ocean, harbors, large lakes; in summer, tundra pools and lakes.

EIDERS AND OTHER DIVING DUCKS

SCOTERS

Heavy, blackish ducks seen in large flocks along ocean coasts. Often fly in thin line formation. Usually silent, but during courtship and mating may utter low whistles, croaks, or grunting noises; wings whistle in flight. **FOOD:** Mainly mollusks, crustaceans.

WHITE-WINGED SCOTER Uncommon to fairly common
Melanitta fusca (see also p. 46)

21 in. (53 cm). The largest of the three scoters; bill is feathered to nostril. On water, white wing patch is often barely visible or fully concealed (wait for bird to flap or fly). *Adult male:* Black, with a "teardrop" of white near eye; bill orange with black basal knob. *Female and juvenile male:* Sooty brown, with white wing patch and two light oval patches on face. *First-year male:* Gradually becomes blackish; bill becomes orange. Underparts bleached white on first-year birds of all three scoters. **SIMILAR SPECIES:** Other scoters. **RANGE:** Rare winter visitor or vagrant to interior states. **HABITAT:** Salt bays, ocean; in summer, lakes.

SURF SCOTER *Melanitta perspicillata* (see also p. 46) Fairly common

19–20 in. (48–51 cm). Medium-sized, the most common scoter along Atlantic Coast. *Adult male:* Black, with bold *white patches* on crown and nape. Sloping bill patterned with orange, black, and white. *Female and juvenile male:* Dusky brown; dark crown; two light spots on each side of head (sometimes obscure). *First-year male:* Gradually becomes blackish; bill becomes orange, swollen. **SIMILAR SPECIES:** Female White-winged Scoter slightly larger overall, has more extensive feathering on bill, more horizontal, oval face patches, and white wing patch (may not show until bird flaps). Black Scoter has rounder head profile (more like Redhead), lacks feathering on bill, and has silvery underside to flight feathers; female has pale cheeks. **RANGE:** Very rare winter visitor or vagrant to interior states. **HABITAT:** Ocean, salt bays; in summer, lakes.

BLACK SCOTER *Melanitta americana* (see also p. 46) Uncommon

18½–19 in. (47–48 cm). The smallest scoter. Bill upturned and not as bulbous as in other scoters. *Adult male:* Entirely black; bright *orange-yellow knob* on bill is diagnostic. In flight, silvery gray underwing more pronounced than in other two scoters. *Female and juvenile male:* Sooty; *entirely light cheeks* contrast with dark cap. *First-year male:* Gradually becomes blackish, especially in head; bill becomes yellow, swollen. **SIMILAR SPECIES:** First-spring male Surf Scoter may lack white head patch and have messy orange coloration on mandible, but note higher-sloping bill. Female and juvenile scoters of other two species have smaller light spots on side of head, not entirely pale cheeks. **RANGE:** Very rare winter visitor to interior states. **HABITAT:** Seacoasts, bays; in summer, tundra and taiga ponds.

SCOTERS
scoters can fly in V formation or in loose clumps
Surf
Black
White-winged
first-winter/ spring male
female
adult male
WHITE-WINGED SCOTER
female
first-winter/ spring male
adult male
SURF SCOTER
first-winter/ spring male
female
adult male
BLACK SCOTER

diving ducks (sea ducks and bay ducks) raft on water, skitter when taking wing

MERGANSERS

Long, slender-bodied, crested diving ducks with spikelike bill, saw-edged mandibles. In flight, bill, head, outstretched neck, and body are on a horizontal axis. Adult sexes not alike; first-year and eclipse males resemble female. **FOOD:** Chiefly fish.

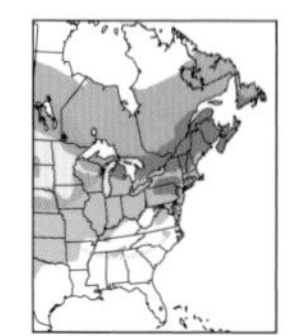

COMMON MERGANSER — Fairly common

Mergus merganser (see also p. 42)

24–25 in. (62–64 cm). Whiteness of adult male and merganser shape (bill, outstretched neck, head, and body held horizontally) identify this species. *Adult male:* Note long whitish body, black back, green-black head; primarily white upperwing. Bill and feet red; breast can be tinged rosy peach. *Female and first-year male:* Gray with rufous head contrasting with white chin and clean white chest; wing patch on trailing edge. First-spring males can show dark green in face. **VOICE:** Male, in display, low staccato croaks. Female, a guttural *karrr.* **SIMILAR SPECIES:** Female Red-breasted Merganser very similar to female Common. Note distinct cut-off of rusty head and neck from breast in Common; this is diffuse in Red-breasted. **HABITAT:** Wooded lakes, ponds, rivers; in winter, open lakes, rivers, rarely coastal bays.

RED-BREASTED MERGANSER — Common

Mergus serrator (see also p. 42)

22½–23 in. (56–58 cm). *Adult male:* Rakish; black head glossed with green and *crested;* breast at waterline dark rusty; *wide white collar* between head and breast; bill and feet red. *Female and first-year male:* Gray, with crested, dull rusty head that *blends* into color of neck; red bill and feet. First-spring male can molt in dark green feathers in face and black feathers in back. **VOICE:** Usually silent. Male, a hoarse croak. Female, *karrr.* **SIMILAR SPECIES:** Male Common Merganser whiter, without collar and breast-band effect; lacks shaggy crest. In female Common, white chin and chest *sharply delineated* from brighter rufous head and pale gray body. Common's bill slightly thicker at base. **HABITAT:** Woodland and coastal lakes, open water; in winter, also bays, tidal channels, nearshore ocean waters.

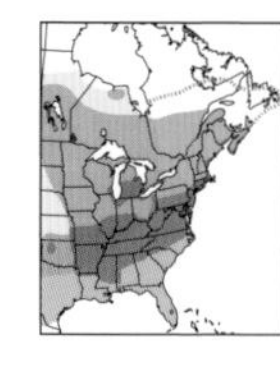

HOODED MERGANSER — Uncommon to fairly common

Lophodytes cucullatus (see also p. 42)

17–18 in. (43–46 cm). *Male:* Note vertical *fan-shaped white crest,* which may be raised or lowered. Breast white, with two black bars on each side. Upperwing has white patch; *flanks rusty brown. Female and first-winter male:* Merganser-like silhouette and spikelike bill; small size, dusky look, and *dark head, bill, and chest.* Note loose *tawny crest.* First-spring male can molt in black and white feathers in head and breast. **VOICE:** In display, low grunting or croaking notes. **SIMILAR SPECIES:** Male Bufflehead smaller and chubbier, with *white* sides. Other female mergansers larger and *grayer,* with rufous head, reddish bill. In flight, wing patch and silhouette separate female Hooded Merganser from female Wood Duck. **HABITAT:** Wooded lakes, ponds, rivers; in winter, also tidal channels, protected bays.

MERGANSERS
mergansers fly with bill, head, body, and tail on the same horizontal axis
saw-edged mandibles of merganser
female
adult male
COMMON MERGANSER
female
adult male
RED-BREASTED MERGANSER
crest down
adult males
crest up
first-winter male
female
HOODED MERGANSER
Common
Red-breasted
Hooded

FLIGHT PATTERNS of DABBLING DUCKS

Note: Only males are described below. Although females are unlike the males in body plumage, their wing patterns are quite similar. The names in parentheses are common nicknames used by hunters.

NORTHERN PINTAIL ("SPRIG") *Anas acuta* p. 26

From below: Needle tail, white breast, thin neck.
Above: Needle tail, neck stripe, single thin white border on speculum.

WOOD DUCK ("WOODY") *Aix sponsa* p. 24

From below: White belly, dusky wings, long square tail.
Above: Stocky; long dark tail, white border on dark wing.

AMERICAN WIGEON ("BALDPATE") *Mareca americana* p. 24

From below: White belly, pointed dark tail.
Above: Large white shoulder patch.

NORTHERN SHOVELER ("SPOONBILL") *Spatula clypeata* p. 28

From below: Dark belly, white breast, white tail, spoon bill.
Above: Large pale bluish shoulder patch, spoon bill.

GADWALL ("GRAYDUCK") *Mareca strepera* p. 26

From below: White belly, white underwing, square white patch on rear edge of wing.
Above: White patch on rear edge of wing.

GREEN-WINGED TEAL ("ROCKET") *Anas crecca* p. 28

From below: Small; light belly, dark head, broad dark borders to underwing.
Above: Small, dark-winged; green speculum.

BLUE-WINGED TEAL ("WHITEFACE") *Spatula discors* p. 28

From below: Small; dark belly, narrow dark borders to underwing.
Above: Small; large chalky blue shoulder patch.
Note: Cinnamon Teal (*S. cyanoptera*) shows similar wing pattern as Blue-winged Teal.

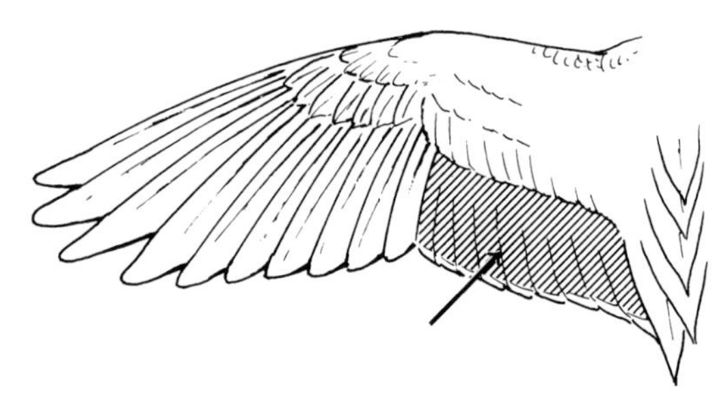

upper wing of a dabbling duck showing the iridescent speculum (secondaries)

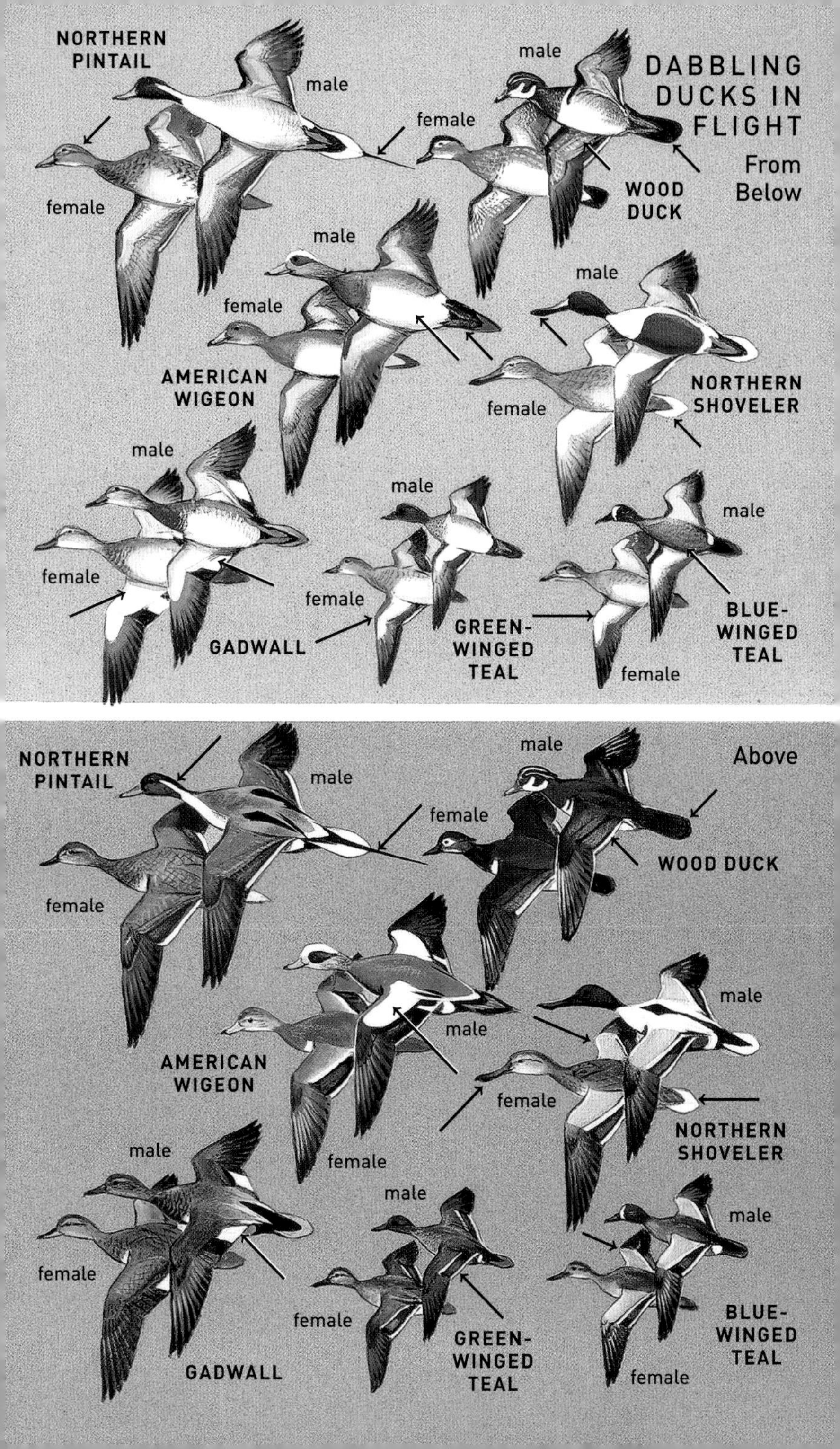
DABBLING DUCKS IN FLIGHT
From Below
NORTHERN PINTAIL
male
female
male
female
WOOD DUCK
male
female
AMERICAN WIGEON
male
female
NORTHERN SHOVELER
male
female
GADWALL
male
female
GREEN-WINGED TEAL
male
female
BLUE-WINGED TEAL
Above
NORTHERN PINTAIL
male
female
male
female
WOOD DUCK
male
female
AMERICAN WIGEON
male
female
NORTHERN SHOVELER
male
female
GADWALL
male
female
GREEN-WINGED TEAL
male
female
BLUE-WINGED TEAL

FLIGHT PATTERNS of DABBLING DUCKS and MERGANSERS

Note: Only males are described below. Although most females are unlike the males, their wing patterns are quite similar. Mergansers have a distinctive flight silhouette. The names in parentheses are common nicknames used by hunters.

MALLARD ("GREENHEAD") *Anas platyrhynchos* p. 26

From below: Dark chest, light belly, white neck ring, white tail.
Above: Dark head, neck ring, two white borders on bluish speculum.

AMERICAN BLACK DUCK ("REDLEG") *Anas rubripes* p. 26

From below: Dark body, white underwing linings.
Above: Dark body, paler head, purplish speculum lacks forward border.

FULVOUS WHISTLING-DUCK *Dendrocygna bicolor* p. 20

From below: Tawny, with blackish underwing linings.
Above: Dark, unpatterned wings; white band on rump.

COMMON MERGANSER ("SAWBILL") *Mergus merganser* p. 38

From below: Merganser shape; outstretched neck, dark head, white body, white underwing linings.
Above: Merganser shape; white chest, large white wing patches.

RED-BREASTED MERGANSER ("SHELDRAKE") *Mergus serrator* p. 38

From below: Merganser shape; outstretched neck, dark chest band, white collar.
Above: Merganser shape; dark chest, large white wing patches.

HOODED MERGANSER ("HOODIE") *Lophodytes cucullatus* p. 38

From below: Merganser shape; dusky underwing linings.
Above: Merganser shape; small white wing patches.

DABBLING DUCKS AND MERGANSERS IN FLIGHT
From Below
AMERICAN BLACK DUCK
MALLARD
male
female
FULVOUS WHISTLING-DUCK
male
COMMON MERGANSER
female
male
female
male
female
RED-BREASTED MERGANSER
mergansers fly on a horizontal axis
female
HOODED MERGANSER

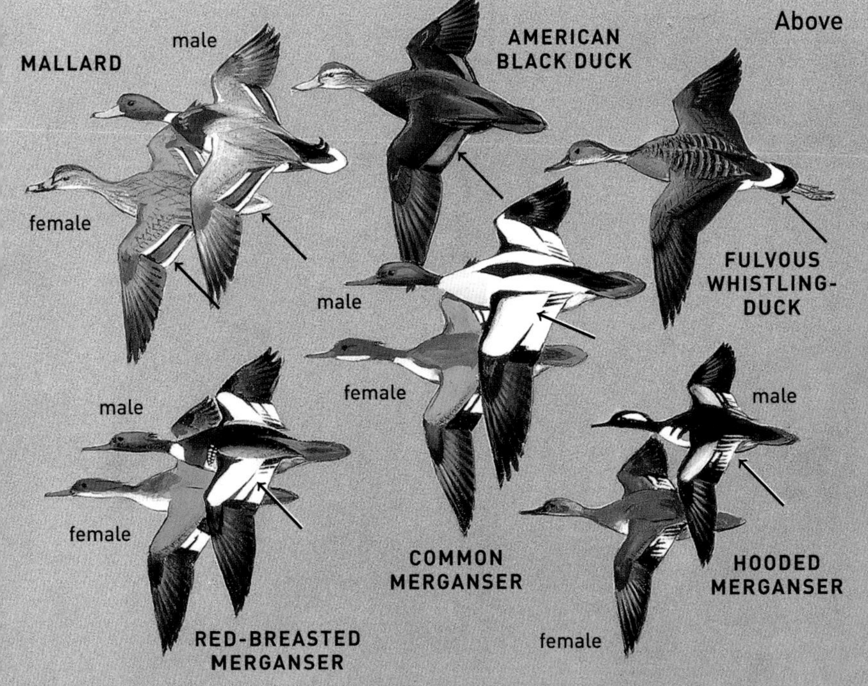

Above
MALLARD
male
female
AMERICAN BLACK DUCK
FULVOUS WHISTLING-DUCK
male
female
COMMON MERGANSER
male
female
RED-BREASTED MERGANSER
male
female
HOODED MERGANSER

FLIGHT PATTERNS of DIVING DUCKS

Note: Only adult males are described below. The first five all have a black chest. The names in parentheses are common nicknames used by hunters.

CANVASBACK ("CANNIE") *Aythya valisineria* p. 30

From below: Black chest, long profile.
Above: White back, long profile. Lacks contrasty wing stripe of next four species.

REDHEAD ("POCHARD") *Aythya americana* p. 30

From below: Black chest, roundish rufous head.
Above: Gray back, broad gray wing stripe.

RING-NECKED DUCK ("BLACKJACK") *Aythya collaris* p. 30

From below: Not safe to tell from scaup from below; gray wing stripe sometimes evident.
Above: Black back, broad gray wing stripe.

GREATER SCAUP ("BROADBILL") *Aythya marila* p. 30

From below: Black chest, white stripe showing through wing.
Above: Broad white wing stripe (extending onto primaries).

LESSER SCAUP ("BLUEBILL") *Aythya affinis* p. 30

Above: Wing stripe shorter than in Greater Scaup.

COMMON GOLDENEYE ("WHISTLER") *Bucephala clangula* p. 32

From below: Dark underwing linings, white wing patches, rounded dark head.
Above: Large white square wing patch, short neck, dark head.

RUDDY DUCK ("STIFFTAIL") *Oxyura jamaicensis* p. 32

From below: Stubby; white face, dark chest, long tail.
Above: Small; dark with white cheeks, long tail.

BUFFLEHEAD ("BUTTERBALL") *Bucephala albeola* p. 32

From below: Like a small goldeneye; note head patch.
Above: Small; large wing patches, white head patch.

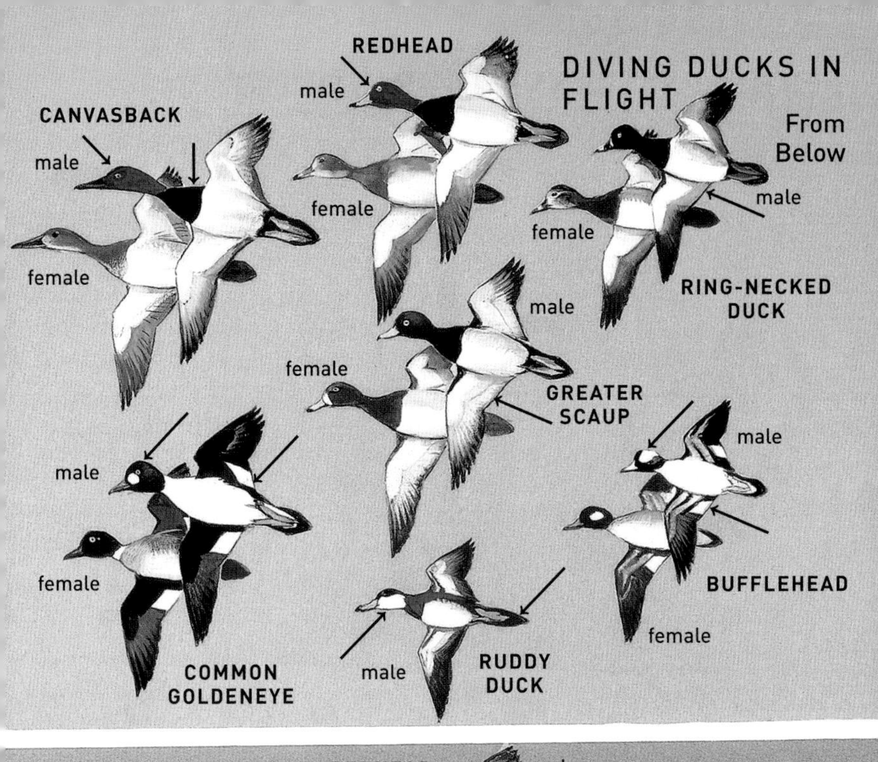
DIVING DUCKS IN FLIGHT
From Below
REDHEAD
male
female
CANVASBACK
male
female
male
female
RING-NECKED DUCK
male
female
GREATER SCAUP
male
female
COMMON GOLDENEYE
male
BUFFLEHEAD
female
male
RUDDY DUCK

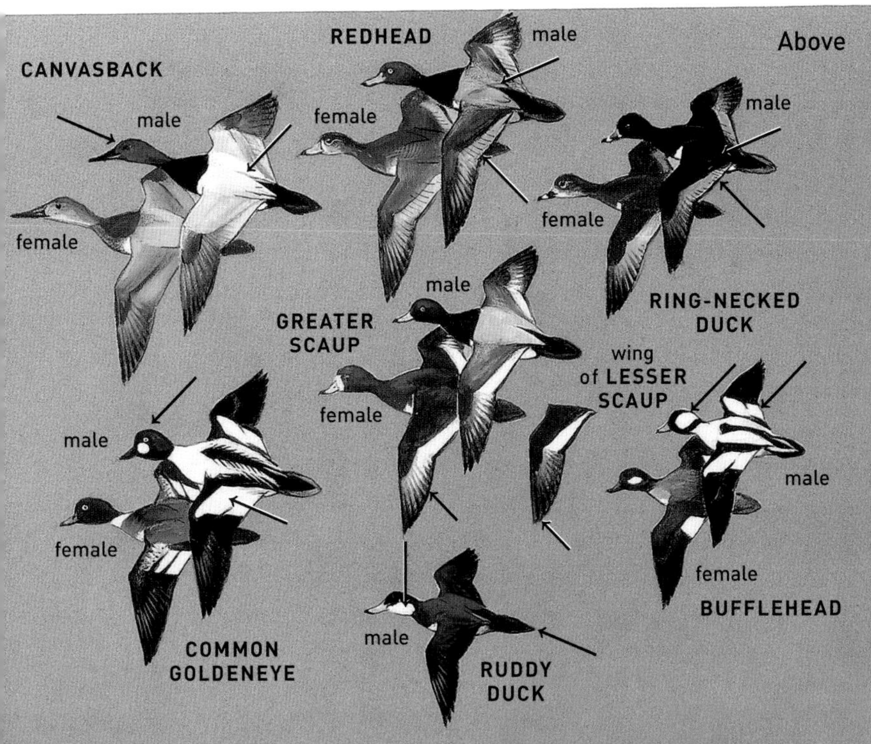
Above
REDHEAD
male
female
CANVASBACK
male
female
male
female
RING-NECKED DUCK
male
GREATER SCAUP
female
wing of LESSER SCAUP
male
female
COMMON GOLDENEYE
male
female
BUFFLEHEAD
male
RUDDY DUCK

FLIGHT PATTERNS of DIVING DUCKS

Note: Only adult males are described below. The names in parentheses are common nicknames used by hunters.

LONG-TAILED DUCK ("KAKAWI") *Clangula hyemalis* p. 34

From below: Dark unpatterned wings, white belly.
Above: Dark unpatterned wings, much white on body.

HARLEQUIN DUCK ("BLUEDUCK") *Histrionicus histrionicus* p. 34

From below: Solid dark below, white head spots, small bill.
Above: Dark with white marks, small bill, long tail.

SURF SCOTER ("SKUNKHEAD") *Melanitta perspicillata* p. 36

From below: Black body, white head patches (not readily visible from below), sloping forehead.
Above: Black body, white head patches, sloping forehead.

BLACK SCOTER ("BUTTERBILL") *Melanitta americana* p. 36

From below: Black plumage, paler flight feathers, rounded forehead.
Above: All-dark plumage. Body slightly smaller and pudgier than Surf Scoter's, rounded forehead.

WHITE-WINGED SCOTER ("WHITEWING") *Melanitta fusca* p. 36

From below: Black body, white wing patches.
Above: Black body, white wing patches.

COMMON EIDER ("IDAH") *Somateria mollissima* p. 34

Above: White back, white forewing, black belly.

KING EIDER ("KING") *Somateria spectabilis* p. 34

Above: Whitish foreparts, black rear parts.

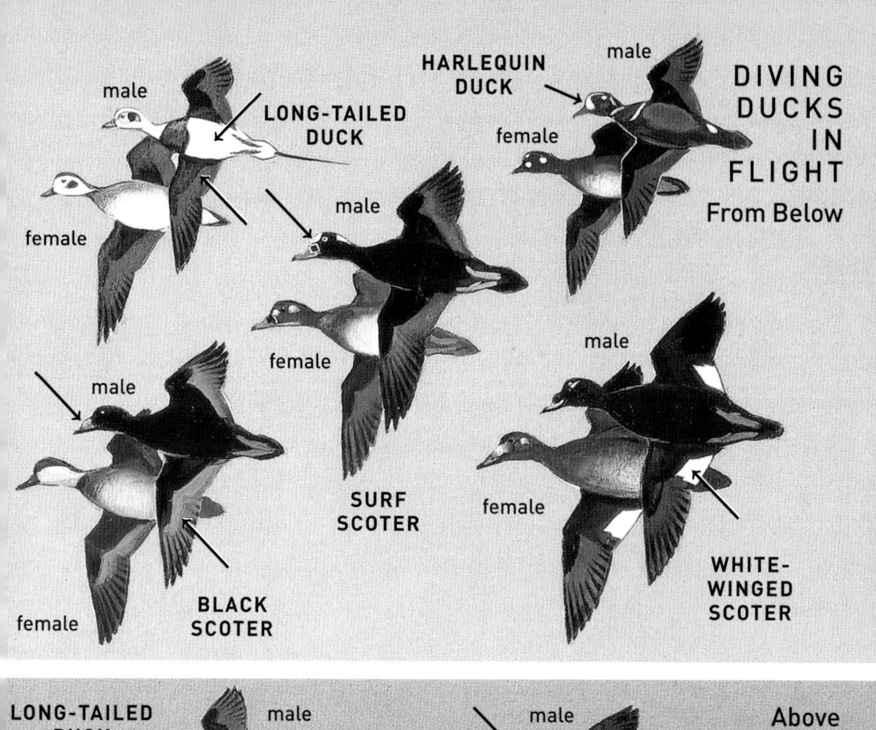
DIVING
DUCKS
IN
FLIGHT
From Below
male
LONG-TAILED
DUCK
female
HARLEQUIN
DUCK
male
female
male
female
SURF
SCOTER
male
female
BLACK
SCOTER
male
female
WHITE-
WINGED
SCOTER

Above
LONG-TAILED
DUCK
male
female
male
female
HARLEQUIN DUCK
male
female
COMMON
EIDER
male
KING EIDER
male
female
BLACK
SCOTER
male
female
SURF
SCOTER
male
female
WHITE-
WINGED
SCOTER

VAGRANT WATERFOWL

PINK-FOOTED GOOSE *Anser brachyrhynchus* Casual vagrant

25–30 in. (65–75 cm). Similar to Greater White-fronted Goose but slightly smaller, head entirely brown, bill mostly dark, tail whiter. **RANGE:** Increasingly observed in the Northeast in winter; accidental farther west. Usually found with Canada Geese.

GARGANEY *Spatula querquedula* Casual vagrant

15–16 in. (38–41 cm). *Male:* Broad white eyebrow stripe, silvery shoulder patch in flight. *Female:* Told from Blue-winged and Cinnamon Teal by bolder face pattern, grayer upperwing patch, and white borders on speculum. **RANGE:** Casual vagrant from Eurasia, primarily along coast.

MASKED DUCK *Nomonyx dominicus* Rare visitor

13–13½ in. (33–34 cm). *Male:* Rusty body with black face and blue bill. *Female:* Similar to female Ruddy Duck, but two distinct face stripes. **RANGE:** Rare visitor from Tropics to TX and FL; accidental elsewhere in East. **HABITAT:** Ponds and marshes with dense vegetation. Often hidden.

TUFTED DUCK *Aythya fuligula* Rare vagrant

16½–17 in. (41–43 cm). *Adult male:* Conspicuous wispy crest; note also black back, white sides, white wing stripe. *Female and first-year male:* Resemble female scaup but develop small tuft and have broad band at bill tip. May or may not have white at base of bill. **RANGE:** Regular visitor from Eurasia to NL; very rare to casual elsewhere in East. **HABITAT:** Sheltered ponds, bays, reservoirs. Usually with scaup.

SMEW *Mergellus albellus* Accidental vagrant

16 in. (41 cm). Smaller and shorter-billed than other mergansers. *Adult male: Very white,* with black-and-white crest; conspicuous black-and-white wings. *Female and first-year male:* Gray with *white cheeks, chestnut cap.* **RANGE:** Accidental. Some birds might be escapees.

COMMON SHELDUCK *Tadorna tadorna* Casual vagrant

23–26 in. (58–67 cm). Plumage unmistakable. Increasing records in ne. Canada and U.S. now regarded as vagrants rather than as escapees.

UNESTABLISHED EXOTIC WATERFOWL

CHINESE GOOSE *Anser cygnoides* Exotic

GRAYLAG GOOSE *Anser anser* Exotic

30–35 in. (75–90 cm). Common domestic species. One shipboard arrival off NL considered an accidental wild vagrant.

BAR-HEADED GOOSE *Anser indicus* Exotic

WHITE-CHEEKED PINTAIL *Anas bahamensis* Provenance in question

(W. Indies) 17 in. (43 cm). Occasional reports from s. FL may include accidental vagrants; scattered records elsewhere are more likely escapees.

MANDARIN DUCK *Aix galericulata* Exotic

RUDDY SHELDUCK *Tadorna ferruginea* Provenance in question

24–26 in. (61–67 cm) Record of six birds in NU may be of vagrants.

VAGRANT WATERFOWL
PINK-FOOTED GOOSE
adult
female
male
GARGANEY
MASKED DUCK
female
adult male
female
adult male
TUFTED DUCK
female
adult male
SMEW
male
COMMON SHELDUCK

UNESTABLISHED EXOTICS
CHINESE GOOSE
WHITE-CHEEKED PINTAIL
male
BAR-HEADED GOOSE
GRAYLAG GOOSE
female
male
MANDARIN DUCK
RUDDY SHELDUCK
adult male

CORMORANTS Family Phalacrocoracidae

Large blackish waterbirds that often stand erect with neck in an S; may rest with wings spread out to dry. Breeding adults have colorful facial skin, throat pouch, and eyes. Bill slender, hook-tipped. Sexes alike. Cormorants swim with bill tilted up at an angle. Silent except for occasional low grunts at nesting colonies. **FOOD:** Fish, crustaceans. **RANGE:** Nearly worldwide.

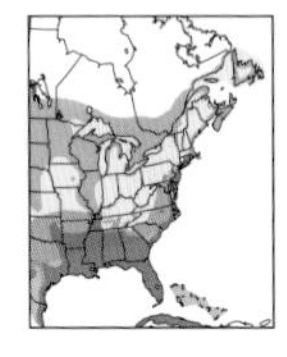

DOUBLE-CRESTED CORMORANT *Phalacrocorax auritus* Common

32–33 in. (81–84 cm). Cormorants found inland or on fresh water are largely this species, except for a few Great Cormorants found in Texas. *Adult:* Glossy black, perches with erect posture. Crest seldom evident. *First-year:* Brownish belly; pale throat and chest can become white by spring. **SIMILAR SPECIES:** Other cormorants, loons. **HABITAT:** Coasts, estuaries, lakes, rivers; nests colonially on rocky islands, structures, or in trees (often with herons).

GREAT CORMORANT *Phalacrocorax carbo* Uncommon

36–37 in. (91–94 cm). *Adult:* Slightly larger than Double-crested Cormorant; note heavier bill and *yellower* throat pouch, bordered by *white throat* strap. In spring/summer, has *white patch* on flanks. *First-year:* Dark breast and *pale belly,* the reverse of first-year Double-crested; also often has suggestion of pale throat patch. **RANGE:** Casual vagrant to Great Lakes and inland. **HABITAT:** Coasts and bays, locally inland on large rivers, lakes. Nests on rocky islands and headlands.

NEOTROPIC CORMORANT *Phalacrocorax brasilianus* Uncommon

25–26 in. (64–66 cm). *Adult and first-year:* Similar to Double-crested Cormorant but smaller, slimmer, and with proportionally *much longer tail.* When breeding (mostly spring but some fall/winter), has white filoplumes on neck. Note smaller throat pouch, in adult with *narrow white border,* forming a point at rear; bare orangey face does not extend to loral area; underparts of first-year not quite as pale. **SIMILAR SPECIES:** Other cormorants. **RANGE:** Casual vagrant well north of range; accidental to NJ. **HABITAT:** Freshwater wetlands, ponds, lakes; lakes near coasts.

DARTERS Family Anhingidae

Represented in N. America by one species. **FOOD:** Fish, small aquatic animals. **RANGE:** N. and S. America, Africa, India, se. Asia, Australia.

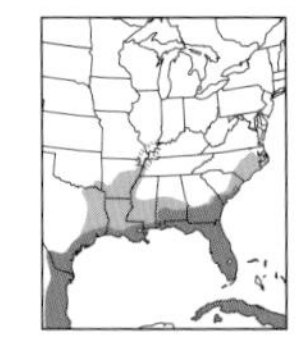

ANHINGA *Anhinga anhinga* Fairly common

34–35 in. (86–89 cm). Similar to a cormorant, but neck snakier, bill more pointed, tail much longer and with corrugations in adult. Note prominent silvery upperwing patch. Male black-bodied; female has buff neck and breast; juvenile like female but abdomen brownish; first-year male slowly acquires black on head and underparts. In flight, flaps and glides with neck extended, long tail spread. Often soars high, hawklike, with wings held flat (arched in cormorants). Perches like a cormorant, often with wings spread or half-spread. May swim submerged, with only head emergent, appearing snakelike. **VOICE:** Occasional grunts and croaks. **SIMILAR SPECIES:** Soaring Double-crested Cormorant can show its tail slightly splayed, recalling Anhinga, but the cormorant's neck is shorter and thicker and tail is shorter. **RANGE:** Rare to casual vagrant well north of range. **HABITAT:** Cypress swamps, rivers, wooded ponds.

CORMORANTS

ANHINGA

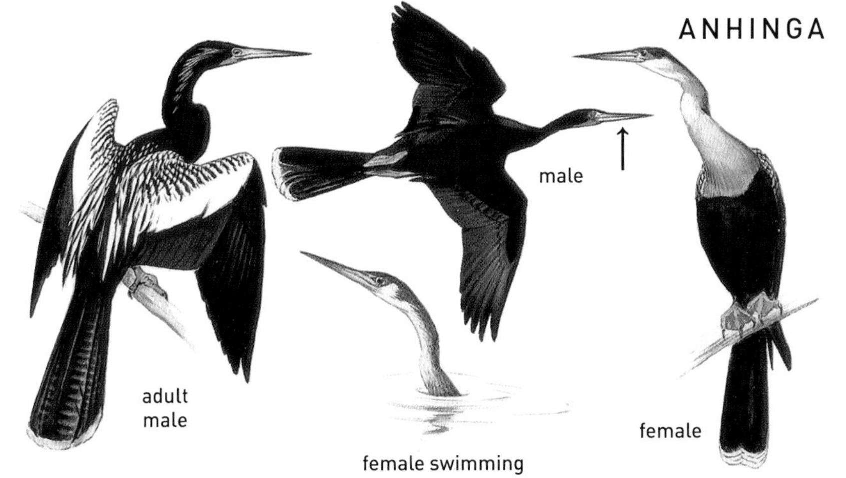

LOONS Family Gaviidae

Large, long-bodied, with daggerlike bills. Airborne, loons are slower and more hunchbacked than most ducks. Sexes alike. Juvenile and first-winter loons are more scaly above than winter adults. **FOOD:** Small fish, crustaceans, other aquatic life. **RANGE:** Northern parts of N. Hemisphere.

RED-THROATED LOON *Gavia stellata* **Common**

25 in. (64 cm). Slimmer head and neck than other loons, and note thin, slightly *upturned bill, often uptilted head.* Flies with neck drooped. *Spring/summer adult:* Plain brown back, gray head, *rufous throat patch. Fall/winter adult and second-year:* Back paler, *spotted white;* extensive white on neck and face includes eye. *Juvenile and first-year:* Face and neck smudgier. **VOICE:** When flying, a repeated *kwuk.* Guttural ptarmigan-like calls and wails on breeding grounds. **SIMILAR SPECIES:** Other loons, Western and Clark's Grebes. **RANGE:** Rare inland. **HABITAT:** Nearshore ocean, bays, estuaries; in summer, tundra lakes.

PACIFIC LOON *Gavia pacifica* **Rare**

25–26 in. (64–66 cm). Smaller than Common Loon, with slightly thinner straight bill. *Spring/summer adult: Pale gray nape;* black throat and foreneck. Back divided into four checkered patches. *Fall/winter adult and second-year:* Note sharp, straight separation of dark and white on neck. Dark feathering around eye. **VOICE:** On breeding gorunds, deep, barking *kwow;* falsetto wails; otherwise silent. **SIMILAR SPECIES:** Winter Red-throated Loon shows more white in face. Face of Common Loon smudgier. **RANGE:** Very rare inland. **HABITAT:** Ocean, large coastal bays; in summer, tundra lakes and sloughs.

ARCTIC LOON *Gavia arctica* **Accidental vagrant**

27–28 in. (69–73 cm). A bit larger than Pacific Loon, with more angular head, larger bill, and whiter sides and rear-flank patches. Juvenile and fall/winter adult with white flanks. **SIMILAR SPECIES:** Red-throated Loon also may have white flanks. **RANGE:** Vagrant along Atlantic Coast in winter. **HABITAT:** Same as Pacific Loon.

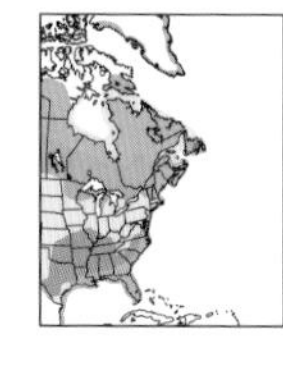

COMMON LOON *Gavia immer* **Common**

31–32 in. (78–81 cm). Large, long-bodied, low-swimming; bill *stout,* daggerlike. In flight shows large, trailing feet. *Spring/summer adult:* Blackish head and bill. Uniformly *checkered back,* broken white necklace. *Fall/winter adult and second-year:* Note *irregular or broken (half-collared) neck pattern. Pale partial eye-ring.* **VOICE:** In breeding locations, falsetto wails, weird yodeling, maniacal quavering laughter; at night, a tremulous *ha-oo-oo.* In flight, a barking *kwuk.* Usually silent in fall and winter. **SIMILAR SPECIES:** Other loons, first-year cormorants. **HABITAT:** In summer, lakes, tundra ponds; in winter, larger lakes, bays, ocean.

YELLOW-BILLED LOON *Gavia adamsii* **Casual vagrant**

34–35 in. (86–89 cm). Similar to Common Loon, but bill *pale ivory* (sometimes with darker base), appears *yellowish* in summer, and slightly uptilted: straight above, angled below. In fall/winter, *paler* and with browner head and neck than Common, usually with small *dark ear patch.* **SIMILAR SPECIES:** Bill of fall/winter Common Loon can be pale, but culmen (upper ridge) is *dark to tip* versus pale in Yellow-billed. **RANGE:** Casual winter vagrant to Great Lakes area; accidental to East Coast. **HABITAT:** Same as Common Loon.

LOONS
first fall/
winter
winter adult
summer
adult
RED-THROATED
LOON
winter adult
first fall/
winter
PACIFIC LOON
summer adult
first fall/
winter
ARCTIC
LOON
summer
adult
winter adult
COMMON
LOON
first fall/
winter
summer adult
winter adult
first fall/winter
summer
adult
YELLOW-BILLED LOON

GREBES Family Podicipedidae

Somewhat ducklike divers with lobed toes, thin necks and bills; tail-less. Sexes alike. Flight labored. **FOOD:** Small fish, other aquatic life. **RANGE:** Worldwide.

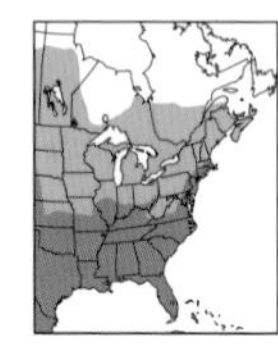

PIED-BILLED GREBE *Podilymbus podiceps* **Fairly common**

13–13½ in. (33–34 cm). Note "chickenlike" bill, puffy white undertail. No wing patch. *Spring/summer: Black throat patch* and *ring* around pale bill. *Fall/winter:* Lacks black bill markings. *Juvenile:* Striped on head. Male's bill thicker than female's. **VOICE:** Song *kuk-kuk-cow-cow-cow-cowp-cowp-cowp;* also a sizzling whinny and sharp *kwah.* **HABITAT:** Ponds, lakes, marshes; in winter, also salt bays and estuaries.

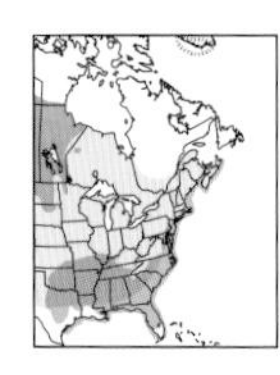

HORNED GREBE *Podiceps auritus* **Fairly common**

13½–14 in. (34–36 cm). *Spring/summer: Golden ear patch* and *chestnut neck. Fall/winter:* Black cap *clean-cut to eye level;* white foreneck, thin straight bill. **VOICE:** A loud *gamp,* trills on breeding grounds; silent otherwise. **SIMILAR SPECIES:** Birds in transitional plumages can have dusky neck and may be confused with Eared Grebe, but note flatter crown, pale lores, and straighter, pale-tipped bill. Red-necked Grebe larger, bill with yellow. **HABITAT:** Lakes, ponds, coastal waters.

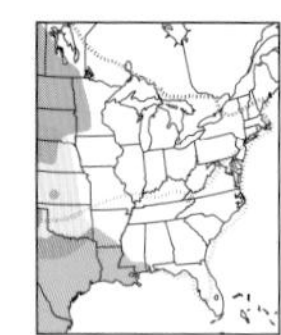

EARED GREBE *Podiceps nigricollis* **Scarce**

12½–13 in. (32–33 cm). Note peaked crown, skinny neck, upturned bill. Floats high in water. *Spring/summer: Wispy golden ear tufts, black neck. Fall/winter:* Dark cap extends *below eye,* face and neck often dusky. **VOICE:** Musical *poo-ee-chk* and a froglike *poo-eep* or *krreep.* **SIMILAR SPECIES:** Horned Grebe. **RANGE:** Rare vagrant to East Coast. **HABITAT:** Prairie lakes, ponds; in winter, open lakes, coastal estuaries.

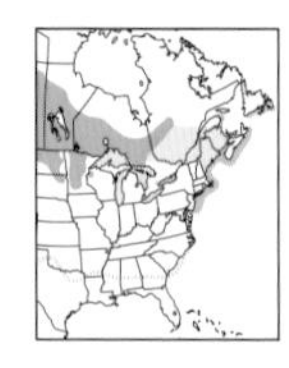

RED-NECKED GREBE *Podiceps grisegena* **Uncommon**

18–19 in. (46–49 cm). A largish grebe. *Spring/summer:* Long *rufous neck, white cheek,* black cap. *Fall/winter:* Grayish brown (including neck); white crescent on face; variable dull *yellowish* base of bill. *First-fall/-winter:* Head pattern less distinct. **VOICE:** A loud braying on breeding grounds. **SIMILAR SPECIES:** Loons, mergansers. **RANGE:** Very rare inland. **HABITAT:** Lakes, ponds; in winter, salt water, estuaries, large lakes.

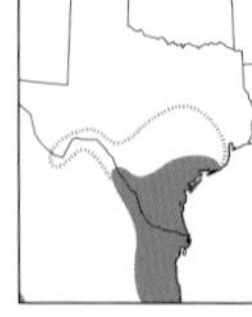

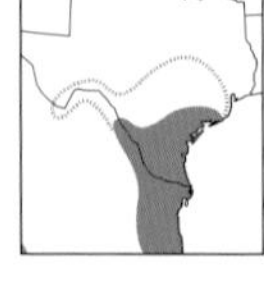

LEAST GREBE *Tachybaptus dominicus* **Uncommon, local**

9½ in. (24 cm). Smaller, darker than Pied-billed Grebe, with white wing patches (usually concealed), puffy undertail, slender *black bill, golden eyes.* **VOICE:** A chattering whinny. **HABITAT:** Ponds, marshes, lake edges.

WESTERN GREBE *Aechmophorus occidentalis* **Uncommon, local**

25 in. (64 cm). A large grebe with long neck. Bill long, greenish yellow with dark ridge. Black of cap extends *below eye.* **VOICE:** A loud, reedy *crik-crick,* often heard year-round. **SIMILAR SPECIES:** Clark's Grebe, Red-throated Loon. **RANGE:** Rare vagrant to Midwest, casual to East Coast. **HABITAT:** Rushy lakes, sloughs; in winter, large lakes, bays, coasts.

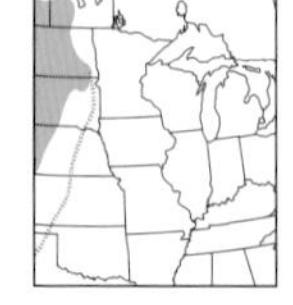

CLARK'S GREBE *Aechmophorus clarkii* **Scarce, local**

25 in. (64 cm). Similar to Western Grebe, but bill *orange-yellow.* Dark eye *surrounded by white* (may be pale gray in first-year and winter plumages). Slightly paler than Western with narrower stripe on back of neck. **VOICE:** A single-noted *creet* or *criik.* **RANGE:** Very rare vagrant to Midwest, accidental to East Coast. **HABITAT:** Similar to Western Grebe.

GREBES
PIED-BILLED GREBE
lobed foot of grebe
fall/winter
spring/ summer
juvenile
spring/ summer
downy young
fall/winter variant
bill of Horned
fall/winter
spring/ summer
HORNED GREBE
fall/winter variant
fall/ winter
spring/ summer
bill of Eared
EARED GREBE
first fall/winter
fall/ winter adult
spring/ summer adult
RED-NECKED GREBE
fall/winter
spring/ summer
LEAST GREBE
male display
WESTERN GREBE
CLARK'S GREBE

AUKS, MURRES, and PUFFINS Family Alcidae

The northern counterparts of penguins, but alcids are smaller and can fly, beating their small narrow wings in a whir, often veering. They have short necks and pointed, stubby, or deep and laterally compressed bills. Alcids swim and dive expertly. Most species nest on sea cliffs or in burrows, often in crowded colonies, and virtually all winter on open ocean. Mostly silent away from breeding grounds. Sexes alike. **FOOD:** Fish, squid, krill, zooplankton. **RANGE:** N. Atlantic, N. Pacific, and Arctic Oceans. Great Auk (*Pinguinus impennis*) formerly bred on rocky islets of ne. Canada and N. Atlantic; became extinct around the late 1840s.

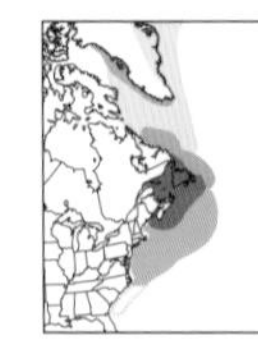

RAZORBILL *Alca torda* Uncommon

17 in. (43 cm). Size of a small duck. *Adult:* Black above, white below; characterized by rather heavy head, thick neck, and flat bill crossed midway by a white mark; male's bill deeper than female's. On water, cocked-up pointed tail is often characteristic. Complete black head in spring/summer replaced by white face and throat in fall/winter. *First-year:* Shows smaller bill, retains white face through first spring/summer; bill develops to adult-sized in second year. **VOICE:** Deep guttural growls; juvenile gives piercing whistle. **SIMILAR SPECIES:** Bill of first-year Razorbill may recall that of a murre but is stubbier and more rounded. See also Long-tailed Duck. **HABITAT:** Nests on rocky offshore islands; forages in coastal waters; winters primarily in open ocean.

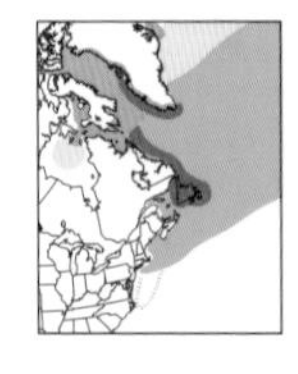

THICK-BILLED MURRE *Uria lomvia* Scarce

18 in. (46 cm). Similar to Common Murre but a bit *blacker above.* Bill slightly shorter, thicker, with *whitish line along gape.* Overall a bit stockier. *Spring/summer adult:* Head and face black, white of foreneck forms inverted V. *Fall/winter adult and first-year:* Face whitish with dark on head extending *well below eye;* no dark line through white ear coverts as in most Common Murres. White bill mark often less evident. Bill also much smaller, shorter, during first year. **VOICE:** Guttural calls and moans, hence the name "murre." Juvenile gives loud whistles. **SIMILAR SPECIES:** Common Murre. **HABITAT:** Nests on coastal cliff ledges. Spends fall/winter season on offshore ocean waters.

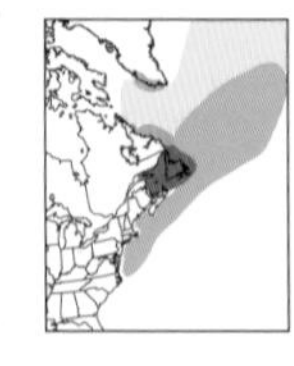

COMMON MURRE *Uria aalge* Uncommon

17–17½ in. (43–45 cm). *Spring/summer adult:* Head, neck, back, and wings dark, *tinged brownish;* underparts, underwing linings, and line on rear edge of wing white. *Dusky markings on flanks* on some birds. *Fall/winter adult and first-year:* Similar, but throat and cheeks white. *Black mark extends from eye to cheek* in most birds. Murres often raft on water, fly in lines, stand erect on sea cliffs. Bridled morph occurs regularly in N. Atlantic. **VOICE:** Similar to Thick-billed Murre. **SIMILAR SPECIES:** Thick-billed Murre, Razorbill, Long-tailed Duck. **HABITAT:** Same as Thick-billed Murre.

ALCIDS (AUKS)
first-year
fall/winter
adult
spring/summer
adult
RAZORBILL
fall/winter
THICK-BILLED
MURRE
COMMON
spring/
summer
adults
chick
COMMON
MURRE
fall/winter
COMMON
THICK-BILLED
bridled
morph
RAZORBILL
spring/
summer
adults
Great Auk
extinct
1844

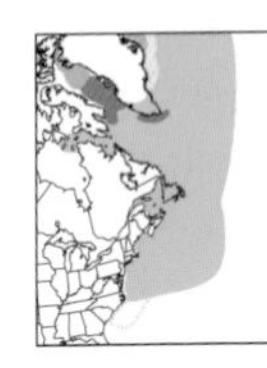

DOVEKIE *Alle alle* Scarce

8–8¼ in. (20–21 cm). By far the smallest alcid in East. Chubby and seemingly neckless, with very stubby bill. In flight, flocks bunch tightly. *Adult:* Black above, white below; black-hooded in spring/summer, white-chested in fall/winter. *First-year:* Similar to fall/winter adult but bill smaller; primaries browner. **VOICE:** A shrill chatter. Noisy on nesting grounds. **SIMILAR SPECIES:** Much smaller and shorter-billed than murres and Razorbill. **RANGE AND HABITAT:** Nests in high Arctic on coastal cliffs. Winters at sea in N. Atlantic.

ANCIENT MURRELET *Synthliboramphus antiquus* Scarce

10 in. (25 cm). In all plumages, *gray back contrasts with black cap.* Bill yellow. **RANGE AND HABITAT:** Casual visitor from AK and BC to lakes, reservoirs, and rivers far inland all the way to Atlantic Coast.

LONG-BILLED MURRELET *Brachyramphus perdix* Accidental vagrant

10–11 in. (25–28 cm). A small alcid with slim bill. Mottled brown with white to whitish throat and pale eye-arcs in spring/summer; blackish above and white below with mottled whitish bar on scapulars in fall/winter and first-summer. **SIMILAR SPECIES:** Fall/winter Dovekie has white hindneck, stubbier bill. **RANGE AND HABITAT:** Casual vagrant from Asia to lakes, reservoirs, and rivers far inland all the way to Atlantic Coast.

BLACK GUILLEMOT *Cepphus grylle* Fairly common

12½–13½ in. (32–34 cm). *Spring/summer adult:* A midsized black alcid with large white wing patch, bright red feet, and pointed bill. Inside of mouth red. *Fall/winter adult:* Pale with whitish underparts and barred back; white wing patch as in summer. *First-year:* Wing patch dingier, mottled. **VOICE:** A wheezy or hissing *peeee;* very high pitched. **SIMILAR SPECIES:** No other Atlantic alcid has white wing patch. White-winged Scoter much larger and chunkier. **VOICE:** A piercing *pseeeei,* other thin notes on breeding grounds. **RANGE:** Accidental vagrant inland. **HABITAT:** Inshore ocean waters; breeds in small groups or singly in holes in ground or under rocks on rocky shores, islands. Less pelagic than other Atlantic alcids.

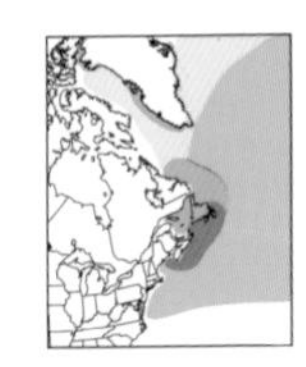

ATLANTIC PUFFIN *Fratercula arctica* Uncommon

12–13 in. (30–33 cm). Colorful triangular bill is most striking feature of this chunky "sea parrot." On the wing, it is a stubby, short-necked, thick-headed bird with buzzy flight. No white border on wing. *Spring/summer adult:* Upperparts black, underbody white, cheeks pale gray; triangular bill bluish and yellow, broadly tipped red. Feet bright orange. *Fall/winter adult:* Cheeks darker gray; bill smaller (summer bill-shield sheds), duller, but still obviously a puffin. *First-year:* Bill much smaller, mostly dark. **VOICE:** Usually silent. When nesting, a low, growling *ow* or *arr.* **SIMILAR SPECIES:** First-years may be mistaken for first-year Razorbill, but note gray cheeks, all-dark underwing. Horned Puffin of Pacific very similar in winter but bill shape differs; malar area, throat, and neck have broader black band; ranges do not currently overlap, but vagrants to opposite coasts might be expected with melting polar cap. **HABITAT:** Very rarely seen from shore except near breeding colonies on rocky islands.

ATLANTIC ALCIDS (AUKS)
DOVEKIE
spring/
summer
adult
fall/
winter
fall/
winter
spring/summer adult
ANCIENT MURRELET
spring/
summer
adult
fall/
winter
spring/
summer
adults
BLACK GUILLEMOT
spring/
summer
adults
LONG-BILLED
MURRELET
fall/winter
fall/
winter
spring/
summer
adult
spring/summer
adults
ATLANTIC
PUFFIN
juvenile
fall/winter
adult
Black
Guillemot
spring/
summer
Dovekie
spring/
summer
Atlantic
Puffin

SHEARWATERS and PETRELS Family Procellariidae

Birds of open sea that bank, or arc, up and down like a roller coaster in strong winds; in calm weather, sit on water or fly with several flaps and then a glide. Have tubelike external nostrils on bill and are thus called "tubenoses." Ages and sexes similar. **FOOD:** Fish, squid, crustaceans, ship refuse. **RANGE:** Oceans of world.

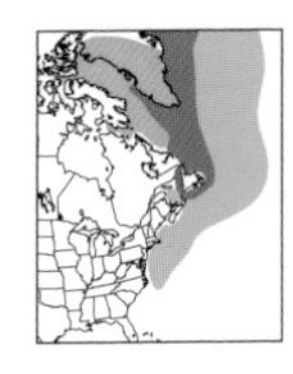

NORTHERN FULMAR *Fulmarus glacialis* **Uncommon to fairly common**
18½–19 in. (47–49 cm). Stockier and with larger head than shearwaters and shorter, rounder wings; flies more horizontally, with quicker wingbeats, less gliding. Forehead rounded; *stubby, yellowish/pinkish, tubenose bill with variable dark band; pale flash or patch* in primaries. Polymorphic. *Light morph:* Gull-like in plumage, some whiter; white wing patches distinct. *Dark morph* (less common in Atlantic): Uniformly smoky gray, reduced whitish wing patches. **VOICE:** A hoarse, grunting *ag-ag-ag-arrr* or *ek-ek-ek-ek-ek.* **SIMILAR SPECIES:** More rounded posture, flight style, and bill distinguish fulmars from gulls and browner (less gray) Sooty Shearwater. **HABITAT:** Open ocean; breeds colonially on sea cliffs.

TRINDADE PETREL *Pterodroma arminjoniana* **Rare**
15½–16 in. (40–41 cm). Known as "Herald Petrel" until recently split from that counterpart Pacific species. Has dark, intermediate, and light morphs. Most in N. America are dark, differing from Sooty Shearwater by *dark* underwing linings, longer tail, slower wingbeat, and jaegerlike pale area at base of primaries. Light morph has dark head, white breast and belly, and prominent white wing patches. Feet and legs black. **RANGE:** Annual in small numbers in Gulf Stream off NC coast May–Sept. Accidental vagrant inland after hurricanes.

BERMUDA PETREL *Pterodroma cahow* **Casual visitor, endangered**
15 in. (38 cm). Also known as "Cahow." One of the world's rarest seabirds. Differs from Black-capped Petrel by *smudgy gray* rump, absence of white collar, smaller size and bill. **RANGE:** Breeds only on islets off Bermuda. Sightings may become more regular in Gulf Stream off NC coast as protection efforts in Bermuda enhance population size.

FEA'S PETREL *Pterodroma feae* **Very rare**
14–15 in. (36–38 cm). Brownish gray above, with M pattern across upperwings. Distinguished from Black-capped Petrel by less contrasty pale *gray rump and tail,* pale gray cowl on head, and *dark underwing.* **RANGE:** Breeds on islands off W. Africa. Rare but regular spring and summer visitor to Gulf Stream off Cape Hatteras, NC; casual elsewhere.

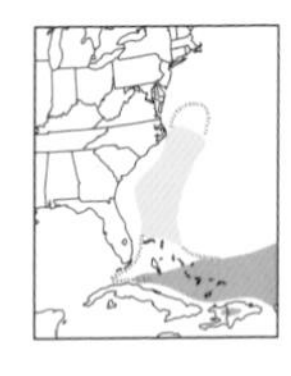

BLACK-CAPPED PETREL *Pterodroma hasitata* **Scarce, endangered**
16 in. (41 cm). Larger than Audubon's and Manx Shearwaters; looks quite similar to Great Shearwater but has thicker bill and characteristic crook-winged shape and flight style of petrels. Note also black cap, white forehead, variable white collar, white rump patch extending to tail. **SIMILAR SPECIES:** See Bermuda Petrel. **RANGE:** Nests on Hispaniola and Cuba. Rarely seen outside Gulf Stream. Casual vagrant inland after hurricanes.

FULMAR AND
ATLANTIC PETRELS
NORTHERN
FULMAR
tubed bill of
fulmar
dark
morph
light
morphs
BERMUDA
PETREL
dark
morph
BLACK-
CAPPED
PETREL
TRINDADE
PETREL
light
morph
FEA'S
PETREL

CORY'S SHEARWATER *Calonectris diomedea* Fairly common

18–20 in. (46–51 cm). A large, pale shearwater; gray-brown head *blends* into white of throat; bill relatively thick, dull *yellow.* Belly all white; rump usually dark with indistinct or no white. **SIMILAR SPECIES:** Great Shearwater has dark cap, black bill, white rump, and dark smudges on belly and underwing. Cory's has more pronounced bend to wing than Great, and wingbeat tends to be slightly slower. Cape Verde Shearwater (*C. edwardsii;* not shown), formerly considered a subspecies of Cory's, is smaller, darker above, and has thinner dark grayish bill with black tip. Casual off Atlantic Coast but should be looked for among Cory's. "Scopoli's" Shearwater of Mediterranean Sea (currently subspecies *C. d. diomedea*), rare to uncommon off cen. Atlantic Coast in summer and fall, could be split and is smaller, smaller billed, slightly paler, and shows more white in undersides of primaries than the more common Atlantic subspecies (*C. d. borealis*) of Cory's that occurs off our coast.

GREAT SHEARWATER *Ardenna gravis* Fairly common

19 in. (48 cm). A large shearwater, dark above and white below, rising above waves on stiff wings off our Atlantic Coast, is likely to be this or Cory's Shearwater. Great Shearwater has dark cap separated by a light band across nape; note also white rump patch and dark smudges on belly and underwing. **SIMILAR SPECIES:** Cory's Shearwater. **RANGE:** Casual vagrant inland after hurricanes.

SOOTY SHEARWATER *Ardenna grisea* Uncommon

17–18 in. (43–46 cm). Appears all dark at a distance; rises over and arcs above waves on narrow, rigid wings. Note *whitish linings* on underwings. Flight rapid and direct, usually low along water surface, birds often following each other in long pathways during migration. **SIMILAR SPECIES:** Dark jaegers (white in primaries, fly differently), dark-morph Northern Fulmar. Short-tailed Shearwater (*Puffinus tenuirostris;* not shown), an accidental to casual vagrant in Atlantic, is very similar to Sooty but smaller, smaller billed, flies more rapidly, and shows darker underwing linings. **RANGE:** Breeds in S. Hemisphere. Undertakes extensive migrations into Atlantic, where it molts.

MANX SHEARWATER *Puffinus puffinus* Uncommon

13½ in. (34 cm). A small black-and-white shearwater; half the bulk of Great Shearwater; shows *completely white undertail coverts* and can have white flank patches on either side of rump. Note dark cap extends below eye; *white extends upward from neck behind ear coverts.* **SIMILAR SPECIES:** See Audubon's Shearwater. Wingbeats quicker than in Great or Cory's Shearwater.

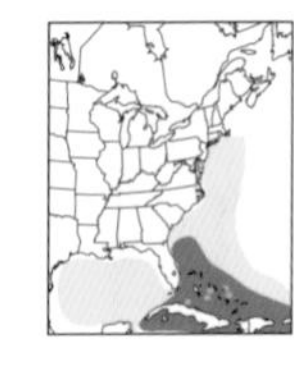

AUDUBON'S SHEARWATER *Puffinus lherminieri* Fairly common

12 in. (30 cm). A very small shearwater, similar to Manx Shearwater but with slightly browner upperparts, *dark undertail.* Wings slightly shorter, *tail longer.* Often has *white markings* around eye. **HABITAT:** Prefers warmer water than Manx Shearwater.

ATLANTIC SHEARWATERS

STORM-PETRELS
Families Oceanitidae and Hydrobatidae

Small seabirds that flutter or bound over open ocean. Nest colonially on islands, returning to burrows at night. Nostrils in a fused tube over top of bill (see p. 60). Usually silent at sea, calling occasionally at feeding frenzies; vocal at breeding colonies. Family Oceanitidae (White-faced and Wilson's Storm-Petrels here) recently split from Hydrobatidae; the former have longer legs used to kick or patter on water while foraging. **FOOD:** Plankton, crustaceans, small fish. **RANGE:** All oceans except Arctic.

WHITE-FACED STORM-PETREL *Pelagodroma marina* **Casual visitor**

7½ in. (19 cm). A medium-large storm-petrel with white head and underparts, two-toned underwing, dark crown and eye patch. Very long legs. When feeding, bounds "kangaroo style" over water on stiff, flat wings. **RANGE:** Se. Atlantic. Casual but probably annual visitor Aug.–Sept. off Atlantic Coast from MA to NC, usually far offshore.

WILSON'S STORM-PETREL *Oceanites oceanicus* **Common**

7¼–7½ in. (18–19 cm). A medium-small storm-petrel with somewhat triangular wings and *white uppertail-covert (often called "rump") patch that wraps around sides;* tail slightly rounded or square-cut, *not forked.* Feet yellow-webbed (hard to see), show *beyond tail* in flight. Direct flight, with short glides, pausing to flutter over water. **SIMILAR SPECIES:** Leach's and Band-rumped Storm-Petrels. **RANGE:** Casual vagrant inland after hurricanes. **HABITAT:** Open ocean. Often follows ships (Leach's does not). May rarely be seen from shore.

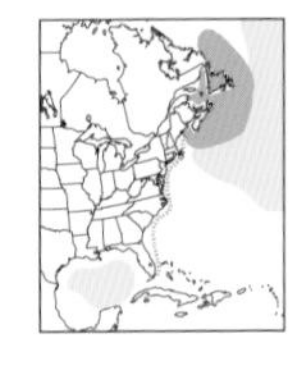

LEACH'S STORM-PETREL *Oceanodroma leucorhoa* **Uncommon**

8 in. (20 cm). Note obscurely divided (double-oval) *white uppertail-covert patch* and moderately forked tail. Pale bar on upperwing often reaches leading edge. In flight, bounds about erratically on fairly long angled wings, suggesting a nighthawk. Breeds in N. Atlantic. *Does not consistently follow ships.* **VOICE:** At night on breeding grounds, nasal chattering notes and long crooning trills. **SIMILAR SPECIES:** Wilson's and Band-rumped Storm-Petrels. Swinhoe's Storm-Petrel (*O. monorhis;* not shown), an accidental vagrant off Atlantic Coast, is similar to Leach's but lacks white in rump; beware also that dark-rumped Leach's occur in Pacific and may occasionally occur in Atlantic as well. **RANGE:** Casual vagrant inland after hurricanes.

BAND-RUMPED STORM-PETREL *Oceanodroma castro* **Scarce**

8½–9 in. (21–23 cm). A "white-rumped" storm-petrel, larger than Wilson's, similar to Leach's. Feet do not project beyond *squarish* tail. Pale bar on upperwing usually much less distinct than in Leach's or Wilson's; *uppertail-covert band more clean-cut than Leach's, less extensive undertail than Wilson's;* bases of outer rectrices white. A stiff-winged flier, with short glides, reminiscent of a shearwater. **RANGE:** Casual vagrant inland after hurricanes.

EUROPEAN STORM-PETREL *Hydrobates pelagicus* **Casual visitor**

6 in. (15 cm). Smaller than Wilson's Storm-Petrel; shorter legs, which do not extend beyond square tail. Yellow on feet, not on webs. Has *whitish underwing* patch. **RANGE:** Nests in ne. Atlantic and Mediterranean. Casual off NC, NS.

WHITE-RUMPED STORM-PETRELS

Leach's

Wilson's

Band-rumped

European

ALBATROSSES Family Diomedeidae

Majestic birds of open ocean, with rigid gliding and banking flight. Much larger than gulls; wings proportionally longer. "Tubenosed" (nostrils in two tubes); bill large, hooked, covered with horny plates. Sexes generally alike. Largely silent at sea. **FOOD:** Cuttlefish, fish, squid, other small marine life; some feeding at night. **RANGE:** Mainly cold oceans of S. Hemisphere.

BLACK-BROWED ALBATROSS Accidental vagrant
Thalassarche melanophris

34–35 in. (86–88 cm); wingspan 7½ ft. (229 cm). Suggests a huge Great Black-backed Gull, but with short blackish tail and very large yellow bill (adult) with hooked tip. Dark eye streak gives it a frowning look. In stiff-winged gliding flight, shows white underwing *broadly outlined* with black. *First-year:* Bill dark. **RANGE:** Accidental off Atlantic Coast.

YELLOW-NOSED ALBATROSS Casual vagrant
Thalassarche chlororhynchos

31–32 in. (79–81 cm); wingspan 7–7½ ft. (213–229 cm). Similar to Black-browed Albatross, but bill *black with yellow ridge* on upper mandible. In flight, underwing whiter, with *narrower* black edging. **RANGE:** Accidental along Atlantic and Gulf Coasts; occasionally seen on shore or up rivers.

TROPICBIRDS Family Phaethontidae

These seabirds resemble (but are unrelated to) large terns with two greatly elongated central tail feathers (adults) and stouter, slightly decurved bills. Fly with shallow wingbeats, rarely glide, dive headfirst, and swim with tail held clear of water. Sexes alike. Largely silent at sea. **FOOD:** Squid, fish, crustaceans. **RANGE:** Tropical oceans.

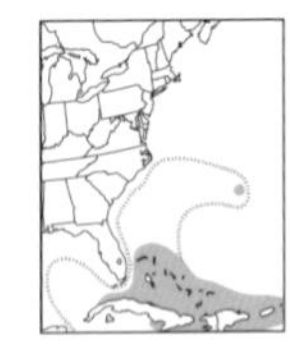

WHITE-TAILED TROPICBIRD *Phaethon lepturus* Rare

15 in. (38 cm), adults to 30 in. (76 cm) with tail-streamers. *Adult:* Distinguished from other tropicbirds by *diagonal black bar* across each wing. Note two extremely long white central tail feathers. Bill orange-red. *Juvenile:* Lacks tail-streamers; has *white,* not black, primary coverts, *coarsely* barred with black above; bill grayish, becoming yellow in first year. **VOICE:** A harsh, ternlike scream. Also *tik-et, tik-et.* **SIMILAR SPECIES:** Red-billed Tropicbird.

RED-BILLED TROPICBIRD *Phaethon aethereus* Rare

18 in. (45 cm), adults to 37 in. (94 cm) with tail-streamers. *Adult:* Slender, white, with *two extremely long white central tail feathers, heavy red bill,* black patch through cheek, black primaries *and primary coverts,* and *finely barred back. Juvenile:* Lacks long tail, has coarser bars on back, grayish-yellow to orangish bill. **SIMILAR SPECIES:** White-tailed Tropicbird. Red-billed slightly larger and larger-billed; has more *finely barred* back than juvenile White-tailed, bright red to slightly orange (not yellow) bill, more black on wing, including *primary coverts.*

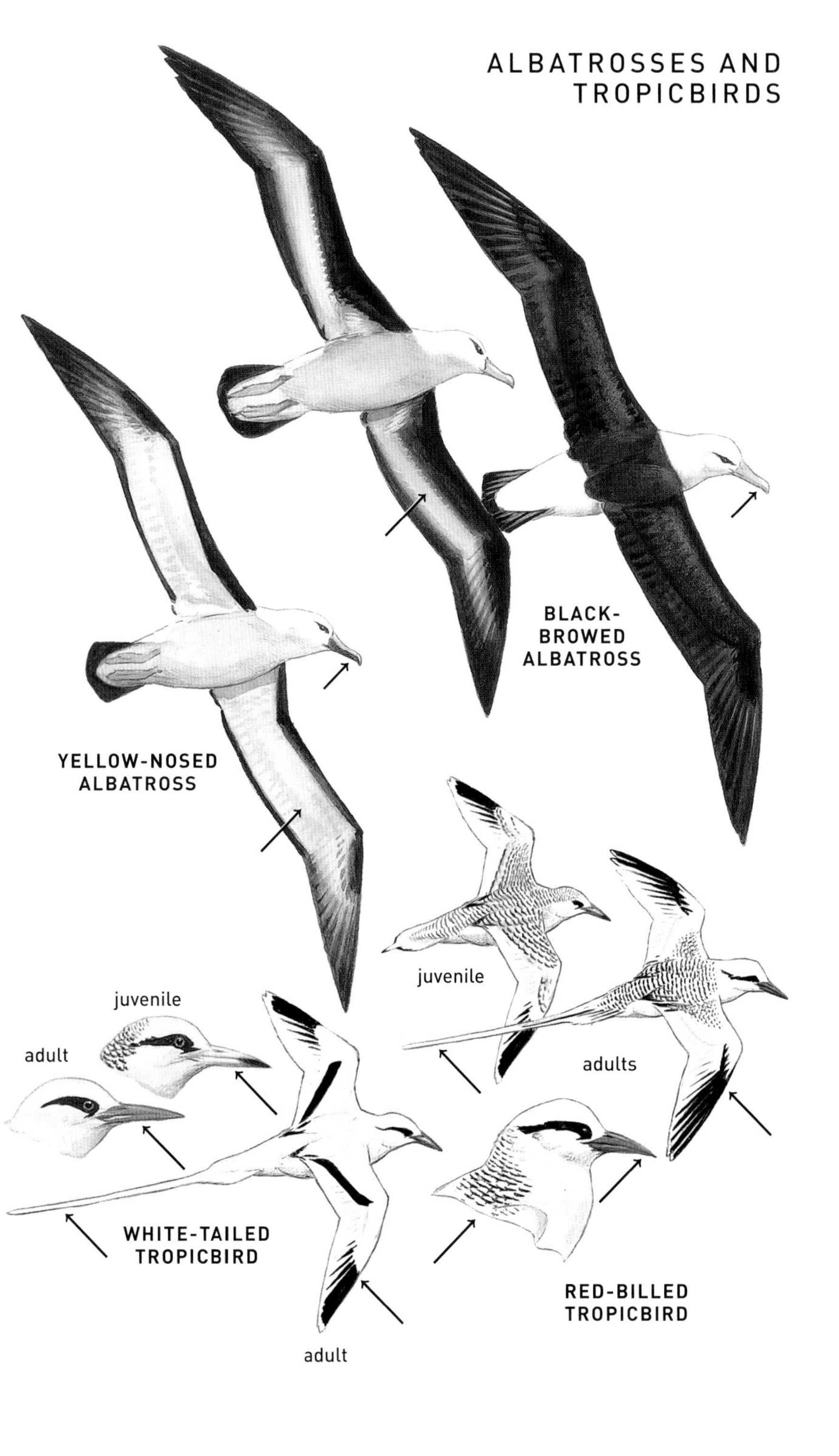
ALBATROSSES AND
TROPICBIRDS
BLACK-
BROWED
ALBATROSS
YELLOW-NOSED
ALBATROSS
juvenile
juvenile
adult
adults
WHITE-TAILED
TROPICBIRD
RED-BILLED
TROPICBIRD
adult

GANNETS and BOOBIES Family Sulidae

Larger and longer necked than most gulls; pointed at all four ends. Sexes largely alike. Boobies sit on buoys, rocks; fish by plunging from air like Brown Pelicans. **FOOD:** Fish, squid. **RANGE:** Gannets live in cold seas (N. Atlantic, S. Africa, Australia), boobies in tropical seas. All nest colonially on islands.

NORTHERN GANNET *Morus bassanus* — Common

37–38 in. (94–97 cm). Goose-sized. Soars over ocean and plunges headlong for fish. Migrates in long lines. Much larger than Herring Gull, with pointed tail, longer neck, larger bill (often pointed toward water). *Adult:* White with extensive black primaries. *Juvenile:* Dusky, but note "pointed-at-both-ends" shape. *Second- and third-year:* Look piebald in transition from juvenile to adult. **VOICE:** In colony, a low barking *arrah.* **SIMILAR SPECIES:** Boobies. In windy conditions, flying gannets may bank, suggesting an albatross. **RANGE:** Scarce winter vagrant to Great Lakes; accidental inland. **HABITAT:** Ocean, but seen regularly from shore. Breeds colonially on sea cliffs.

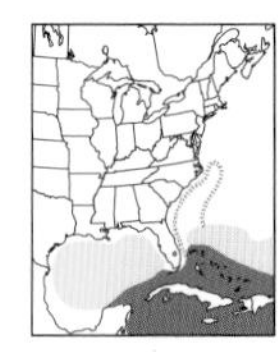

MASKED BOOBY *Sula dactylatra* — Scarce, local

31–32 in. (79–81 cm). *Adult:* White; smaller than Northern Gannet, with *black tail,* black along *entire rear edge* of wing, and black in *face.* Yellowish bill; dark bluish facial skin; *feet dark olive to slate.* Mostly white underwing. *Juvenile:* Variably mottled with dark on upperwing and head, *white hindcollar.* **VOICE:** Usually silent at sea. **SIMILAR SPECIES:** Other boobies, first-year Northern Gannet.

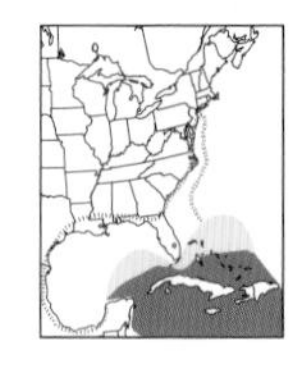

BROWN BOOBY *Sula leucogaster* — Scarce, local

29–30 in. (74–76 cm). *Adult:* Chocolate brown with *white belly in clean-cut contrast* to dark breast. White underwing linings contrast with dark flight feathers. *Feet yellowish. Juvenile:* Underparts mostly dark, with little or no contrast between breast and belly; bill grayish. *Second-year:* White lower breast and belly mottled brown. **SIMILAR SPECIES:** First-year Northern Gannet lacks clean-cut breast contrast; shows some white patches or mottling above; feet dark (not yellowish). First-year Red-footed Booby (which has dark tail) more buffy overall with dark underwing; has blackish bill that becomes tinged pinkish then lilac at base; feet pinkish to pale reddish. First-year Masked Booby resembles adult Brown Booby, but brown of head not sharply demarcated from paler underparts and set off from back by white breast sides and nape collar.

RED-FOOTED BOOBY *Sula sula* — Rare visitor

27–28 in. (69–71 cm). The smallest booby. *Adult:* Feet *bright red,* tail *white.* Polymorphic. *White morph:* Gannetlike; white, with black tip and trailing edge of wing (as in Masked Booby, but tertials white), tail white. *Dark morph:* Brown back and wings, paler head; white tail and belly; in flight, *underwing dark,* thin dark trailing edge on upperwing. *Juvenile:* Brownish overall with *dark underwing,* blackish bill that becomes pink with dark tip by second year, *pink feet* that become red by second year. **SIMILAR SPECIES:** Juvenile Brown Booby is darker than juvenile Red-footed; has a more conical and paler grayish bill, yellowish feet. **RANGE:** Nests in Tropics. Very rare, mostly young birds, at Dry Tortugas, FL; casual elsewhere in FL.

GANNET AND BOOBIES

FRIGATEBIRDS Family Fregatidae

Primarily black tropical seabirds with extremely long wings. Bill long, hooked; tail deeply forked. Normally do not swim. **FOOD:** Fish, jellyfish, squid, chicks of other seabirds. Food snatched from water or ground in flight, scavenged, or pirated from other seabirds. **RANGE:** Pantropical oceans.

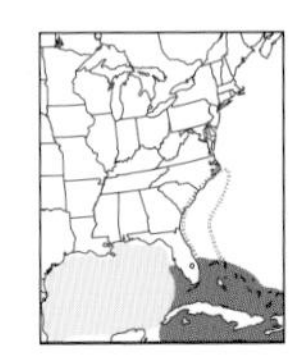

MAGNIFICENT FRIGATEBIRD *Fregata magnificens* **Uncommon, local**

36–46 in. (91–117 cm); wingspan 7–8 ft. (215–245 cm). Large, mostly black, with extremely long angled wings and *scissorlike* tail (often folded in a *point*). Soars with extreme ease. Bill long, hooked; orbital skin bluish. *Male:* All black, with *red throat pouch* (inflated like a balloon in display). *Female:* White breast, dark head. *Juvenile:* Head and breast white. Can take up to 10 years to develop adult plumages. **VOICE:** Voiceless at sea. A gargling whinny during display. **SIMILAR SPECIES:** Lesser Frigatebird (*F. ariel;* not shown), accidental across N. America, is smaller; adult has white spur on axillars; female has red orbital skin; juvenile has russet head. **RANGE:** Accidental to scarce vagrant away from coast. **HABITAT:** Tropical oceans; coastal habitats; breeds in mangroves.

PELICANS Family Pelecanidae

Huge waterbirds with long flat bills and great throat pouches (flat when deflated). Neck long, body robust. Sexes alike. Flocks fly in lines, Vs, or kettles, alternating several flaps with a glide. In flight, head is hunched back on nape, with long bill resting on breast. Pelicans swim buoyantly. **FOOD:** Mainly fish, crustaceans. **RANGE:** N. and S. America, Africa, s. Eurasia, E. Indies, Australia.

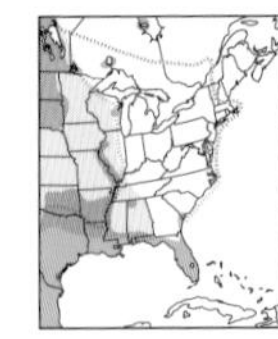

AMERICAN WHITE PELICAN **Fairly common**
Pelecanus erythrorhynchos

62 in. (157 cm). Huge; wingspan 8–9½ ft. (244–290 cm). White, with black primaries and a great orange-yellow bill and throat pouch. Adult in breeding condition has keratinous appendage or "centerboard" on ridge of bill that develops in spring, drops off in fall. *First-year:* Dusky wash on head, neck; dark mottling to upperwing coverts; second-year birds intermediate. This pelican does not plunge from air like Brown Pelican but scoops up fish while swimming, often working in groups. Flocks often circle high in air on thermals. **VOICE:** In colony, a low groan. Young utter whining grunts. **SIMILAR SPECIES:** Swans have no black in wings. Wood Stork and Whooping Crane fly with neck and long legs extended. Snow Goose much smaller; noisy. **RANGE:** Rare vagrant to Southeast and Gulf coasts; casual to Northeast. **HABITAT:** Lakes, marshes, estuaries.

BROWN PELICAN *Pelecanus occidentalis* **Common**

48–50 in. (122–127 cm); wingspan 6½ ft. (198 cm). An unmistakable, ponderous dark waterbird. *Adult:* Much white and buff on head and front of neck. Dark chestnut brown on back of neck and reddish throat when breeding. *First-year:* Duskier brown overall, with dark head, paler underparts. *Second-year:* Intermediate. Large size, head and bill shape, and powerful slow flight (a few flaps and a glide) indicate a pelican; dark color and habit of *plunging bill-first* proclaim it as this species. Lines or broken Vs of pelicans glide low over water, wingtips almost touching. **VOICE:** Adults silent (rarely a low croak). Nestlings squeal. **RANGE:** Casual to accidental vagrant inland. **HABITAT:** Salt bays, beaches, ocean; more rarely inland lakes. Perches on posts, piers, rocks, buoys, beaches.

MAGNIFICENT FRIGATEBIRD
FRIGATEBIRD AND PELICANS
adult male display
adult female
adult male
first-years
spring/ summer adult
first- year
adults
fall/winter adult
AMERICAN WHITE PELICAN
spring/summer adults
adult
BROWN PELICAN
juvenile/ first-year

SKUAS and JAEGERS Family Stercorariidae

Falconlike seabirds that harass gulls, terns, and shearwaters, forcing them to disgorge or drop their food. Light, intermediate, and dark morphs exist in at least two species; all have flash of white in primaries. Adult jaegers have two projecting central tail feathers, which differ in shape and length among species and ages. In juveniles and molting birds, these feathers can be shorter or lacking, or sometimes blunter tipped than in adults. Separating jaegers in most plumages can be very difficult. Skuas are larger, powerful birds that lack elongated tail feathers and are broader winged. Sexes alike. **FOOD:** In Arctic, lemmings, eggs, young birds. At sea, food taken from other birds or from water. **RANGE:** Seas worldwide, breeding in subpolar regions. In e. N. America, all five species can occur as rare to accidental vagrants inland.

SOUTH POLAR SKUA *Stercorarius maccormicki* Scarce

21 in. (53 cm). Skuas are near size of a large gull, but stockier, with deep-chested, hunchbacked look. Dark, with short, slightly wedge-shaped tail and *conspicuous white wing patch at base of primaries visible on both upperwing and underwing.* South Polar Skua is slightly slimmer in build and bill, is colder and grayer brown, and averages paler nape than Great Skua. *Adult:* Has *pale head and underparts* contrasting with darker wings; older adults can be much paler than shown. *Juvenile and first-year:* Darker and more uniform, similar to but often with a paler nape than like-aged Great Skuas. **SIMILAR SPECIES:** Great Skua. Dark jaegers (particularly Pomarine Jaeger) may lack elongated tail feathers, but skuas larger, their wings wider, and they have more striking white wing patches. **HABITAT:** In our area, open ocean; rarely seen from or close to shore.

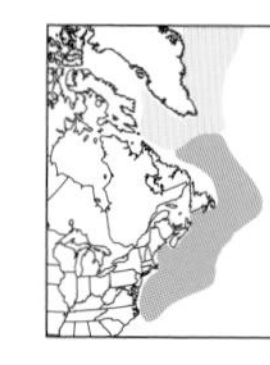

GREAT SKUA *Stercorarius skua* Scarce

22–23 in. (56–58 cm). Note conspicuous white wing patch visible on both upper- and underwing. Near size of large gull, but stockier. Flight strong and swift; harasses other seabirds. *Adult:* Dark brown with rusty and streaked upperparts and short, slightly wedge-shaped tail. *Juvenile and first-year:* Darker brown, less rusty, and with fewer streaks; found only at sea post-fledging. **SIMILAR SPECIES:** Dark jaegers may lack distinctive tail-feather extensions. However, skuas' wings wider, less falconlike, white wing patches more striking both above and below, and flight more powerful. Very much like South Polar Skua but averages larger and heavier-billed. Note *warmer brown color, dark cap, often less distinct pale nape,* and more *streaked appearance* to upperparts. **VOICE:** A soft, nasal *kare, kare* on breeding grounds. **HABITAT:** Rocky islands in subarctic regions for breeding; otherwise, open ocean, seldom close to shore.

SKUAS AND JAEGERS

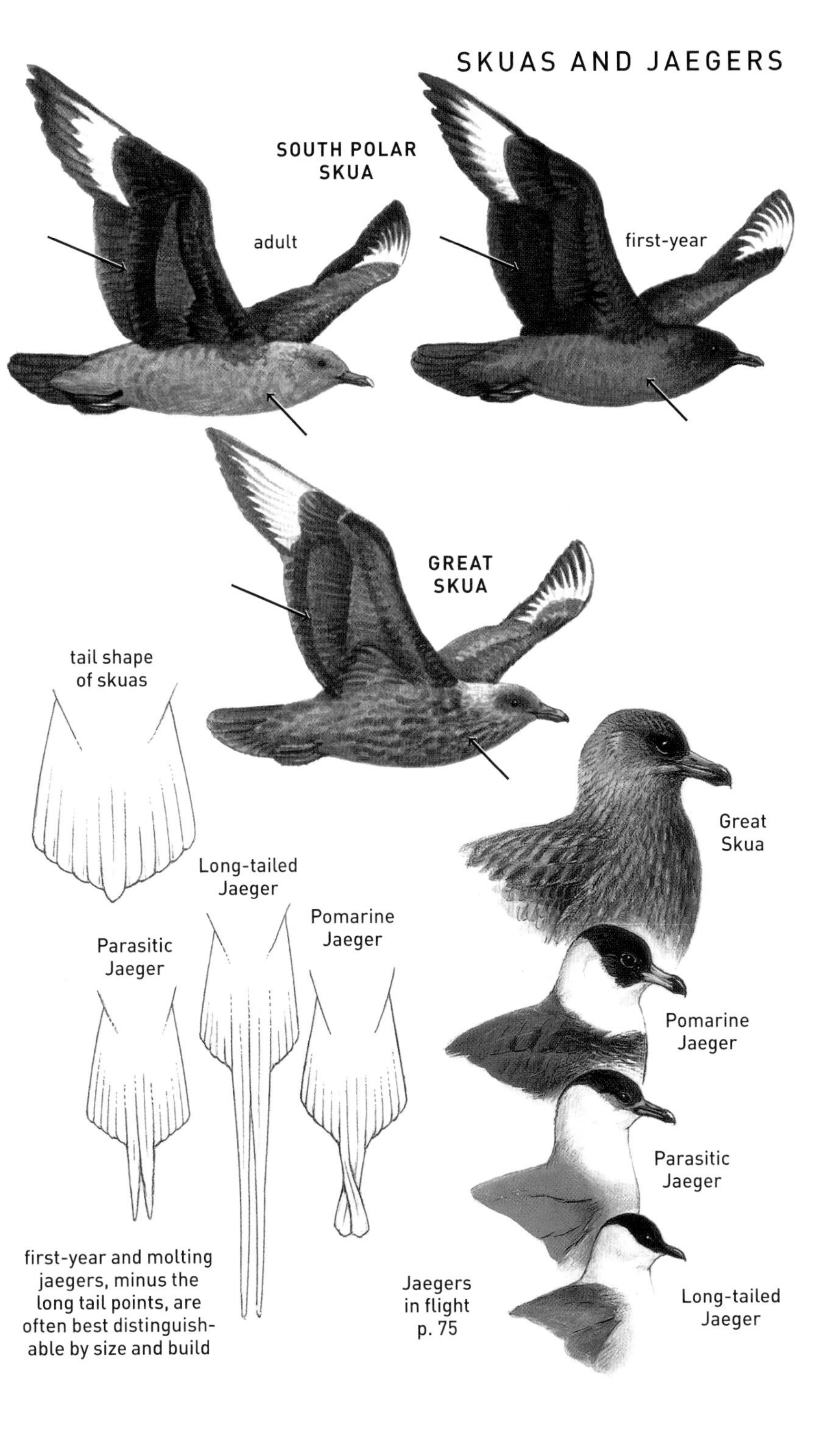

PARASITIC JAEGER *Stercorarius parasiticus* **Uncommon**

17–19 in. (44–49 cm). The jaeger most frequently seen from shore. Flies with strong, falconlike wing strokes. Smaller and less chesty than Pomarine Jaeger; larger and with longer bill than Long-tailed Jaeger. Typically chases larger terns and medium-sized gulls. Like other jaegers, shows variable white wing-flash. *Spring/summer adult:* Dark crown and pale underparts (light morph) to completely dark brown (dark morph). *Sharp central tail feathers* project up to 3½ in. (9 cm). *Juvenile and first-year:* Juvenile jaegers have heavy barring, especially on underwing. Juvenile Parasitic often with *more distinct white patch on upperwing;* dark morph usually *warmer brown* than other juvenile jaegers. *Second-year:* All three jaeger species retain partial barring on underwing and elsewhere. *Winter adult* (not seen in our area): Can lack dark crown and has barring on back and flanks. **SIMILAR SPECIES:** Pomarine and Long-tailed Jaegers. **HABITAT:** Primarily ocean, regularly seen from shore; in summer, tundra.

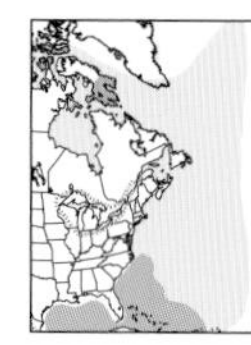

POMARINE JAEGER *Stercorarius pomarinus* **Uncommon**

19–21 in. (48–53 cm). Like Parasitic Jaeger, but slightly heavier with more gull-like flight style. Typically chases larger gulls, shearwaters. *Adult: Broad and twisted* central tail feathers are blunt-tipped and project 2–7 in. (5–18 cm); bill heavy and *pink-based.* In light morph, dark cap extends *farther down* sides of head and near bill base. Dark morph averages sootier than dark morph Parasitic. *Juvenile:* Compared with juvenile Parasitic, lacks warm tones, and very short central tail feathers are blunt-tipped. Look for white-based primary coverts, creating *double white flash* on underwing. Second-year and adult winter plumages as described under Parasitic Jaeger. **SIMILAR SPECIES:** Plumages of Pomarine and Parasitic Jaegers are so variable that the two species are often best distinguished by structural features and behavior. Large molting Pomarines can resemble skuas, but bills slenderer, white wing-flashes not as extensive, especially from above. **HABITAT:** Open ocean, seen from shore in small numbers; in summer, tundra.

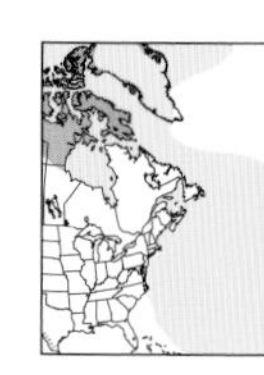

LONG-TAILED JAEGER *Stercorarius longicaudus* **Scarce**

17–22 in. (44–56 cm). The smallest, slimmest jaeger, with buoyant, ternlike flight and small, short bill. Typically chases smaller terns and gulls. *Adult:* Less variable than other jaegers, virtually all being light morph, paler and grayer above, with distinctly *two-toned upperwing* in flight; *long attenuated tail-streamers* project 8–15 in. (20–38 cm); black cap *sharply defined; no breast-band;* almost *no white in wings. Juvenile:* Polymorphic. All have very *limited white on upperwing* (two or three primary shafts), *stubby bill,* and longer, blunter tipped central tail feathers than juvenile Pomarine Jaegers. White patch on underwing variable. Light-morph juvenile has *pale grayish head and breast* and extensively *white belly.* Dark morph cold gray-brown; often shows pale nape and *pale lower breast patch.* **HABITAT:** Open ocean; tundra in summer. Most pelagic of the jaegers; seldom if ever seen from shore.

JAEGERS
Parasitic
Pomarine
Long-tailed
light
morph
adults
PARASITIC
JAEGER
dark
morph
intermediate
morph
dark-morph
juvenile
Parasitic
first-year
adults
POMARINE
JAEGER
dark
morph
Pomarine
light
morph
adult
LONG-
TAILED
JAEGER
light-morph
juvenile
Long-tailed
dark-morph
juvenile
Long-tailed

GULLS Family Laridae

Long-winged swimming birds with superb flight. Most are more robust, wider winged, and longer legged than terns, and most have larger and slightly hooked bills. Tails square or rounded (terns usually have forked tail). Gulls seldom dive (most terns hover, then plunge headfirst). **FOOD:** Omnivorous; marine life, plant and animal food, refuse, carrion. **RANGE:** Nearly worldwide.

AGING GULLS

It is often important to determine the age of a gull before identifying it. Knowing what a gull looks like in both its adult and first-year plumages is helpful in placing the bird to species in its intermediate (second- and third-year) stages. The sequence of plumages in gulls can be divided into three groups based on the age at which "adult" plumage is reached. Generally, this equates to the size of the gull, but note that maturation of plumage in all species is variable, with some individuals reaching full adult plumage a year before or after that described below. Most (but not all) gulls also have differing plumages in fall/winter and spring/summer, which become more distinct in each successive age class. When learning gulls, it is helpful to first focus on the size and structure of easier-to-identify adults, then consider size and structure of younger birds.

SEQUENCE OF PLUMAGES IN SMALL GULLS

In the top panel of the opposite page, the Bonaparte's Gull illustrates the transition of plumages directly from first-year to adult, usually without a very distinctive second-year plumage. Species in the East in this category include Bonaparte's, Black-headed, Little, Ross's, Sabine's, and Ivory Gulls. Adult Bonaparte's is also an example of a gull that has a distinctive spring/summer plumage for breeding.

SEQUENCE OF PLUMAGES IN MEDIUM-SIZED GULLS

In the middle panel of the opposite page, the Ring-billed Gull illustrates the transition of plumages from first-year to adult, with a distinctive second-year plumage that then generally transitions into adult. Species in the East in this category are mostly medium-sized gulls, including Ring-billed, Laughing, Franklin's, and Mew Gulls and Black-legged Kittiwake. Of these, Laughing and Franklin's Gulls have distinctive spring/summer plumages in their second year and as adults, whereas in the others, winter plumages have slight dusky streaks to the head, lost for breeding.

SEQUENCE OF PLUMAGES IN LARGE GULLS

In the bottom panel of the opposite page, the Herring Gull illustrates the transition of plumages from first-year to adult, including distinctive second- and third-year plumages. Species in this category in the East are most of the larger gulls, including California, Herring, Lesser Black-backed, Great Black-backed, Slaty-backed, Glaucous, and Iceland Gulls. Plumages tend to be similar between winter and summer, although some species have streaking on the head in winter, which is lost in summer.

Caution: There is extensive variation in plumage within species (particularly the second- and third-year plumages), resulting from variation in molt extents and timing, and plumage wear and bleaching. In addition, hybridization is a regular phenomenon among most large species, although not so much in the East. Even expert birders leave some gulls unidentified.

BONAPARTE'S GULL

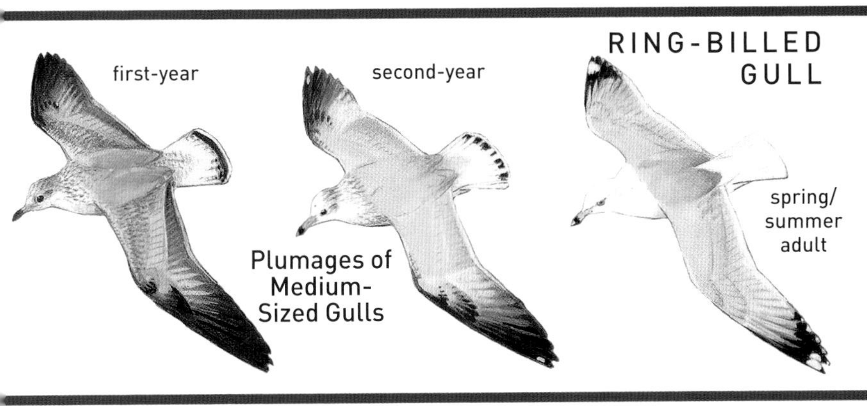

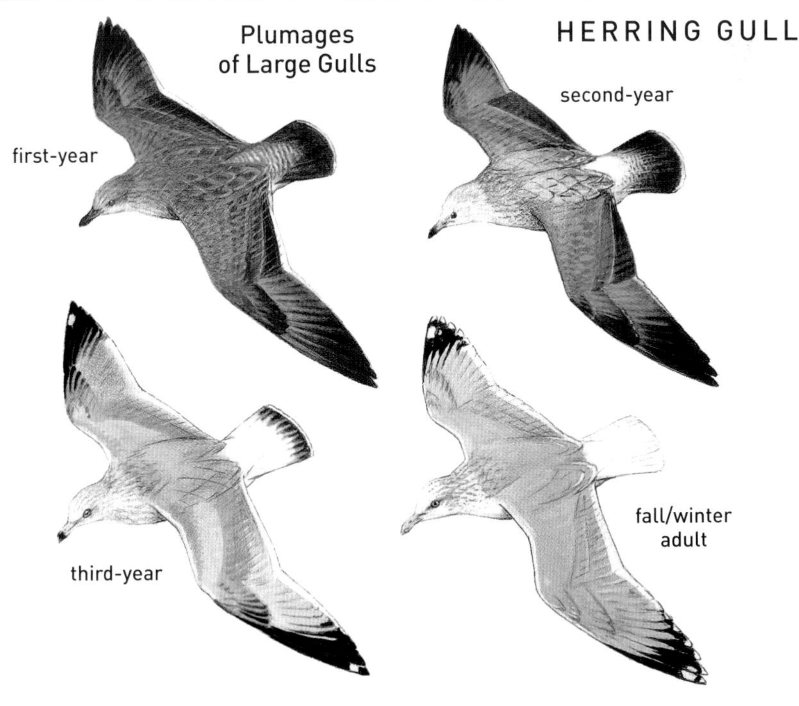

LAUGHING GULL *Leucophaeus atricilla* **Common**

16–16½ in. (41–42 cm). A medium-small coastal gull named for its call. *Adult: Dark mantle blends into black wingtips.* Bold white trailing edge to dark wing. Head *black* in spring/summer; pale in fall/winter with dark gray smudge across eye and nape. Bill longish, often with slight droop to tip; reddish when breeding, mostly dark when not breeding. *Juvenile and first-winter:* See p. 84. **VOICE:** A nasal *ha-a* and strident laugh, *ha-ha-ha-ha-ha-haah-haah-haah,* etc. **SIMILAR SPECIES:** Franklin's Gull slightly smaller, shorter billed, has broader white eye-arcs and *different wingtip pattern.* **RANGE:** Rare inland. **HABITAT:** Salt marshes, coastlines, parks, farm fields.

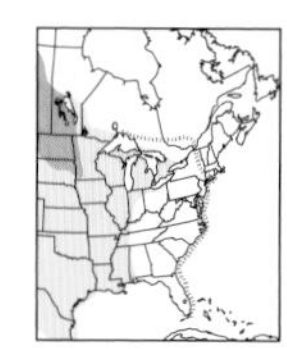

FRANKLIN'S GULL *Leucophaeus pipixcan* **Fairly common**

14½–15 in. (37–38 cm). *Adult:* Note *white band* near wingtip, separating black from gray. In spring/summer, head black; breast often has rosy bloom; bill red. In fall/winter, head paler but with dark cheeks and nape forming partial hood; bill mostly dark. *First-year:* See p. 84. **VOICE:** A shrill *kuk-kuk-kuk;* also mewing, laughing cries. **SIMILAR SPECIES:** Laughing and Bonaparte's Gulls. **RANGE:** Casual vagrant to East Coast. **HABITAT:** Prairies, inland marshes, lakes; in winter, coasts, ocean, primarily in S. America.

SABINE'S GULL *Xema sabini* **Rare**

13½–14 in. (34–36 cm). A small, *ternlike* gull with slightly *forked tail. Adult:* Note *bold upperwing pattern* of black outer primaries and *triangular white wing patch.* Bill black with *yellow tip;* legs dark. In winter, head whitish with dusky wash. *Juvenile:* See p. 84. **VOICE:** Grating or buzzy ternlike calls given on breeding grounds. **SIMILAR SPECIES:** Black-legged Kittiwake. **RANGE:** Casual vagrant inland. **HABITAT:** Ocean; nests on tundra pools.

BLACK-HEADED GULL *Chroicocephalus ridibundus* **Uncommon**

15¾–16 in. (40–41 cm). *Adult:* Similar in pattern to Bonaparte's Gull and often associates with it, but slightly larger; mantle slightly paler; shows much *blackish gray on underside of primaries;* bill *dark red,* not black. In fall/winter, head pale with black ear spot. *First-year:* See p. 84. **VOICE:** Harsh *kerrr.* **HABITAT:** Same as Little and Bonaparte's Gulls; also beaches, lawns.

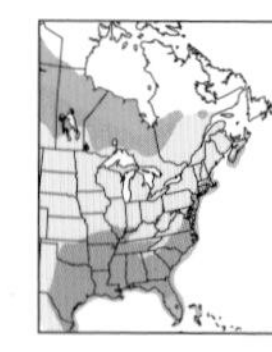

BONAPARTE'S GULL *Chroicocephalus philadelphia* **Common**

13–13½ in. (33–34 cm). A petite, almost ternlike gull. *Adult:* Note *wedge of white* on *fore edge* of wing. Legs red to pinkish; bill small, black. Head blackish in spring/summer, whitish with *black ear spot* in fall/winter. *First-year:* See p. 84. Also see Sequence of Plumages in Small Gulls, p. 76. **VOICE:** A nasal, grating *cheeer* or *cherr*. Some calls ternlike. **SIMILAR SPECIES:** Black-headed and Little Gulls. **HABITAT:** Ocean, bays, lakes, sewage-treatment ponds; in summer, muskeg.

LITTLE GULL *Hydrocoloeus minutus* **Rare**

11 in. (28 cm). The smallest gull; usually associates with Bonaparte's Gull. *Adult:* Note *blackish undersurface* of *rather rounded wing* and absence of black above. Legs red. Head black and bill dark red in spring/summer; head *dark-capped,* with *black ear spot,* and bill black in fall/winter. *First-year:* See p. 84. **VOICE:** A series of one- or two-syllable *key* notes. **SIMILAR SPECIES:** Bonaparte's Gull. **HABITAT:** Lakes, rivers, bays, coastal waters, sewage-treatment ponds; often with Bonaparte's Gulls.

SMALL HOODED GULLS
Adults
fall/
winter
LAUGHING
GULL
breeding adults
flying
spring/
summer
FRANKLIN'S
GULL
fall/
winter
spring/
summer
SABINE'S
GULL
fall/
winter
spring/
summer
BLACK-
HEADED
GULL
fall/
winter
spring/
summer
BONAPARTE'S
GULL
fall/
winter
spring/
summer
LITTLE
GULL
fall/
winter
spring/
summer

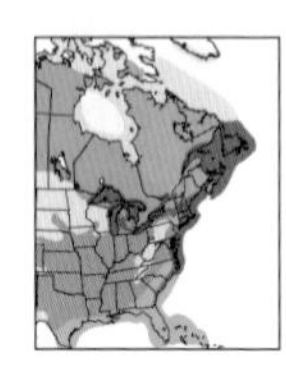

HERRING GULL *Larus argentatus* Common

24–25 in. (61–64 cm). A widespread, fairly large gull. *Adult: Pale gray* mantle, *pinkish* legs, *pale eye.* Outer primaries contrastingly *black* with moderately extensive white spots or "mirrors." Bill yellow with red spot on lower mandible. In fall/winter, head and neck streaked or mottled with brownish; bill and legs duller. *First- and second-years:* See p. 86. Also see Sequence of Plumages in Large Gulls, p. 76. **VOICE:** A loud *hiyak . . . hiyak . . . hyiah-hyak* or *yuk-yuk-yuk-yuk-yuckle-yuckle.* Anxiety call *gah-gah-gah.* **SIMILAR SPECIES:** "Thayer's" Gull (subspecies of Iceland Gull) and California Gull; first-years of these species can be similar (see p. 86). **HABITAT:** Ocean, coasts, bays, beaches, lakes, dams, piers, farmland, dumps.

CALIFORNIA GULL *Larus californicus* Scarce, local

21–21½ in. (53–55 cm). Between Ring-billed and Herring Gulls in size, but proportionally longer winged. *Adult:* Note *yellow to greenish legs,* darker mantle and *darker eye,* bill with *both red and black spots.* Has more white in wingtips than Ring-billed. In fall/winter, head streaked or mottled brownish, bill and legs slightly duller, the latter often *grayish green. First- and second-years:* See p. 86. **VOICE:** Like Herring Gull but higher, more hoarse. **RANGE:** Casual vagrant to Gulf Coast; accidental to Atlantic Coast. **HABITAT:** Ocean and coasts, lakes, farms, dumps, urban centers.

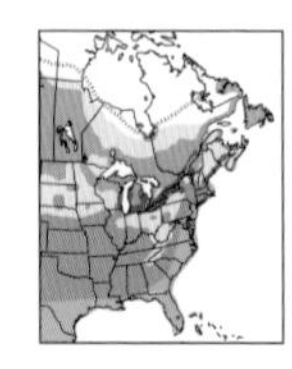

RING-BILLED GULL *Larus delawarensis* Common

17–17½ in. (43–45 cm). Smaller than Herring Gull, more delicate and buoyant; somewhat dovelike. *Adult:* Shows *pale eye* and *light gray mantle* (similar to Herring but paler than California and Mew Gulls); *legs yellow or greenish yellow.* Note *complete black ring* encircling bill. In fall/winter shows some fine dark streaking on head, and bill and legs become duller. *First-year:* See p. 84. Also see Sequence of Plumages in Medium-sized Gulls, p. 76. **VOICE:** Higher pitched than Herring Gull. **SIMILAR SPECIES:** Mew Gull has smaller bill that lacks bold blackish ring, darker mantle, dark eye, and in fall/winter, more extensive dark mottling on head and neck. **HABITAT:** Lakes, bays, coasts, dumps, plowed fields, sewage outlets, fast-food restaurants; rarer on open ocean than other gulls.

MEW GULL *Larus canus* Very rare vagrant

16–17 in. (41–43 cm). Called "Common Gull" in Europe. *Adult:* Slightly smaller than Ring-billed Gull, with *dainty, short, unmarked greenish-yellow bill.* Darkish eye. Mew shows *medium gray mantle,* noticeably darker than Ring-billed's, and larger white "mirrors" in its black wingtips than either California or Ring-billed Gull. In fall/winter, head streaked and bill duller. *First-year:* See p. 84. **VOICE:** A low, mewing *queeu* or *meeu* in winter. **SIMILAR SPECIES:** First-year Ring-billed Gull, adult Black-legged Kittiwake. **HABITAT:** In winter, ocean, coastlines, parks, wet fields.

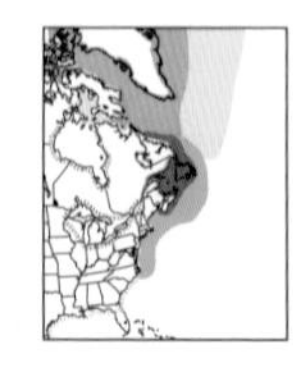

BLACK-LEGGED KITTIWAKE *Rissa tridactyla* Uncommon

16–17 in. (41–43 cm). A small, buoyant oceanic gull. *Adult:* Wingtips lack white spots; *solid black,* almost *straight across,* as if dipped in ink. Bill slightly curved, without angle to lower mandible as in other gulls, and is pale yellow, unmarked. Legs and feet *black. Eyes dark.* Fall/winter (not shown) similar, but rear nape has dusky band. *First-year:* See p. 84. **VOICE:** At nesting colony, a raucous *kaka-week* or *kitti-waak.* **SIMILAR SPECIES:** Mew, Ring-billed, and Sabine's Gulls. **RANGE:** Rare to casual vagrant inland. **HABITAT:** Chiefly oceanic; rarely on beaches. Nests on sea cliffs.

GULLS
Adults
fall/winter
HERRING GULL
spring/summer
fall/winter
CALIFORNIA GULL
spring/summer
RING-BILLED GULL
fall/winter
spring/summer
fall/winter
spring/summer
MEW GULL
BLACK-LEGGED KITTIWAKE
spring/summer

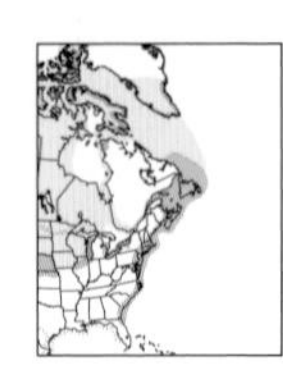

ICELAND GULL *Larus glaucoides* Uncommon

22–24 in. (56–61 cm). Two distinct subspecies. Eastern-breeding "Kumlien's" Gull (*L. g. kumlieni*) is a pale ghostly gull, slightly smaller than Herring Gull. *Adult:* Mantle pale gray; primaries, which extend *well beyond tail,* whitish with gray or dark markings, variable in hue, with large white "mirrors" (not black with white mirrors as in Herring Gull). Western-breeding "Thayer's" Gull (*L. g. thayeri;* scarce vagrant to East Coast), formerly a separate species, is slightly larger and larger-billed than "Kumlien's" and has blacker primary tips like Herring Gull. Differs from Herring by having *darker eye,* smaller bill (often tinged greenish in winter), *thinner trailing edge of black* on *grayish* underside of primaries, and *more extensive white mirrors* to outer primaries. Variation in wingtip pattern between "Kumlien's" and "Thayer's" breeding in high Arctic may be nearly continuous; thus often paler than in Herring Gull. *First- and second-years:* See p. 86. **VOICE:** Similar to Herring Gull but higher pitched; rarely heard away from breeding grounds. **SIMILAR SPECIES:** Glaucous Gull is similar to "Kumlien's" but has larger bill, shorter primary extension. Herring Gull larger and has paler back and eye than "Thayer's." **RANGE:** Rare to casual well inland. **HABITAT:** Ocean, coastlines, freshwater outflows, dumps.

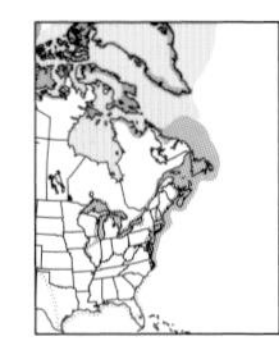

GLAUCOUS GULL *Larus hyperboreus* Uncommon

27–28 in. (68–72 cm). A large, chalky white gull with pinkish legs. *Adult:* Note "frosty" wingtips. Has pale gray mantle and *unmarked white outer primaries. Light eye.* Head slightly streaked and bill duller in fall/winter. *First- and second-years:* See p. 86. **VOICE:** Much like Herring Gull. **SIMILAR SPECIES:** "Kumlien's" Iceland Gull similar but smaller; bill also smaller, head rounder, and wings proportionally longer (extending well beyond tail when sitting). Spring/summer adult Iceland has narrow red eye-ring (Glaucous has yellow), but this is hard to see. **RANGE:** Rare inland and casual vagrant well south of normal winter range. **HABITAT:** Mainly coastal; inland, at large lakes and dumps.

GREAT BLACK-BACKED GULL *Larus marinus* Common

29–30 in. (73–76 cm). The largest gull in the world, with broad wings and heavy body and bill. *Adult:* Black back and wings, snow-white underparts, no head streaking in winter. Legs and feet *pale* pinkish. *First- and second-years:* See p. 86. **VOICE:** A harsh, deep, seal-like *kyow* or *owk.* **SIMILAR SPECIES:** Lesser Black-backed Gull; see also Slaty-backed Gull. **HABITAT:** Mainly coastal waters, estuaries, dumps; occasionally well inland on large lakes and rivers.

LESSER BLACK-BACKED GULL *Larus fuscus* Uncommon

21–22½ in. (53–57 cm). A Eurasian species that has increased along Atlantic Coast and in Great Lakes; may breed somewhere in Holarctic. Similar to Great Black-backed Gull but smaller (usually smaller than Herring Gull) and slimmer, with longer wings and smaller, slimmer bill. Distinguished by yellowish (not pink) legs and slate gray (not black) mantle. Extensive head and neck streaking or mottling in fall/winter. Pale eye. Oblong red spot on bill. *First- and second-years:* See p. 86. **VOICE:** A harsh *kyah.* **SIMILAR SPECIES:** Great Black-backed and Slaty-backed Gulls. **RANGE:** Rare vagrant well inland across N. America. **HABITAT:** Same as Herring Gull.

PALE AND DARK-BACKED GULLS
Adults
pale
extreme
ICELAND
GULL
"Kumlien's"
(typical)
"Thayer's"
GLAUCOUS
GULL
fall/
winter
spring/
summer
GREAT
BLACK-BACKED
GULL
LESSER
BLACK-BACKED
GULL

FIRST-YEAR and SECOND-YEAR SMALL GULLS

First-year and second-year gulls can be difficult to identify. They are usually darkest the first year and become lighter and more adultlike during the second and third years. Body and bill size and structure are useful for identification.

LAUGHING GULL *Leucophaeus atricilla* **Adult, p. 78**

Most reach adult plumage by third year. *Juvenile:* Dark brown with black tail, white rump, and *broad white* trailing edge of wing. *First-year:* Neck and back extensively smudged with gray. *Second-year:* Similar to fall/winter adult, but upperwing tips darker, some black in tail.

FRANKLIN'S GULL *Leucophaeus pipixcan* **Adult, p. 78**

Most reach adult plumage by third year. *First-year:* Similar to first-year Laughing Gull but more petite and with *smaller and straighter bill, blackish extensive half-hood,* outermost tail feather *white,* paler underside to primaries. *Second-year:* Close to second-year Laughing but with blackish half-hood, pale underside to primaries.

BLACK-HEADED GULL *Chroicocephalus ridibundus* **Adult, p. 78**

Most reach adult plumage by second year. *First-year:* Similar to first-year Bonaparte's Gull but slightly larger; bill longer, *orange to red* at base, black at tip; *sooty underwing;* dusky trailing edge to upperwing.

BONAPARTE'S GULL *Chroicocephalus philadelphia* **Adult, p. 78**

Most reach adult plumage by second year. *First-year:* Note dark ear spot, narrow black tail band, dark trailing edge to wings, and white in outer primaries. Pale underwing. See also p. 76.

LITTLE GULL *Hydrocoloeus minutus* **Adult, p. 78**

Most reach adult plumage by second year. *First-year:* Small with bolder *black and white M pattern* across back and wings, *dusky cap.*

SABINE'S GULL *Xema sabini* **Adult, p. 78**

Most reach adult plumage by second year. *Juvenile:* Dark grayish brown and white-scaled back, but with adult's bold *triangular wing pattern. First-spring:* Black hood partial.

BLACK-LEGGED KITTIWAKE *Rissa tridactyla* **Adult, p. 80**

Most reach adult plumage by third year. *First-year: Dark bar on nape, black M across back and wings. Second-year:* Like adult but with more black to upperwing.

MEW GULL *Larus canus* **Adult, p. 80**

Most reach adult plumage by third year. *First-year and second-year:* Smaller than Ring-billed Gull, darker, with smaller bill, and darker tail.

RING-BILLED GULL *Larus delawarensis* **Adult, p. 80**

Most reach adult plumage by third year. *First-year: Bicolored (pinkish-based) bill,* mostly whitish underneath and on rump and upper tail, *pale gray back.* Well-defined subterminal tail band; contrasty wing pattern. *Second-year:* Like adult but upperwing has more brown and black and tail has some black. See also p. 76.

SMALL GULLS
First- and Second-Years
juvenile
first-winter
juvenile
first-winter
LAUGHING GULL
first-year
first-year
FRANKLIN'S GULL
first-year
BLACK-HEADED GULL
BONAPARTE'S GULL
first-year
first-year
LITTLE GULL
SABINE'S GULL
juvenile
BLACK-LEGGED KITTIWAKE
first-year
first-year
MEW GULL
first-year
RING-BILLED GULL

FIRST-YEAR, SECOND-YEAR, and THIRD-YEAR LARGE GULLS

LESSER BLACK-BACKED GULL *Larus fuscus* **Adult, p. 82**

Most reach adult plumage by fourth year. *First-year:* Like other first-year gulls but with broader tail band, darker wings and back, more heavily streaked breast; white tail base; black bill. *Second- and third-years:* Follow patterns of other large gulls; some show more adultlike plumage in second spring/summer.

HERRING GULL *Larus argentatus* **Adult, p. 80**

Most reach adult plumage by fourth year. *First-year:* Extremely variable; brownish overall, with brownish-black wingtips and dark brown tail; only all-brown gull commonly seen in East. *Pale area on inner primaries visible in flight.* Bill variably all dark or paler at base. *Second- and third-years:* Head and underparts become whiter; back pale gray; bill pink-based, then yellow with dark tip. See also p. 76.

GREAT BLACK-BACKED GULL *Larus marinus* **Adult, p. 82**

Most reach adult plumage by fourth or fifth year. *First-year:* Larger and with more contrasting salt-and-pepper pattern than first-year Herring Gull; whiter on head, rump, and underparts. Bill entirely black. *Second-year:* Mantle becomes blacker during winter; bill becomes paler at base and often tipped yellow. *Third-year and some fourth-years:* Adultlike, but secondaries and wing coverts brownish or washed brown; wingtips darker, with smaller white mirrors; tail with black; bill variably black and yellowish.

CALIFORNIA GULL *Larus californicus* **Adult, p. 80**

Most reach adult plumage by fourth year. *First-year:* Like Herring Gull but slender, with slimmer, distinctly bicolored bill. Upperparts with distinct checkering; wings lack paler inner primaries. *Second-year:* Legs and bill often dull gray-green. Larger than first-winter Ring-billed Gull; upperparts, eye, and tail darker. *Third-year:* Like adult but with more black in wing, some black usually in tail, bill can have black ring (but red spot also present).

GLAUCOUS GULL *Larus hyperboreus* **Adult, p. 82**

Most reach adult plumage by fourth year. *First-year:* Pale tan, becoming white by late winter; primaries white with small black marks when fresh. Brownish mottling in wing coverts and tail when fresh. Bill *pale pinkish* with *sharply demarcated* dark tip. *Second-year:* Paler gray back. First- and second-year plumages often bleached pure white by spring. *Third-year:* Like adult, but bill usually retains dark tip or smudge.

ICELAND GULL *Larus glaucoides* **Adult, p. 82**

Most reach adult plumage by fourth year. Plumages of "Kumlien's" subspecies similar to Glaucous Gull's, but size and structure differ as in adult (p. 82); bill usually mostly dark, only rarely as sharply demarcated as in Glaucous. Most birds have at least a hint of tail band as well as some dark in outer primaries, both lacking in Glaucous. First-year "Thayer's" similar to juvenile Herring Gull but paler; primaries paler; *bill entirely or almost entirely blackish, more petite; underside of primaries pale. Third-year:* In both subspecies, similar to adults, but tail and bill usually have some dusky; white mirrors to outer primaries smaller.

LARGE GULLS

First-Years, Second-Years, Third-Years

RARE GULLS

BLACK-TAILED GULL *Larus crassirostris* **Accidental vagrant**

18–18½ in. (46–47 cm). Size and shape of California Gull, with slightly longer bill. *Adult:* Has red tip to black-banded bill, slate gray mantle, wide black subterminal band on tail. *First-year:* Very dark with bright pink-based bill. **RANGE:** Accidental vagrant from e. Asia, with widely scattered records across much of N. America.

YELLOW-LEGGED GULL *Larus michahellis* **Casual vagrant**

24–24½ in. (61–63 cm). A native of s. Europe; very similar to Herring Gull, but bill slightly stouter, adult's mantle slightly *darker gray,* and head only *finely streaked on crown* in fall/winter. *Yellow legs* of adult usually distinctive. **RANGE:** Casual vagrant to NL and Atlantic Coast.

SLATY-BACKED GULL *Larus schistisagus* **Accidental vagrant**

25–26 in. (64–67 cm). A dark-backed Asian gull. *Adult:* Blackish-backed with *pale "staring" eye,* deep pinkish feet; white subterminal bars or "tongues" form *thin white bar* crossing dark outer primaries. *First-year:* Similar to Herring Gull but dumpier, legs darker purplish, bill stout and black, inner primaries not as pale. *Second- and third-years:* Follow plumage and bill color changes of Kelp Gull, but legs dark pink, eye paler. **RANGE:** Casual vagrant across N. America to East Coast. **HABITAT:** Seacoasts, beaches, dumps.

KELP GULL *Larus dominicanus* **Casual vagrant**

22–25 in. (56–64 cm). A black-backed, stocky gull from S. America. *Adult:* Black back, reduced mirrors in primaries (typically a *square patch on outermost primary* only), *very stout bill, bright greenish yellow legs.* Younger plumages and bill colors parallel those of other large dark-backed gulls. **SIMILAR SPECIES:** Lesser Black-backed Gull usually smaller and has slimmer bill; juvenile and first-year Kelp Gulls have darker legs, blacker base to tail in flight. **RANGE:** Casual vagrant to Gulf Coast; accidental elsewhere throughout N. America. Small numbers bred or hybridized with Herring Gulls in 1990s in LA.

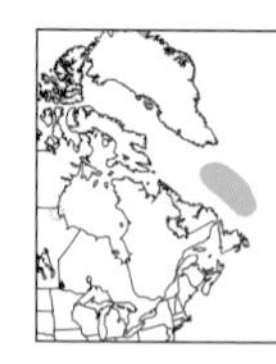

ROSS'S GULL *Rhodostethia rosea* **Very rare**

13–13½ in. (33–35 cm). A rare Arctic gull of drift ice. Note *wedge-shaped tail, medium gray underwing linings,* and *small black bill. Spring/summer: Rosy* blush on underparts, *fine black collar. Fall/winter:* Less rosy, lacks black collar. *First-year:* Similar in pattern to first-year Black-legged Kittiwake or Little Gull, but intermediate in size and note *longer wedge-shaped tail* with black tip; lacks dark nape of young kittiwake. **RANGE:** Vagrant well south of normal winter range. **HABITAT:** Arctic waters, tundra in summer.

IVORY GULL *Pagophila eburnea* **Very rare**

17 in. (43 cm). A declining species of Arctic pack ice; those that wander south are usually first-years. Bill dark greenish *with yellow tip. Adult:* Small, all white, with *black legs.* Pigeonlike in size and head shape; wings long, flight ternlike. *First-year:* Dark *smudge on face, black spots* above, wing and tail feathers tipped black. **RANGE AND HABITAT:** Open Arctic waters near pack ice; vagrants well south of winter range found on coasts, lakes.

RARE GULLS

YELLOW-
LEGGED
GULL
BLACK-
TAILED GULL
SLATY-BACKED
GULL
fall/winter
adult
adult
fall/winter
adult
juvenile
KELP
GULL
second-
year
adult
fall/
winter
spring/
summer adult
first-year
adult
ROSS'S
GULL
first-year
spring/
summer
adults
IVORY
GULL

TERNS Subfamily Sterninae

Graceful waterbirds. Bill pointed, often tilted down. Most whitish with black cap in summer, white forehead in fall/winter. Sexes alike. Terns hover and plunge but rarely swim. **FOOD:** Small fish, insects. **RANGE:** Worldwide.

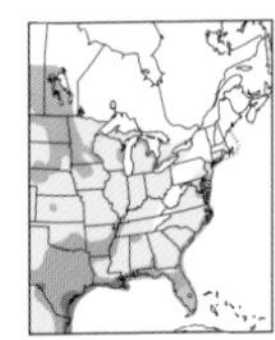

FORSTER'S TERN *Sterna forsteri* **Common**

14½ in. (37 cm). Similar to Common Tern, but *paler wingtips than rest of wing,* tail grayer, adults with orangey bill. In fall/winter has isolated *black mask, usually not connecting around nape.* Juvenile has upperpart fringing washed cinnamon. **VOICE:** A harsh, nasal *za-a-ap* and nasal *kyarr.* **HABITAT:** Fresh and salt marshes, lakes, bays, beaches. Nests in marshes.

COMMON TERN *Sterna hirundo* **Common**

14 in. (36 cm). *Spring/summer adult:* Pearl gray mantle and black cap; bill red with black tip; feet orange-red. Similar to Forster's and Arctic Terns but has *dark wedge on upperwing primaries. Grayer below* than Forster's; bill and legs smaller than in Forster's, *larger than in Arctic. Fall/winter adult and first-year:* Cap, nape, and bill blackish. *Juvenile:* Upperparts washed brownish and marked dark. **VOICE:** A drawling *kee-arr* (downward inflection); also *kik-kik-kik;* a quick *kirri-kirri.* **HABITAT:** Ocean, bays, marshes, beaches. Nests colonially on beaches, small islands.

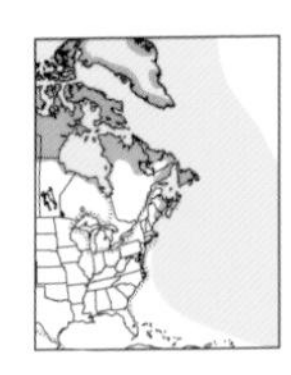

ARCTIC TERN *Sterna paradisaea* **Uncommon**

15 in. (38 cm). A pelagic tern when away from nesting grounds. Similar to Common Tern, but bill and neck shorter, head rounder. *Legs shorter. Bill smaller.* From below, note translucent effect of primaries and *narrow black trailing edge. Spring/summer adult:* Bill usually *blood red* to tip, extensive wash of *gray below,* setting off white cheeks. *Fall/winter and juvenile:* Like Common, but black on head more extensive, shoulder bar *weaker,* structural differences noted above. **VOICE:** *Kee-yak,* less slurred and higher than Common's *kee-arr.* **RANGE:** Rare vagrant on coasts, casual inland. **HABITAT:** Open ocean; in summer, taiga lakes, tundra.

ROSEATE TERN **Uncommon, local, endangered/threatened**
Sterna dougallii

15½ in. (39 cm). Similar to Common Tern but bill longer, thinner, black year-round, paler overall; *at rest, tail extends well beyond wingtips;* also has *shallower wingbeats* and *different call.* In spring, may acquire rosy blush to breast and dark red on base of bill. *Adult winter, first-year, and juvenile:* Similar to respective Common Terns, but in juvenile, back has pattern of *coarse black crescents* (can be darker than shown); forehead darker. **VOICE:** A rasping *ka-a-ak;* a soft two-syllable *chu-ick* or *chiv-ick.* **HABITAT:** Salt bays, estuaries, ocean. Northeastern populations endangered, FL populations threatened.

LEAST TERN *Sternula antillarum* **Locally common**

9 in. (23 cm). A *very small* tern, with quick wingbeats. *Spring/summer adult:* Dark-tipped *yellow bill, yellow legs and feet,* and *white forehead. Long black wedge on outer wing. First-year:* Dark bill, dark carpal (shoulder) bar. *Juvenile:* Upperpart feathers fringed cinnamon. **VOICE:** A sharp, repeated *kit;* a harsh, squealing *zree-eek* or *k-zeek;* also a rapid *kitti-kitti-kitti.* **SIMILAR SPECIES:** Forster's Tern much bigger. **RANGE:** Rare vagrant away from breeding areas. **HABITAT:** Beaches, bays, large rivers, sandbars. Populations breeding in Mississippi R. endangered.

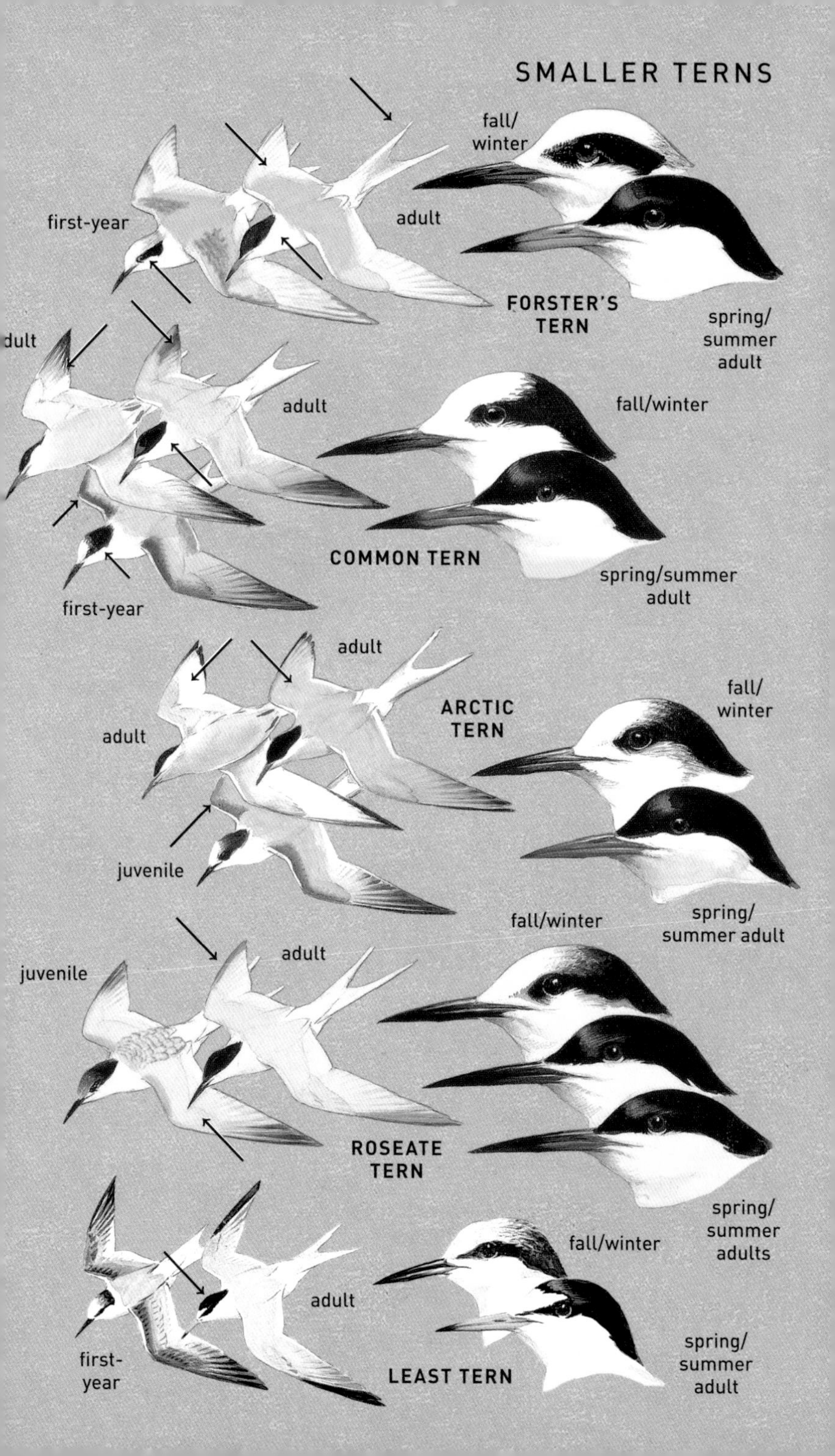
SMALLER TERNS
fall/
winter
first-year
adult
FORSTER'S
TERN
spring/
summer
adult
dult
adult
fall/winter
COMMON TERN
spring/summer
adult
first-year
adult
ARCTIC
TERN
fall/
winter
adult
juvenile
fall/winter
spring/
summer adult
adult
juvenile
ROSEATE
TERN
spring/
summer
adults
fall/winter
adult
first-
year
LEAST TERN
spring/
summer
adult

GULL-BILLED TERN *Gelochelidon nilotica* Uncommon

14 in. (36 cm). Note *stout black* bill. Stockier and paler than Common Tern; tail much less forked; feet *black. Fall/winter adult:* Head white with smudgy dark ear patch, pale dusky on nape. *First-year:* Similar to fall/winter adult; carpal bar dusky; crown mostly pale in spring/summer. *Juvenile:* Crown and upperparts washed pale brown; wing coverts grayish. This tern plucks food from water's surface and often hawks for insects over marshes and fields, swooping (rarely diving) after prey. **VOICE:** *Kay-weck, kay-weck;* also a throaty, rasping *za-za-za.* **SIMILAR SPECIES:** Sandwich Tern, Forster's Tern in winter, small gulls. **RANGE:** Accidental vagrant inland. **HABITAT:** Marshes, fields, coastal bays.

SANDWICH TERN *Thalasseus sandvicensis* Fairly common

15–15½ in. (38–40 cm). Larger than Common Tern. Note *long black bill with yellow tip* "as though dipped in mustard." Outer primaries variably dark from above, tipped dusky from below. Legs black. *Adult:* All-black cap in spring/summer, white forehead in fall/winter; feathers on back of crown form shaggy crest. *First-year:* Like winter adult, tertials with dark centers, cap mostly white in spring/summer. *Juvenile:* Upperpart feathers tipped black; *bill can be mostly black or mostly yellow.* **VOICE:** A grating *kirr-ick* (higher than Gull-billed Tern's *kay-weck*). **SIMILAR SPECIES:** Gull-billed Tern has stout black bill. **RANGE:** Scarce vagrant inland after hurricanes. **HABITAT:** Coastal waters, jetties, beaches. Often seen with Royal Tern.

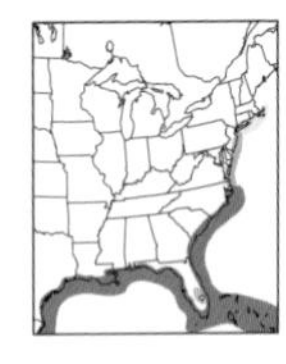

ROYAL TERN *Thalasseus maximus* Common

20 in. (51 cm). A large tern, slimmer than Caspian Tern, with medium-large *orange* bill. Tail forked. Keeps solid black crown for short time in spring; for most of year has *much white on forehead,* black crown feathers forming a crest. Dusky upperside and *pale underside to primaries,* opposite of Caspian. *First-year:* Like winter adult, tertials with dark centers, cap mostly white, and outer primaries darker in spring/summer. *Juvenile:* Upperpart feathers have neat black crescents; wing coverts streaked black. **VOICE:** A sonorous *karr-rik;* also *kaak* or *kak.* **SIMILAR SPECIES:** Caspian Tern. **RANGE:** Accidental vagrant inland. **HABITAT:** Ocean, coasts, beaches, salt bays. More closely tied to coastal waters than Caspian, which is regular inland.

CASPIAN TERN *Hydroprogne caspia* Uncommon

21 in. (53 cm). Large size and *stout reddish bill with small dark mark near tip* set Caspian apart from all other terns. Tail of Caspian *shorter;* head and bill larger, crest shorter and less shaggy. Royal Tern's forehead usually *clear white* in fall/winter adults, whereas Caspian has *gray-streaked* forehead. Caspian shows obvious *grayish black on undersurface of primaries, but pale upper surface. First-year:* Rare in our area, but cap like winter adult's. *Juvenile:* Upperpart feathers boldly marked gray and black; wing coverts have dusky markings. **VOICE:** A raspy, low *kraa-uh* or *karr,* also a repeated *kak;* juvenile gives a whistled *wheee-oo.* **SIMILAR SPECIES:** Caspian ranges inland, Royal usually does not. **HABITAT:** Large lakes, rivers, coastal waters, beaches, bays.

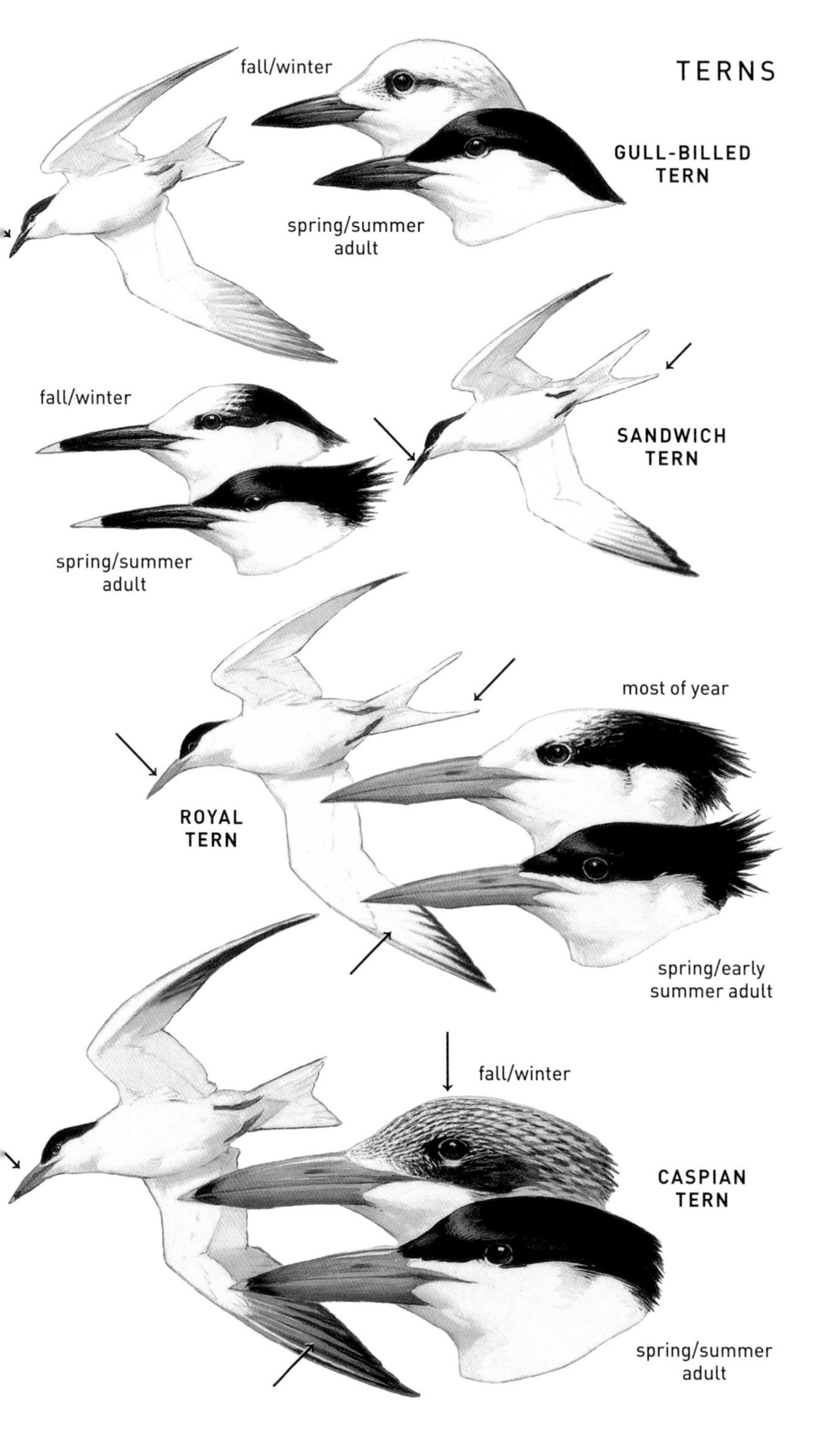
TERNS
fall/winter
GULL-BILLED TERN
spring/summer adult
fall/winter
SANDWICH TERN
spring/summer adult
most of year
ROYAL TERN
spring/early summer adult
fall/winter
CASPIAN TERN
spring/summer adult

BLACK TERN *Chlidonias niger* — Uncommon

9½–9¾ in. (24–25 cm). *Spring/summer adult:* Head and underparts primarily *black; back, wings, and tail dark gray;* underwing linings whitish. *Fall/winter adult and first-year:* Pied head, smudgy dark crown, and hind collar. *Juvenile:* Upperpart feathers fringed brown. **VOICE:** A sharp *kik, keek,* or *klea.* **SIMILAR SPECIES:** White-winged Tern (*C. leucopterus;* not shown), a casual vagrant in East, has mostly white upperwing and black underwing lining in spring/summer adult; paler and lacks dark breast mark in first-year and winter plumages. **HABITAT:** Freshwater marshes, lakes; in migration, also coastal waters, including open ocean.

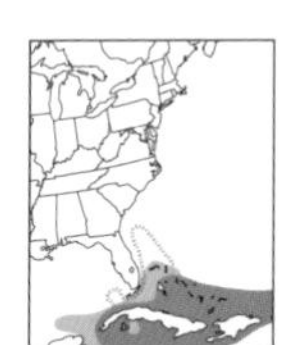

BROWN NODDY *Anous stolidus* — Uncommon, local

15–15½ in. (38–40 cm). A brown tern with *whitish cap.* Long, wedge-shaped tail. First-year has duller cap. **VOICE:** A ripping *karrrrk* or *arrr-rowk;* a harsh *eye-ak.* **SIMILAR SPECIES:** Black Noddy occurs occasionally with Brown Noddies at Dry Tortugas, FL. **RANGE:** Scarce vagrant along coast after hurricanes. **HABITAT:** Warm ocean waters.

BLACK NODDY *Anous minutus* — Very rare

13½ in. (34 cm). Slightly smaller and sootier colored than Brown Noddy, with thinner and proportionally *longer bill, shorter forked tail,* and more *sharply defined white cap.* **VOICE:** A variety of chatters, croaks, and bill rattles. **SIMILAR SPECIES:** Brown Noddy. **RANGE:** Rare spring visitor to Dry Tortugas, FL; casual visitor to TX. **HABITAT:** Tropical islands.

SOOTY TERN *Onychoprion fuscatus* — Uncommon, local

16 in. (41 cm). *Adult:* Cleanly patterned, black above and white below. Patch on forehead white; bill and feet black. *Juvenile:* Dark brown; back spotted with white; underwing lining and vent grayish. **VOICE:** A nasal *wide-a-wake.* **SIMILAR SPECIES:** Juvenile larger than Black Tern and with upperparts spotted white. **RANGE:** Rare vagrant inland and to Northeast coast after hurricanes. **HABITAT:** Warm ocean waters.

BRIDLED TERN *Onychoprion anaethetus* — Uncommon, local

15 in. (38 cm). *Adult:* Resembles Sooty Tern, but back brownish, not blackish; *note whitish collar separating black cap from back;* white forehead patch *extending behind eye. Juvenile and first-year:* Head whiter; upperparts barred white. **VOICE:** Sometimes gives a soft, nasal *wheeep.* **SIMILAR SPECIES:** Sooty Tern. **RANGE:** Scarce vagrant inland and to Northeast after hurricanes. **HABITAT:** Warm, offshore ocean waters.

SKIMMERS Subfamily Rynchopinae

Slim, short-legged relatives of gulls and terns. Scissorlike red bill; *lower mandible longer than upper.* **FOOD:** Small fish, crustaceans. **RANGE:** Coasts, ponds, marshes, beaches, rivers of warmer parts of world.

BLACK SKIMMER *Rhynchops niger* — Fairly common

18–18½ in. (46–47 cm). Very long wings; skims low, with stiff wingbeats, dipping lower mandible in water for prey. *Adult:* Black above (nape becomes white in fall/winter); white face and underparts. Bright red bill (tipped with black); *lower mandible juts up to a third beyond upper.* Reddish legs. *Juvenile:* Upperpart feathers paler, broadly fringed whitish, bill smaller, bill and legs duller. **VOICE:** Soft, short, barking notes. Also *kaup, kaup.* **RANGE:** Accidental vagrant inland. **HABITAT:** Bays, marshes, beaches, protected ocean waters.

DARK TERNS
AND SKIMMER
BLACK TERN
fall/
winter
fall/winter
spring/
summer
adult
spring/
summer
adults
fall/
winter
White-
winged Black
Tern for
comparison
BROWN
NODDY
BLACK
NODDY
SOOTY
TERN
juvenile
adult
BRIDLED
TERN
adult
juvenile
adult
BLACK SKIMMER
adult

SHOREBIRDS

Many shorebirds (or "waders," as they are called in the Old World) are real puzzlers to the novice, and to many experienced birders as well! There are 12 plovers in our area, and nearly 60 sandpipers and their allies. Many species have up to four different plumages: spring/summer adult (Apr.–Sept.), winter adult and first-winter (Oct.–Mar.), first-summer (Apr.–Sept.), and juvenile (July–Sept.). Being able to properly age many species is an important part of correctly identifying them. Noting size, shape, and feeding style is also a critical part of the identification process.

Plovers are usually more compact and thicker necked than most sandpipers, with a pigeonlike bill and larger eyes. They run in short starts and stops.

Sanderlings run.
Phalaropes swim
and spin.
Does it teeter like a
Spotted Sandpiper?
Is it slim like a
yellowlegs?
Or does it nod
like a Solitary
Sandpiper?
Or is it squat
like a turnstone?
Does it probe with a sewing-machine
motion like a dowitcher?

PLOVERS Family Charadriidae

Largely non-wading birds, more compactly built and thicker necked than most sandpipers, with shorter, pigeonlike bills and larger eyes. Call notes are distinctive and assist identification. Unlike most sandpipers, plovers run in short starts and stops, often on dry mud and in fields. Sexes alike or differ slightly. **FOOD:** Small marine life, insects, some vegetable matter. **RANGE:** Nearly worldwide.

BLACK-BELLIED PLOVER — Common

Pluvialis squatarola (see also p. 120)

11½ in. (29 cm). A large plover, recognized as one by stocky shape, hunched posture, and short, pigeonlike bill. *Spring/summer adult:* Has *black face and breast* (duller and mottled white in female) and pale speckled back. *Fall/winter adult, first-winter, and juvenile:* Look tan-gray to grayish white (juvenile scalier backed). *First-spring/-summer:* Variable between winter and summer. In any plumage, note *black wingpits* and white rump and tail in flight. **VOICE:** A plaintive slurred whistle, *tlee-oo-eee* or *whee-er-eee* (middle note lower). **SIMILAR SPECIES:** American Golden-Plover slightly smaller and slimmer, smaller billed, buffier or more golden on at least some feathering, has more distinct supercilium, and *lacks pattern of white in wings and tail.* **RANGE:** Uncommon to rare inland. **HABITAT:** Mudflats, marshes, beaches, rocks, short-grass habitats; in summer, tundra.

AMERICAN GOLDEN-PLOVER — Uncommon

Pluvialis dominica (see also p. 120)

10¼–10½ in. (26–27 cm). Size of Killdeer. Shows distinct wingtip extension, primaries extending well beyond tail tip when standing. *Spring/summer adult and first-summer:* Dark, spangled above with *whitish and pale yellow spots;* underparts black (slightly mottled white in female). *Broad white stripe* runs over eye and down sides of neck and breast. *Winter adult and first-winter:* Gray-brown, darker above than below, with distinct pale supercilium, dark crown. *Juvenile:* Similar to winter plumages, but back slightly brighter golden, more scaled. **VOICE:** A whistled *queedle* or *que-e-a* (dropping at end). **SIMILAR SPECIES:** Black-bellied Plover. See Pacific Golden-Plover and European Golden-Plover (p. 128). **HABITAT:** Prairies, mudflats, shores, short-grass pastures, sod farms; in summer, tundra.

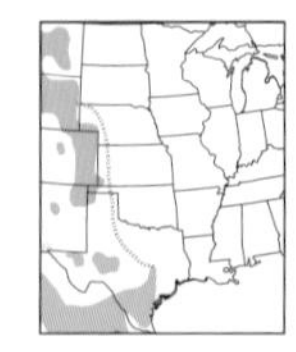

MOUNTAIN PLOVER *Charadrius montanus* — Scarce, local

9 in. (23 cm). *Spring/summer adult and first-summer:* White forehead and face, black forecrown and loral stripe, and brownish rufous back. *Winter adult and first-winter:* May be told from golden-plovers by tan-brown back devoid of mottling and by pale tan, unmarked breast; juvenile (not shown) is scalier backed. Has pale blue-gray legs, light wing stripe, and dark tail band. **VOICE:** A low whistle, variable. **SIMILAR SPECIES:** Black-bellied Plover, golden-plovers, Buff-breasted Sandpiper. **RANGE:** Casual vagrant well east of range; accidental to East Coast. **HABITAT:** Plowed fields, short-grass plains, dry sod farms.

PLOVERS
fall/winter
fall/winter
juvenile
spring/
summer adult
BLACK-BELLIED
PLOVER
juveniles
AMERICAN
GOLDEN-PLOVER
fall/winter
spring/
summer
MOUNTAIN
PLOVER
fall/
winter
spring/
summer

COMMON RINGED PLOVER *Charadrius hiaticula* **Rare, local**

7½ in. (19 cm). A Eurasian species, very similar to Semipalmated Plover; distinguished by voice, slightly longer bill, darker cheeks. Male averages bolder supercilium, wider breast-band. Less extensive webbing between toes is difficult to see. **VOICE:** A softer, more minor *poo-eep* or *too-li.* **RANGE:** Breeds in e. Canadian Arctic; winters in Old World. Accidental migrant in East. **HABITAT:** Same as Semipalmated Plover.

SEMIPALMATED PLOVER **Common**

Charadrius semipalmatus (see also p. 120)

7¼ in. (18 cm). Half the size of Killdeer, with *single dark breast-band. Adult:* Bill orangey with black tip or (in winter) nearly all dark. Male brighter and with more blackish than female; spring/summer brighter than fall/winter. *Juvenile:* Like winter female but back slightly scaly. **VOICE:** A plaintive, upward-slurred *chi-we* or *too-li.* **SIMILAR SPECIES:** Darker above than Piping and Snowy Plovers; the latter also has thinner bill, darker legs. **HABITAT:** Shores, tidal flats; in summer, tundra.

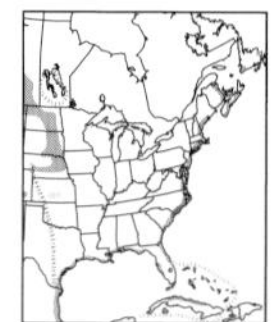

PIPING PLOVER **Uncommon, endangered/threatened**

Charadrius melodus (see also p. 120)

7¼ in. (18 cm). Quite pallid in color, like dry sand. Legs yellow or orange. *Spring/summer male:* Bill has yellow-orange base, black tip; black band on upper breast can be complete or incomplete. *Female and juvenile:* Black on collar less distinct or lacking, bill dark. Note tail pattern. **VOICE:** A plaintive whistle: *peep-lo* (first note higher). **SIMILAR SPECIES:** Snowy and Semipalmated Plovers. **RANGE:** Rare in interior East. **HABITAT:** Sandy beaches, dry mudflats; in summer, also lakeshores and river islands. Midwestern populations considered endangered, others threatened.

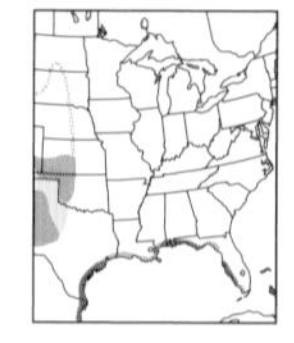

SNOWY PLOVER *Charadrius nivosus* (see also p. 120) **Uncommon**

6¼–6½ in. (16–17 cm). A pale, flatter-headed plover of beaches and alkaline flats. Note *slim black bill,* dark (sometimes pale) legs. *Male:* Has *dark ear patch,* paler and better marked in summer than winter. *Female and juvenile:* Duller, lack black in winter. **VOICE:** A musical whistle, *pe-wee-ah* or *o-wee-ah;* also a low *prit.* **SIMILAR SPECIES:** Piping Plovers are rounder headed, have brighter orange legs, and paler rump and uppertail coverts in flight. **RANGE:** Casual to accidental vagrant in East away from Gulf Coast. **HABITAT:** Beaches, sandy flats, alkaline lakeshores.

WILSON'S PLOVER *Charadrius wilsonia* (see also p. 120) **Uncommon**

7¾–8 in. (19–20 cm). Larger than Semipalmated Plover, with *wider breast-band* and longer, *heavier black bill.* Legs pinkish gray. Male has black breast-band in summer; female and first-year male browner. **VOICE:** An emphatic *whit!* or *wheet!* **RANGE:** Casual vagrant to midwestern states and well north of range. **HABITAT:** Open beaches, tidal flats, sandy islands.

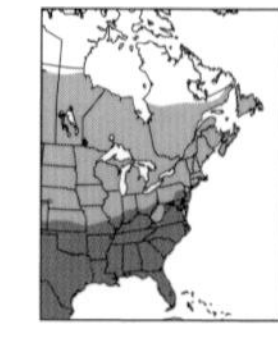

KILLDEER *Charadrius vociferus* (see also p. 120) **Common**

10½ in. (27 cm). The common, noisy plover of farm country and ball fields. Note *two black breast-bands* (chick has only one band and might be confused with Wilson's Plover). In flight or distraction display (near nest), shows *rusty orange rump,* longish tail, white wing stripe. Sexes similar. **VOICE:** A loud, insistent *kill-deeah,* repeated. Also a plaintive *dee-ee* (rising), *dee-dee-dee,* etc. **SIMILAR SPECIES:** Smaller banded plovers have single breast-band. **HABITAT:** Fields, airports, lawns, riverbanks, mudflats, shores.

BANDED PLOVERS
spring/ summer adult male
COMMON RINGED PLOVER
fall/winter
SEMIPALMATED PLOVER
spring/ summer adult male
spring/ summer adult male
PIPING PLOVER
fall/ winter
SNOWY PLOVER
spring/ summer adult male, breast-band may be unbroken
spring/summer adult male
adult
first-year female
WILSON'S PLOVER
KILLDEER
adult and first-years
chick
Adult Males
Common Ringed
Piping
Killdeer
Wilson's
Semipalmated
Snowy

OYSTERCATCHERS Family Haematopodidae

Large shorebirds with long, laterally flattened, chisel-tipped, red bills. Sexes alike. **FOOD:** Mollusks, crabs, marine worms. **RANGE:** Widespread on coasts of world; inland in some areas of Europe and Asia.

AMERICAN OYSTERCATCHER *Haematopus palliatus* **Fairly common**

17½–18½ in. (44–47 cm). A noisy, thickset, black-headed shorebird with dark back, white belly, and large white wing and tail patches. Outstanding feature is large straight red bill, flattened laterally. Legs pale pink. *Juvenile and first-year:* Bill dark-tipped; upperpart feathers fringed pale when fresh (juvenile). **VOICE:** A piercing *wheep!* or *kleep!;* a loud *pic, pic, pic.* **RANGE:** Accidental vagrant inland. **HABITAT:** Coastal beaches, tidal flats.

STILTS and AVOCETS Family Recurvirostridae

Tall, slim waders with very long legs and very slender bills (bent upward in avocets). Sexes fairly similar. **FOOD:** Insects, crustaceans, other aquatic life. **RANGE:** N., Cen., and S. America, Africa, s. Eurasia, Australia, Pacific region.

BLACK-NECKED STILT *Himantopus mexicanus* **Fairly common**

14 in. (36 cm). A large, extremely slim wader; black above (female and juvenile have browner backs), white below. Note *extremely long, dark pinkish legs,* needlelike bill. In flight, black *unpatterned* wings contrast strikingly with white rump, tail, and underparts. **VOICE:** A sharp yipping: *kyip, kyip, kyip.* **SIMILAR SPECIES:** Fall/winter American Avocet. **RANGE:** Rare vagrant well east of range and to East Coast. **HABITAT:** Marshes, mudflats, pools, shallow lakes (fresh and alkaline), flooded fields.

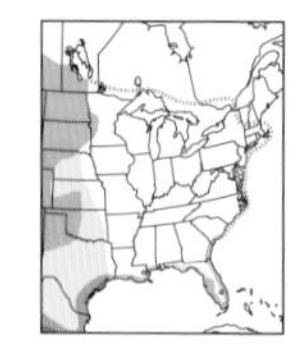

AMERICAN AVOCET *Recurvirostra americana* **Fairly common**

18 in. (46 cm). A large, slim shorebird with very slender, *upturned bill,* more upturned in female. This and striking white-and-black pattern make this bird unique. In spring/summer, head and neck pinkish tan or orangey buff; in fall/winter, this color replaced by pale gray. Avocets feed with scythelike sweep of head and bill. **VOICE:** A sharp *wheek* or *kleet,* excitedly repeated. **RANGE:** Rare vagrant to East Coast. **HABITAT:** Mudflats, shallow lakes, marshes, prairie ponds.

OYSTERCATCHER, STILT, AND AVOCET

SANDPIPERS, PHALAROPES, and ALLIES
Family Scolopacidae

Small to large shorebirds. Bills more slender than those of plovers. Sexes mostly similar, except in phalaropes. **FOOD:** Insects, crustaceans, mollusks, worms, etc. **RANGE:** Cosmopolitan.

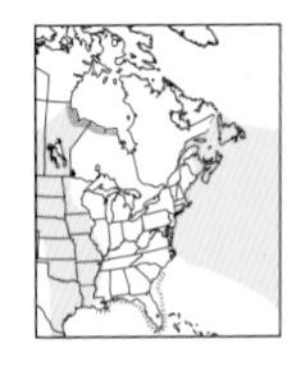

HUDSONIAN GODWIT *Limosa haemastica* (see also p. 122) **Uncommon**
15–15½ in. (38–39 cm). Long, *slightly upturned* bill; *blackish underwing linings;* black tail *ringed broadly with white. Spring/summer:* Male ruddy-breasted, female duller. *Fall/winter:* Gray-backed, pale-breasted; juvenile with more patterned scaly back. **VOICE:** *Tawit!* (or *godwit!*); higher pitched than Marbled Godwit call. **SIMILAR SPECIES:** Bar-tailed and Black-tailed Godwits (see p. 128). **RANGE:** Rare migrant throughout most of East. **HABITAT:** Mudflats, prairie pools; in summer, marshy taiga and tundra.

MARBLED GODWIT *Limosa fedoa* (see also p. 122) **Uncommon**
17½–18½ in. (44–46 cm). *Buff-brown* with *cinnamon* underwing linings. Spring/summer adults have more barring underneath than fall/winter birds and juveniles. **VOICE:** An accented *kerwhit! (godwit!);* also *raddica, raddica.* **SIMILAR SPECIES:** When head tucked in, difficult to tell from Long-billed Curlew except slightly smaller and thinner, leg color blackish (more blue-gray in the curlew), supercilium averages more distinct. See Bar-tailed Godwit (p. 128). **RANGE:** Rare migrant or vagrant inland. **HABITAT:** Prairies, pools, shores, mudflats, beaches.

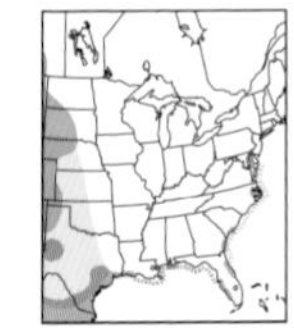

LONG-BILLED CURLEW **Uncommon to rare**
Numenius americanus (see also p. 122)
22–24 in. (55–60 cm). Note *very long, sickle-shaped bill* (4–8½ in.; 10–21 cm). Larger than Whimbrel and warmer colored overall; lacks distinct dark crown stripes. From below has *cinnamon underwing linings.* Ages and sexes rather similar; female larger with longer bill. **VOICE:** A loud *cur-lee* (rising inflection) or *curlew;* rapid, whistled *kli-li-li-li;* on breeding grounds, a longer drawn-out *curleeeeeeeeuuu.* **SIMILAR SPECIES:** See Marbled Godwit. Whimbrel smaller, grayer (lacks cinnamon tones), has shorter and blacker bill. **RANGE:** Rare to casual migrant or winter vagrant to East Coast. **HABITAT:** High plains, rangeland; in winter, cultivated land, mudflats, beaches, salt marshes.

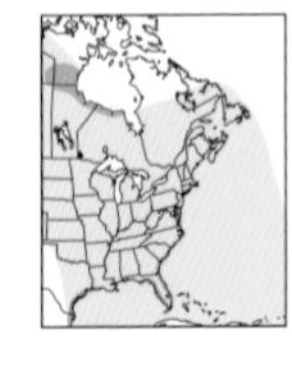

WHIMBREL *Numenius phaeopus* (see also p. 122) **Fairly common**
17–18 in. (43–46 cm). A large gray-brown shorebird with *decurved bill.* Much grayer brown than Long-billed Curlew; bill shorter (2¾–4 in.; 7–10 cm); crown *striped;* in flight, uppersurface of outer two primaries has distinct white shafts. Ages and sexes similar through year. Casual vagrant "Eurasian" subspecies (*N. p. phaeopus*) has white wedge up lower back (p. 128). **VOICE:** Five to seven short, rapid whistles: *chee-chee-chee-chee-chee-chee.* **SIMILAR SPECIES:** Long-billed Curlew. **RANGE:** Uncommon to rare inland. **HABITAT:** Mudflats, beaches; in summer, tundra.

LARGE SANDPIPERS
fall/winter
spring/summer
fall/winter
HUDSONIAN GODWIT
spring/
summer male
MARBLED
GODWIT
LONG-BILLED
CURLEW
WHIMBREL

RUDDY TURNSTONE **Fairly common**
Arenaria interpres (see also p. 120)
9½ in. (24 cm). A squat, robust, *orange-legged* shorebird, with *harlequin pattern. Spring/summer:* Russet back and curious face and breast pattern make this bird unique, but in flight it is even more striking. *Fall/winter and juvenile:* Duller, but retain body feathers and striking upperpart and wing patterns. **VOICE:** A staccato *tuk-a-tuk* or *kut-a-kut;* also a single *kewk.* **RANGE:** Uncommon to rare inland. **HABITAT:** Beaches, mudflats, rocky shores, jetties; in summer, tundra.

PURPLE SANDPIPER *Calidris maritima* (see also p. 126) **Uncommon**
9 in. (23 cm). A stocky, dark sandpiper seen on rocks, jetties, and breakwaters along our n. Atlantic Coast in winter. *Fall/winter:* Slate gray with white belly. At close range, note short yellow-orange legs, dull orangish base of bill, white eye-ring. *Spring/summer:* Much browner, more heavily streaked above and below with purplish sheen to some back feathers. **VOICE:** A low, scratchy *weet-wit* or *twit.* **SIMILAR SPECIES:** Fall/winter Dunlin, also found roosting on jetties, has plain brown back and breast, black bill and legs. **RANGE:** Casual vagrant inland and to Gulf Coast. **HABITAT:** Wave-washed rocks, jetties, rarely sandy shoreline. Often quite tame. In summer, coastal tundra.

SANDERLING *Calidris alba* (see also p. 126) **Common**
8 in. (20 cm). A plump, active sandpiper of outer beaches, where it chases retreating waves like a wind-up toy. Note bold *white wing stripe* in flight. *Spring/summer:* Bright rusty about head, back, and breast (male averages brighter than female). *Fall/winter:* The palest sandpiper; snowy white underparts, plain pale gray back, *black shoulders. Juvenile:* Has salt-and-pepper pattern on back and breast sides. **VOICE:** A short *kip* or *quit.* **SIMILAR SPECIES:** Semipalmated and Western Sandpipers (p. 108), Red-necked Stint (p. 130). **RANGE:** Uncommon to rare inland. **HABITAT:** Beaches, mudflats, lakeshores; when nesting, stony tundra.

DUNLIN *Calidris alpina* (see also p. 126) **Common**
8½–8¾ in. (22–23 cm). Larger than a peep (p. 108), with *longish, droop-tipped bill.* Black legs. *Spring/summer: Rusty red above,* with *black patch on belly. Fall/winter:* Unpatterned gray or gray-brown above, with *grayish wash across breast. Juvenile* (this plumage rarely seen away from nesting areas): Rusty above, with buffy breast and suggestion of belly patch. **VOICE:** A nasal, rasping *cheezp* or *treezp.* **SIMILAR SPECIES:** Winter Sanderling and (smaller) Semipalmated and Western Sandpipers have clean white breast; Sanderling also paler above and has straighter bill. See also Purple Sandpiper, Curlew Sandpiper (p. 130). **HABITAT:** Tidal flats, beaches, muddy pools; in summer, moist tundra.

RED KNOT *Calidris canutus* (see also p. 126) **Uncommon, threatened**
10½ in. (27 cm). Larger than Sanderling. Stocky, with medium-length, straight bill and short legs. *Spring/summer:* Face and underparts *pale robin red;* back mottled with black, gray, and russet. *Fall/winter:* A dumpy wader with washed-out gray look and mottled flanks; medium bill, *pale rump* in flight, greenish legs. *Juvenile:* Has *pale feather edgings* above and pale buff wash on breast. **VOICE:** A low, mellow *tooit-wit* or *wah-quoit;* also a short, low *tchrrt.* **SIMILAR SPECIES:** Dowitchers. **RANGE:** Uncommon to rare inland. **HABITAT:** Tidal flats, sandy beaches, shores; tundra when breeding. Populations breeding in Canadian Arctic and migrating through East (subspecies *C. c. rufa*) considered threatened.

fall/winter
RUDDY TURNSTONE
spring/summer male
PURPLE SANDPIPER
spring/summer
fall/winter
juvenile
SANDERLING
fall/winter
spring/summer
juvenile
fall/winter
DUNLIN
spring/summer
fall/winter
juvenile
RED KNOT
spring/summer
fall/winter

PEEPS

Collectively, the three common small sandpipers of N. America are nicknamed "peeps" (other slightly larger *Calidris* sandpipers also are referred to sometimes as "peeps"). In Old World, similar small peeps are called "stints."

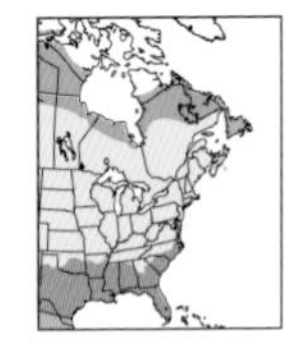

LEAST SANDPIPER *Calidris minutilla* (see also p. 126) **Common**

6 in. (15 cm). Distinguished from the other two common peeps by its slightly smaller size, *browner* upperparts and breast, and *yellowish or greenish*—not blackish—legs (but which might appear dark if caked in mud). *Bill slighter, finer, and slightly drooped at tip.* Plumage variable. *Adult:* Mostly brownish with some rufous and black in back (spring/summer) or brownish gray (fall/winter). *Juvenile:* Much brighter, with extensive rufous on upperparts and buff wash across breast. **VOICE:** A thin *krreet, kree-eet.* **SIMILAR SPECIES:** Western and Semipalmated Sandpipers have blackish legs, thicker-based bill, paler upperparts, and different voice; whitish breast in fall/winter plumage. **HABITAT:** Mudflats, marshes, rain pools, shores, flooded fields; in summer, taiga wetlands.

SEMIPALMATED SANDPIPER **Common**
Calidris pusilla (see also p. 126)

6¼ in. (16 cm). A small black-legged peep with *straight,* somewhat *bulbous-tipped bill* of short length (female's bill longer than male's). *Spring/summer:* Gray-brown above, many birds with a tinge of russet to cheeks and back; dark streaks on breast. *Fall/winter:* Uniformly plain gray across upperparts (rarely seen in our area). *Juvenile:* Breast washed with buff and with fine streaks on sides; scaly upperpart pattern rather uniform, with pale feather edges tinged buff (sometimes reddish) when fresh. **VOICE:** Call a *chit* or *chirt* (lacks *ee* sound of Least and Western Sandpipers). **SIMILAR SPECIES:** Most Western Sandpipers have *longer bill, slightly drooped* at tip. Spring/summer Western more rufous above, more heavily streaked below, particularly on flanks. Juvenile Western has rusty scapulars forming a diagonal bar on grayer back and slightly paler face. Not all birds distinguishable in winter plumages. Least Sandpiper smaller, browner, thinner billed; has *yellowish or greenish* legs; in fall/winter plumage has darker breast. See also Red-necked and Little Stints (p. 130). **HABITAT:** Mudflats, marshes, shores, beaches; in summer, tundra.

WESTERN SANDPIPER *Calidris mauri* **Uncommon**

6½ in. (17 cm). Very similar to Semipalmated Sandpiper. Legs black. Bill averages thicker at base and longer than Semipalmated's and *droops near tip. Spring/summer: Heavily spotted* on breast and flanks; *rusty scapulars, crown, and ear patch. Fall/winter:* Gray or gray-brown above, unmarked whitish below. *Juvenile:* Buff wash on breast; scaly upperparts, like juvenile Semipalmated but with distinct rusty scapular bar. **VOICE:** A distinct high-pitched *jeet* or *cheet,* unlike lower, soft *chirt* of Semipalmated. **SIMILAR SPECIES:** Semipalmated and Least Sandpipers, Dunlin. Many male Westerns may be particularly difficult to separate from female Semipalmateds; see also Voice. Semipalmated rarely winters in our area, but Western regularly does. **HABITAT:** Shores, beaches, mudflats, marshes; in summer, tundra.

SMALL "PEEP" SANDPIPERS

WHITE-RUMPED SANDPIPER — Uncommon

Calidris fuscicollis (see also p. 126)

7½ in. (19 cm). Larger than Semipalmated Sandpiper, smaller than Pectoral Sandpiper. The only peep with completely *white rump.* At rest, this long-winged bird has *tapered* look, with *wingtips extending well beyond tail.* Distinct pale supercilium. *Spring/summer:* Some rusty on crown, face, back. *Dark streaks and chevrons on sides extend to flanks.* Base of lower mandible bright reddish orange. *Juvenile:* Spangled upperparts scalloped rufous and white, fine streaks to buff-washed breast; bold *white eyebrow.* Winter birds are grayer and plainer; not seen in our area. **VOICE:** A high, thin, mouselike *jeet,* like two flint pebbles scraping. **SIMILAR SPECIES:** Long wings and very attenuated look shared only by Baird's Sandpiper among other peeps, but Baird's buffier brown overall, has scalier back and dark center to rump, lacks bold supercilium and dark streaks on flanks, and has much lower pitched call. **RANGE:** Rare along East Coast. **HABITAT:** Prairie pools, shores, mudflats, marshes; in summer, tundra.

BAIRD'S SANDPIPER *Calidris bairdii* (see also p. 126) — Uncommon

7½ in. (19 cm). Larger than Semipalmated and Western Sandpipers, with more *long-winged, tapered look* (wings extend ½ in., 1 cm, beyond tail tip). Legs *black. Spring/summer:* Grayish white upperparts with black centers to back feathers; white throat; black breast streaking heaviest to sides. *Juvenile:* Head and breast washed buff, throat and breast finely streaked; back feathers with *dark centers and rich buff to buff-orange fringing,* creating highly scaled appearance. *Fall/winter:* Browner and duller than juvenile; not found in our area. **VOICE:** Call a low *kreep* or *kree;* a rolling trill. **SIMILAR SPECIES:** White-rumped and Pectoral Sandpipers. Buff-breasted Sandpiper buffier below, without streaks, and has *yellow* (not *black*) legs. **RANGE:** Scarce migrant or vagrant to East Coast. **HABITAT:** Pond margins, grassy mudflats, shores, upper beaches; in summer, tundra.

PECTORAL SANDPIPER — Fairly common

Calidris melanotos (see also p. 124)

8¼–8¾ in. (21–23 cm). Medium-sized (but variable; male larger than female); plump-bodied, but neck longer than in smaller peeps. Note that heavy breast streaks end rather *abruptly,* like a bib. Dark back with two white stripes. Wing stripe faint or lacking; crown variably rusty. Legs usually dull yellowish. Bill may be pale yellow-brown at base. On breeding grounds, males display by expanding breast, exposing black-based feathers. *Juvenile:* Similar to adult, but brighter rufous present on upperparts and crown, buffier wash on breast under streaking. **VOICE:** A low, reedy *churrt* or *trrip, trrip.* **SIMILAR SPECIES:** Sharp-tailed Sandpiper (p. 130). Baird's and Least Sandpipers smaller, usually lack sharp breastband; legs of Baird's black. **HABITAT:** In migration, prairie pools, sod farms, muddy shores, fresh and tidal marshes; in summer, tundra.

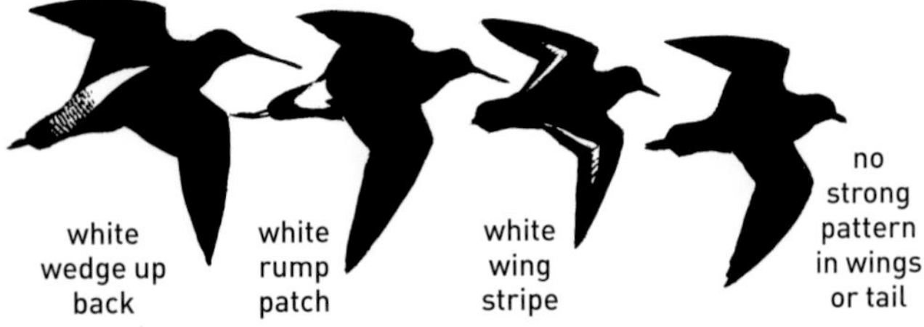

SANDPIPERS
juvenile
spring/
summer
WHITE-RUMPED
SANDPIPER
juvenile
spring/
summer
BAIRD'S
SANDPIPER
breeding
male
juvenile
spring/
summer
PECTORAL
SANDPIPER

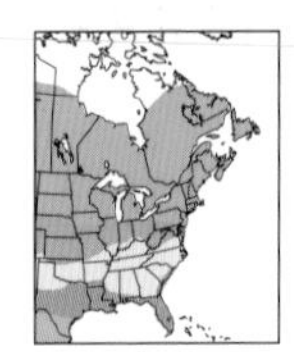

SPOTTED SANDPIPER Fairly common

Actitis macularius (see also p. 126)

7½ in. (19 cm). A widespread sandpiper along shores of small freshwater lakes and streams. Usually solitary. Teeters rear body up and down nervously. Note moderately *long tail,* prominent wing stripe. *Spring/summer:* Note *round breast spots. Fall/winter and juvenile:* No spots; brown above, with white line over eye (juvenile lightly scaled above). Dusky smudge encloses white wedge near shoulder. Flight distinctive: wings beat in a *shallow arc,* giving a stiff, bowed appearance. **VOICE:** A clear *peet* or *peet-weet!* or *peet-weet-weet-weet-weet.* **SIMILAR SPECIES:** Solitary Sandpiper. **HABITAT:** Pebbly shores, ponds, streamsides, marshes; in winter, also seashores, rock jetties.

STILT SANDPIPER Uncommon

Calidris himantopus (see also pp. 116 and 124)

8½ in. (22 cm). A tall sandpiper with slight *droop* to tip of bill, legs long and greenish yellow. Feeds like a dowitcher (sewing-machine motion) but *tilts tail up* more than a dowitcher while probing. *Spring/summer:* Heavily marked below with *transverse bars;* back brown with black mottling. Note *rusty cheek patch. Fall/winter:* Yellowlegs-like but unmarked gray above, dark-winged and *white-rumped;* note also more *greenish legs* and *white eyebrow. Juvenile:* Brownish-buff wash to breast, upperpart feathers brown with even, pale edgings. **VOICE:** A single *whu* (like Lesser Yellowlegs but lower, hoarser). **SIMILAR SPECIES:** Yellowlegs. Dowitchers pudgier, have longer, yellowish-based, less drooped bills, and in flight show white wedge up back. See also juvenile Wilson's Phalarope (see pp. 116 and 124) and Curlew Sandpiper (p. 130). **HABITAT:** Shallow pools, mudflats, marshes; in summer, tundra.

UPLAND SANDPIPER Uncommon

Bartramia longicauda (see also p. 124)

12 in. (30–31 cm). A "pigeon-headed" brown sandpiper; larger than Killdeer. Short bill, *small head,* shoe-button eye, thin neck, and *long tail* are helpful points. Often perches with erect posture on fenceposts and poles; on alighting, holds wings elevated. Ages and sexes similar through year. **VOICE:** A mellow, whistled *kip-ip-ip-ip,* often heard at night. Song a weird windy whistle: *whoooleeeeee, wheelooooooooooo.* **SIMILAR SPECIES:** Buff-breasted Sandpiper smaller, richer buff and unmarked below. See also Eskimo Curlew (p. 128). **HABITAT:** Grassy prairies, open meadows, fields, airports, sod farms.

BUFF-BREASTED SANDPIPER Scarce

Calidris subruficollis (see also p. 124)

8¼ in. (21 cm). No other small shorebird is as *rich buffy* below (paling to whitish on undertail coverts). A docile, buffy bird, with erect stance, small head, short bill, and yellowish legs. Dark eye stands out on plain face. In flight or in "display," buff body plumage contrasts with underwing, which is *white* with marbled tip and distinct dark crescent at base of primaries. Ages and sexes similar through year. **VOICE:** A low, trilled *pr-r-r-reet;* sharp *tik.* **SIMILAR SPECIES:** Baird's, Pectoral, and Upland Sandpipers. Juvenile Ruff (p. 130). **RANGE:** Rare inland away from Mississippi Valley. **HABITAT:** Dry dirt, sand, and short-grass habitats, including drying lakeshores, pastures, sod farms; in summer, drier tundra ridges.

fall/winter
flies with
stiff wings
spring/
summer
SPOTTED SANDPIPER
juvenile
fall/
winter
spring/
summer
STILT
SANDPIPER
UPLAND
SANDPIPER
BUFF-BREASTED
SANDPIPER

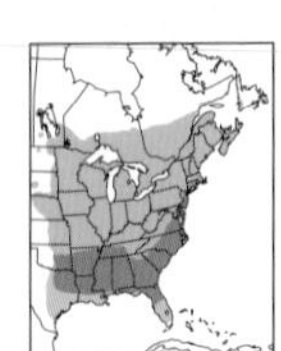

AMERICAN WOODCOCK

Fairly common but secretive

Scolopax minor (see also p. 124)

11 in. (28 cm). A woodland-loving shorebird. Near size of Northern Bobwhite, with extremely long bill and large bulging eyes placed high on head. Rotund, almost neckless, with leaflike brown camouflage pattern, broadly barred crown. When flushed, produces whistling sound with wings. Ages and sexes similar through year. **VOICE:** At dusk in spring, a nasal *beezp* (suggesting nighthawk). Aerial "song" a chipping trill made by wings as bird ascends, changing to a bubbling twittering on descent. **HABITAT:** Wet thickets, moist woods, brushy swamps. Spring courtship by male is a crepuscular display ("sky dance") high over semiopen fields, pastures.

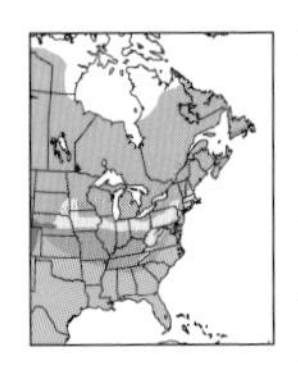

WILSON'S SNIPE *Gallinago delicata* (see also p. 124)

Fairly common

10¼–10½ in. (26–27 cm). A tight-sitting bog, marsh, and wet-field prober; on nesting grounds may be seen standing on posts. Note *extremely long bill.* Brown, with *buff stripes on back* and a *striped head.* Ages and sexes similar through year. Flies off in *zigzag,* showing *short rusty orange tail.* **VOICE:** When flushed, a rasping *scaip.* Song a measured *chip-a, chip-a, chip-a,* etc. In high aerial display, a winnowing *huhuhuhuhuhuhu.* **SIMILAR SPECIES:** Dowitchers. **HABITAT:** Marshes, bogs, ditches, wet fields and meadows.

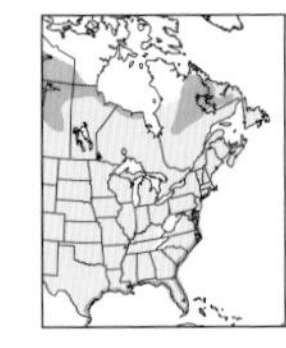

SHORT-BILLED DOWITCHER

Common

Limnodromus griseus (see also p. 126)

11–11¼ in. (27–28 cm). A snipelike bird of open mudflats. Note long bill, sewing-machine feeding motion, and in flight, *long white wedge up back. Spring/summer:* Underparts rich rusty with some barring on flanks. *Fall/winter:* Gray. *Juvenile:* Brighter upperparts, buff wash to neck and breast; *patterned tertial feathers* (fringes broken orange and black). Western birds (uncommon in East) brighter in spring. **VOICE:** A staccato, muted *tu-tu-tu;* pitch of Lesser Yellowlegs. **SIMILAR SPECIES:** Bill length overlaps with that of Long-billed Dowitcher, but use this with caution, especially among groups of birds; shorter bill also results in more angled back when feeding, on average. In fall/winter, differences in call notes and habitat are usually the best way to distinguish the dowitchers; Long-billed migrates later in fall than Short-billed. See also Stilt Sandpiper, Red Knot. **RANGE:** Uncommon to rare inland. **HABITAT:** More frequent on large tidal mudflats than Long-billed Dowitcher. In summer, taiga and tundra.

LONG-BILLED DOWITCHER

Uncommon

Limnodromus scolopaceus (see also p. 126)

11½ in. (29 cm). Shows more round-bodied profile than Short-billed Dowitcher; dark tail bars average wider; bill averages longer (see Short-billed). *Spring/summer:* Underparts *evenly bright rusty to lower belly* (white or very pale in most Short-billeds), with dark spotting on neck and barring on sides. Dark bars on tail broader, giving tail a darker look. *Fall/winter:* Averages darker than Short-billed, with smoother gray breast and darker centers to scapulars. *Juvenile:* Gray tertials *with unpatterned solid pale fringe;* unlike "tiger barring" edges of Short-billed. **VOICE:** A single sharp, high *keek,* occasionally given in twos or threes but differs in quality from Short-billed call. **SIMILAR SPECIES:** See Short-billed Dowitcher. **HABITAT:** Shallow pools, marshes, mudflats during migration; when breeding, tundra. More partial to fresh water than Short-billed, especially in winter, but some overlap.

AMERICAN
WOODCOCK
SHORT-
BILLED
DOWITCHER
winnowing
display
flight
WILSON'S
SNIPE
juvenile
eastern
spring/
summer
fall/winter
snipe
Short-billed
western
spring/
summer
probing
juvenile
spring/
summer
Long-
billed
LONG-BILLED
DOWITCHER
probing
fall/
winter

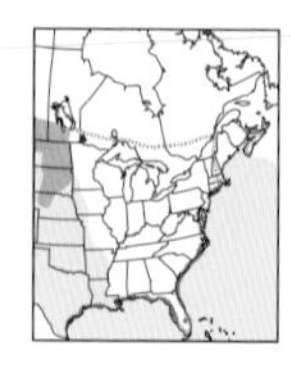

WILLET *Tringa semipalmata* (see also p. 122) **Fairly common**

15–16 in. (38–41 cm). Stockier than Greater Yellowlegs; has grayer look, heavier bill, blue-gray legs. In flight, note *striking black-and-white wing pattern.* At rest, this large wader is rather nondescript: gray above, mottled or barred below in spring/summer, unmarked in fall/winter. *Juvenile:* Browner above with light buff spots and bars. **VOICE:** A musical, repetitious *pill-will-willet* (in breeding season); a loud *kay-ee* (second note lower). Also a rapidly repeated *kip-kip-kip,* etc. In flight, *kree-ree-ree.* **SIMILAR SPECIES:** Greater Yellowlegs; see also dowitchers. **RANGE:** Uncommon to rare inland, away from breeding range. **HABITAT:** Marshes, wet meadows, mudflats, beaches.

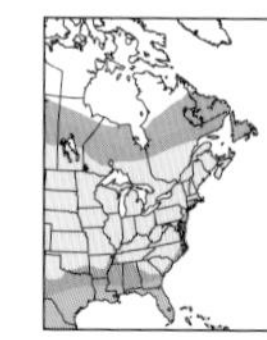

GREATER YELLOWLEGS **Common**

Tringa melanoleuca (see also p. 124)

14 in. (36 cm). Note *bright yellow legs* (shared with next species). A slim gray sandpiper; back checkered with gray, black, and white. Often teeters body. In flight, appears *dark-winged* (no stripe), with *whitish rump and tail.* Bill long, *slightly upturned, paler at base.* Spring/summer adults blacker above, more barred on breast; fall/winter birds grayer above, whiter below; juveniles pale brownish gray, evenly scaled above. **VOICE:** A three-note strident whistle, *dear! dear! dear!* or *teer-teer-turr* with emphasis on first note. **SIMILAR SPECIES:** Lesser Yellowlegs, Willet, Spotted Redshank (p. 128). **HABITAT:** Marshes, mudflats, streams, ponds, flooded fields; in summer, wooded muskeg, spruce bogs.

LESSER YELLOWLEGS *Tringa flavipes* (see also p. 124) **Common**

10½ in. (27 cm). Like Greater Yellowlegs but smaller (obvious when both species are together). Lesser's shorter, slimmer, all-dark bill is *straight* and about *equal to length of head;* Greater's appears slightly uptilted, paler based, and longer than bird's head. Readily separated by voice. Age and seasonal differences similar to Greater. **VOICE:** *Yew* or *yu-yu* (usually one or two notes); less forceful than three-syllable call of Greater. **SIMILAR SPECIES:** Solitary and Stilt Sandpipers, Wilson's Phalarope. Both yellowlegs species may swim briefly, like a phalarope. **HABITAT:** Marshes, mudflats, ponds, flooded fields; in summer, open, moist boreal woods and taiga.

SOLITARY SANDPIPER *Tringa solitaria* (see also p. 124) **Uncommon**

8½ in. (22 cm). Note *dark wings* and conspicuous *white sides of tail* (crossed by bold black bars). A dark-backed sandpiper, whitish below, with *light eye-ring* and greenish legs. Nods like a yellowlegs. Usually alone, seldom in groups. Ages and sexes fairly similar. **VOICE:** *Peet!* or *peet-weet-weet!* (more strident than Spotted Sandpiper call). **SIMILAR SPECIES:** Lesser Yellowlegs has bright yellow legs, white rump, lacks bold eye-ring. Spotted Sandpiper teeters tail (not head), has different wing and tail patterns. **HABITAT:** Streamsides, wooded swamps and ponds, ditches, freshwater marshes.

STILT SANDPIPER *Calidris himantopus* **See p. 112**

Fall/winter: Long yellow-green legs, slight droop to bill, white rump; distinct light supercilium.

WILSON'S PHALAROPE *Phalaropus tricolor* **See p. 118**

Fall/winter: Straight needle bill, clear white underparts, pale gray back, dull yellow legs.

SANDPIPERS
spring/summer
WILLET
fall/winter
LESSER
YELLOWLEGS
GREATER
YELLOWLEGS
SOLITARY
SANDPIPER
STILT
SANDPIPER
(p. 112) for
comparison
fall/winter
WILSON'S
PHALAROPE
(p. 118) for
comparison
fall/
winter

PHALAROPES

Shorebirds with lobed toes; more at home wading or swimming than on land. Phalaropes often spin like tops, rapidly dabbling for plankton and other marine invertebrates and insects. Female slightly larger and, in spring/summer, more colorful than male. **RANGE:** Circumpolar, with one species confined to Americas.

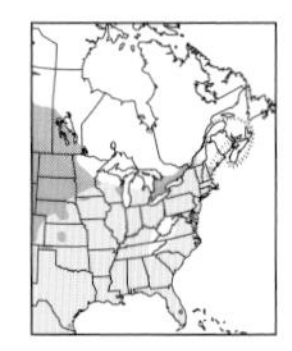

WILSON'S PHALAROPE — Uncommon

Phalaropus tricolor (see also pp. 116 and 124)

9¼ in. (23½ cm). A trim phalarope, plain-winged (no stripe) with white rump. In addition to spinning in water, also feeds by dashing about on shorelines. *Spring/summer:* Female unique, with *broad black face and neck stripe blending into cinnamon.* Male duller. *Fall/winter* (not found in our area): Whiter below, with no breast streaking. *Juvenile:* Buff and brown pattern above, buffy wash on breast. **VOICE:** A low, nasal *wurk;* also *check, check, check.* **SIMILAR SPECIES:** Other two phalaropes in fall have white wing stripe, dark rump, and bold dark patch through eye. See also yellowlegs (p. 116) and Stilt Sandpiper (p. 112), which may swim for brief periods of time. **RANGE:** Rare vagrant to East Coast. **HABITAT:** Shallow lakes, freshwater marshes, pools, shores, mudflats.

RED-NECKED PHALAROPE — Common

Phalaropus lobatus (see also p. 126)

7¾ in. (20 cm). Found primarily at sea, although not as pelagic as Red Phalarope. Note dark patch through eye and needlelike black bill. *Spring/summer:* Female gray above with *rufous chestnut on neck,* white throat and eyebrow. Male duller but similar in pattern. *Fall/winter:* Both sexes dark gray above with whitish streaks, white below; rare in our area in this plumage. *Juvenile:* Has distinct buff stripes on back. **VOICE:** A sharp *kit* or *whit,* similar to call of Sanderling. **SIMILAR SPECIES:** Red Phalarope. **RANGE:** Rare to casual inland. **HABITAT:** In migration, nearshore ocean, bays, ponds; in summer, tundra.

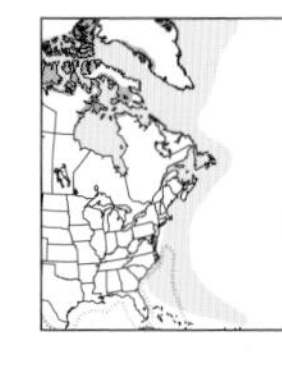

RED PHALAROPE *Phalaropus fulicarius* (see also p. 126) — Uncommon

8¼–8½ in. (21–22 cm). Seagoing habits and buoyant swimming at sea. *Spring/summer:* Female has deep *reddish underparts, white face,* mostly yellow bill. Male duller. *Fall/winter:* Both sexes plain pale gray above, white below; *dark patch* through eye. Bill mostly dark with *yellow base. Juvenile:* Has peach-buff wash on neck. **VOICE:** *Whit* or *kit,* higher than Red-necked Phalarope call. **SIMILAR SPECIES:** Red-necked Phalarope slightly slimmer, has more needlelike bill; juvenile darker gray above with thin pale back stripes. Thicker yellow-based bill of Red Phalarope visible at closer range. In our area, most or all phalaropes observed in winter are Reds. **RANGE:** Uncommon offshore; casual inland. **HABITAT:** More strictly pelagic (less coastal) than Red-necked in migration and winter. In summer, tundra.

PHALAROPES
spring/
summer
female
fall/winter
WILSON'S
PHALAROPE
fall/winter
juvenile
spring/
summer
male
phalaropes
spin
fall/
winter
RED-NECKED
PHALAROPE
juvenile
spring/
summer
female
fall/winter
spring/
summer
male
spring/summer
female
fall/winter
RED
PHALAROPE
juvenile
fall/winter
spring/summer male
lobed foot of
phalarope

PLOVERS and TURNSTONE in FLIGHT

Learning their distinctive flight calls can help substantially with identification.

PIPING PLOVER *Charadrius melodus* p. 100

Pale sand color above, wide black tail spot, whitish rump.
Call a plaintive whistle, *peep-lo* (first note higher).

SNOWY PLOVER *Charadrius nivosus* p. 100

Pale sand color above; tail with dark center, white sides; rump not white.
Call a musical whistle, *pe-wee-ah* or *o-wee-ah.*

SEMIPALMATED PLOVER *Charadrius semipalmatus* p. 100

Mud brown above; dark tail with white borders.
Call a plaintive upward-slurred *chi-we* or *too-li.*

WILSON'S PLOVER *Charadrius wilsonia* p. 100

Similar in pattern to Semipalmated Plover; larger with big bill.
Call an emphatic whistled *whit!* or *wheet!*

KILLDEER *Charadrius vociferus* p. 100

Tawny orange rump, longish tail.
Noisy; a loud *kill-deeah* or *killdeer;* also *dee-dee-dee,* etc.

BLACK-BELLIED PLOVER *Pluvialis squatarola* p. 98

Spring/summer adult: Black below, silvery white above, white undertail coverts.
Fall/winter, juvenile, and some first-summer birds: Pale grayish above and below.
Year-round: Black wingpits, white in wing, white rump and tail base.
Call a plaintive, slurred whistle, *tlee-oo-eee* or *whee-er-ee.*

AMERICAN GOLDEN-PLOVER *Pluvialis dominica* p. 98

Spring/summer adult: Black below, black undertail coverts.
Fall/winter, juvenile, and some first-summer birds: Speckled brown and buff above, grayish below.
Year-round: Underwing grayer than Black-bellied Plover's; no black in wingpits.
Call a querulous, whistled *queedle* or *que-e-a.*

RUDDY TURNSTONE *Arenaria interpres* p. 106

Harlequin pattern in face distinctive; bold white upperpart patterns.
Call a low, chuckling *tuk-a-tuk* or *kut-a-kut.*

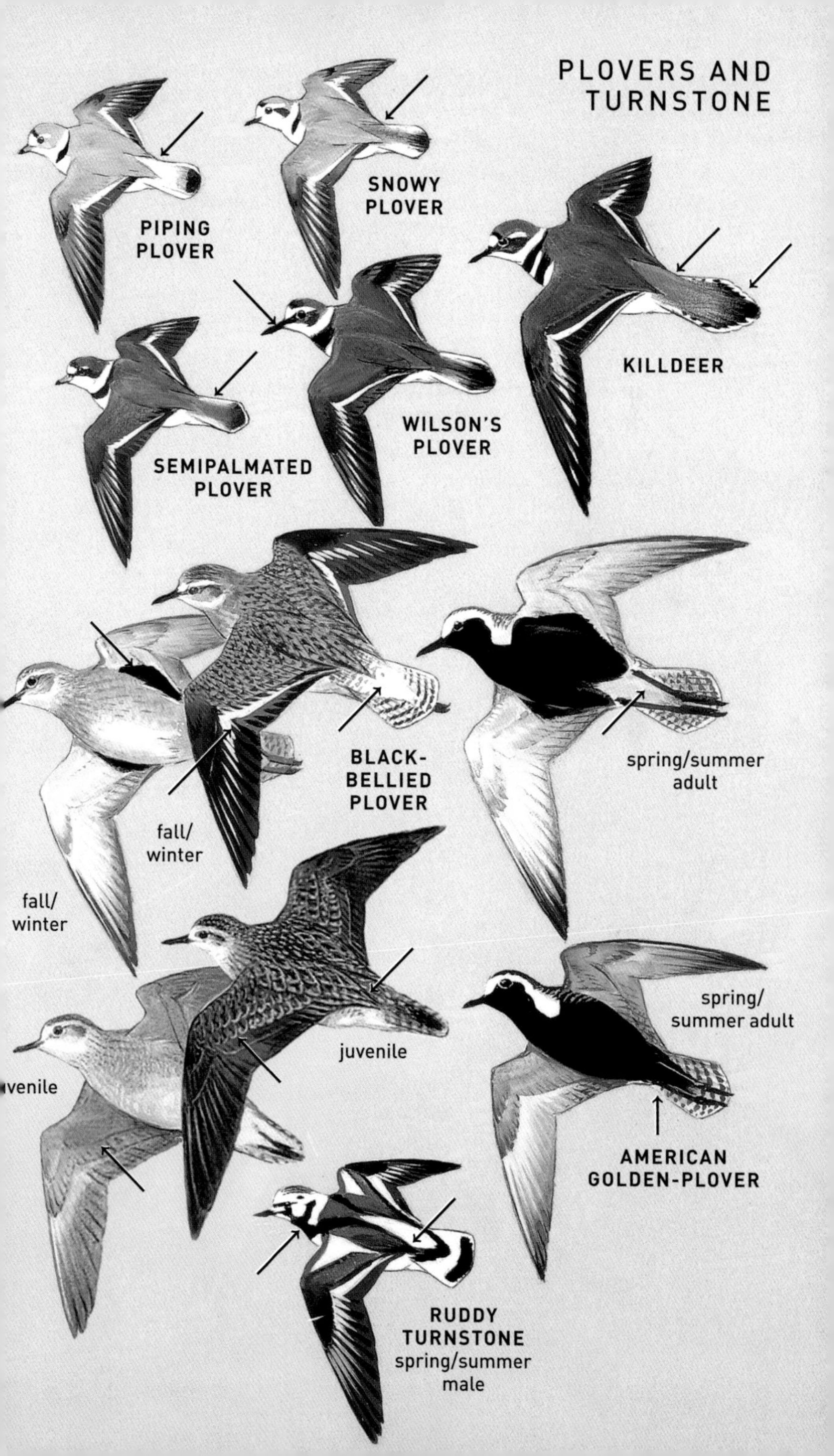
PLOVERS AND TURNSTONE
PIPING PLOVER
SNOWY PLOVER
KILLDEER
SEMIPALMATED PLOVER
WILSON'S PLOVER
BLACK-BELLIED PLOVER
fall/winter
fall/winter
spring/summer adult
juvenile
venile
spring/summer adult
AMERICAN GOLDEN-PLOVER
RUDDY TURNSTONE
spring/summer male

LARGE WADERS in FLIGHT

Learn to know their flight calls, which are distinctive.

HUDSONIAN GODWIT *Limosa haemastica* p. 104

Upturned bill, white wing stripe, ringed tail. Blackish underwing linings. Flight call *tawit!*, higher pitched than Marbled Godwit.

WILLET *Tringa semipalmata* p. 116

Contrasty black, gray, and white wing pattern from both above and below. Flight call a whistled one- to three-note *kree-ree-ree.*

MARBLED GODWIT *Limosa fedoa* p. 104

Long upturned bill, tawny brown color, cinnamon underwing linings. Flight call an accented *kerwhit!* (or *godwit!*).

WHIMBREL *Numenius phaeopus* p. 104

Decurved bill, gray-brown overall color, distinctly striped crown. Grayer than next species; lacks cinnamon underwing linings; bill darker, to blackish.
Flight call five to seven short, rapid whistles: *chee-chee-chee-chee-chee-chee.*

LONG-BILLED CURLEW *Numenius americanus* p. 104

Very long, sicklelike bill (longer in female than male); no head striping. Bright cinnamon underwing linings. Juvenile's bill shorter, but note head patterns.
Flight call a rapid, whistled *kli-li-li-li.* Also a husky *curr-liew* (second note rising).

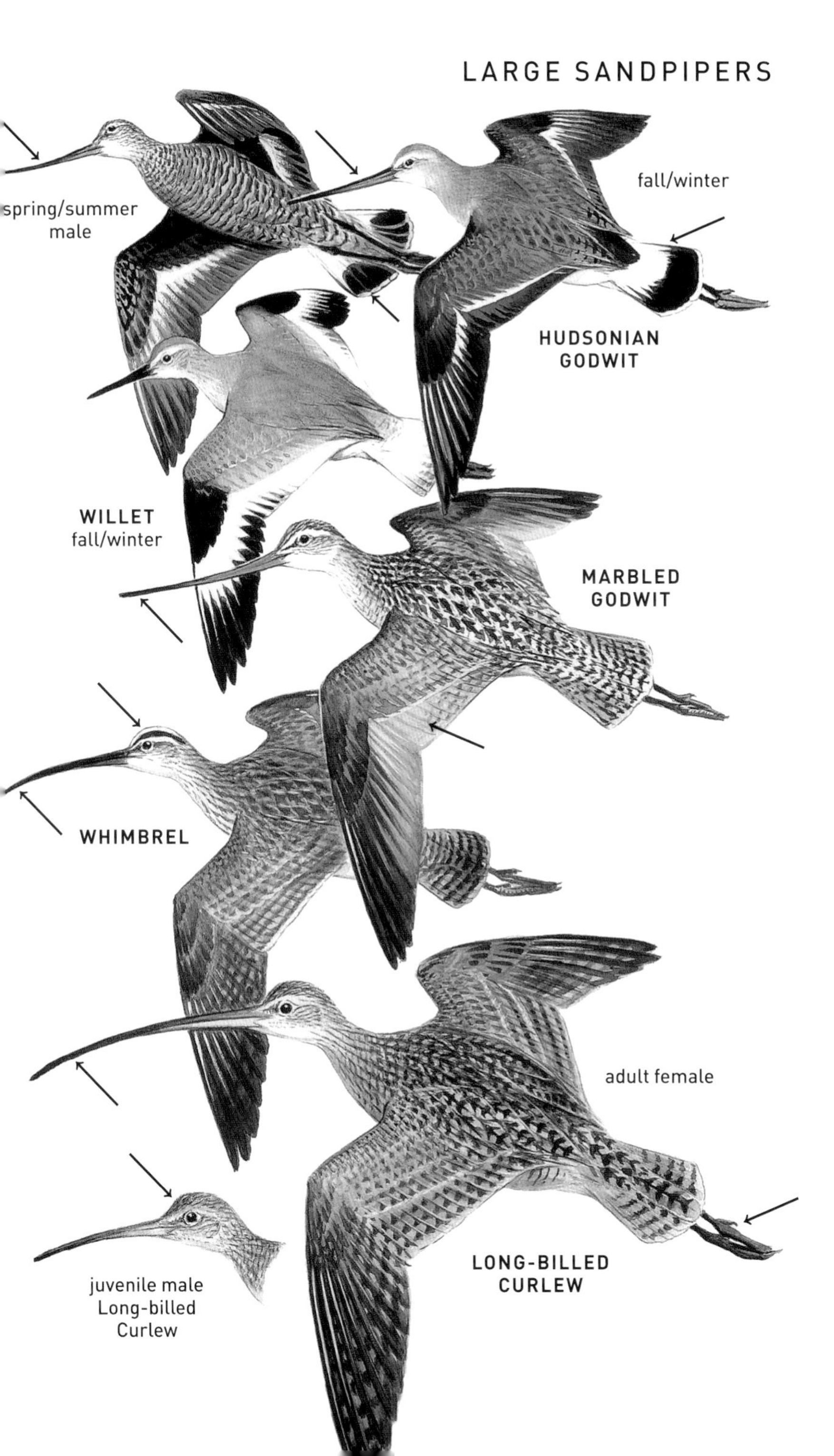
spring/summer
male
fall/winter
HUDSONIAN
GODWIT
WILLET
fall/winter
MARBLED
GODWIT
WHIMBREL
adult female
LONG-BILLED
CURLEW
juvenile male
Long-billed
Curlew

SNIPELIKE WADERS and SANDPIPERS in FLIGHT

These species and those on the next plate show their basic flight patterns. Most of these have unpatterned wings, lacking a pale stripe. All are shown in full color on other plates. Learning their distinctive flight calls helps with identifications.

WILSON'S SNIPE *Gallinago delicata* p. 114

Long bill, pointed wings, rusty orange tail, zigzag flight.
Flight call, when flushed, a distinctive rasping *scaip.*

AMERICAN WOODCOCK *Scolopax minor* p. 114

Long bill, rounded wings, chunky shape. Wings whistle in flight.
At dusk, aerial flight "song."

SOLITARY SANDPIPER *Tringa solitaria* p. 116

Very dark unpatterned wings (underwing dark also; pale in yellowlegs), conspicuous bars on white sides of tail.
Flight call *peet!* or *peet-weet-weet!* (higher than Spotted Sandpiper).

LESSER YELLOWLEGS *Tringa flavipes* p. 116

Similar to Greater Yellowlegs but smaller, with smaller bill.
Flight call *yew* or *yu-yu* (rarely three), softer than Greater's call.

GREATER YELLOWLEGS *Tringa melanoleuca* p. 116

Plain unpatterned wings, whitish rump and tail, long bill.
Flight call a distinctive and forceful three-note whistle, *dear! dear! dear!*

WILSON'S PHALAROPE *Phalaropus tricolor* p. 118

Fall/winter: Suggests Lesser Yellowlegs; smaller, whiter, bill needlelike. Differs in posture and behavior from Stilt Sandpiper.
Flight call a low, nasal *wurk.*

STILT SANDPIPER *Calidris himantopus* p. 112

Suggests Lesser Yellowlegs, but legs greenish yellow, bill longer and drooped. Differs in posture and behavior from Wilson's Phalarope.
Flight call a single *whu,* lower than Lesser Yellowlegs call.

UPLAND SANDPIPER *Bartramia longicauda* p. 112

Brown; small head, long tail.
Often flies "on tips of wings," like Spotted Sandpiper.
Flight call a mellow, whistled *kip-ip-ip-ip.*

BUFF-BREASTED SANDPIPER *Calidris subruficollis* p. 112

Buff below; white underwing linings with distinct "comma" marks; plain upperparts.
Flight call a low, trilled *pr-r-r-reet;* usually silent.

PECTORAL SANDPIPER *Calidris melanotos* p. 110

Like an oversized Least Sandpiper. Wing stripe faint or lacking.
Flight call a low, reedy *churrt* or *trrip, trrip.*

SNIPELIKE WADERS AND SANDPIPERS IN FLIGHT
WILSON'S SNIPE
AMERICAN WOODCOCK
SOLITARY SANDPIPER
GREATER YELLOWLEGS
LESSER YELLOWLEGS
WILSON'S PHALAROPE
fall/winter
STILT SANDPIPER
fall/winter
UPLAND SANDPIPER
BUFF-BREASTED SANDPIPER
PECTORAL SANDPIPER

SANDPIPERS and PHALAROPES in FLIGHT

DOWITCHERS *Limnodromus* spp. p. 114
Long bill, long wedge of white up back.
Flight calls of Long-billed and Short-billed (see text).

DUNLIN *Calidris alpina* p. 106
Fall/winter: Slightly larger than peeps, darker than Sanderling.
Flight call a nasal, rasping *cheezp* or *treezp.*

RED KNOT *Calidris canutus* p. 106
Fall/winter: Washed-out gray look, pale rump.
Flight call a low *knut.*

PURPLE SANDPIPER *Calidris maritima* p. 106
Slaty color.
Flight call a low *weet-wit* or *twit.*

WHITE-RUMPED SANDPIPER *Calidris fuscicollis* p. 110
White rump; but beware partial or poor views.
Flight call a mouselike squeak, *jeet.*

CURLEW SANDPIPER *Calidris ferruginea* p. 130
Fall/winter: Suggests Dunlin, but rump white.

RUFF *Calidris pugnax* p. 130
If seen well, oval white patch on each side of dark tail distinctive.

SPOTTED SANDPIPER *Actitis macularius* p. 112
Shallow wing stroke gives stiff, bowed effect; longish tail.
Flight call a clear *peet* or *peet-weet.*

SANDERLING *Calidris alba* p. 106
The most contrasting wing stripe of any small shorebird.
Flight call a sharp, metallic *kip* or *quit.*

RED PHALAROPE *Phalaropus fulicarius* p. 118
Fall/winter: Paler above and plumper than Red-necked Phalarope; bill slightly thicker, yellow-based.

RED-NECKED PHALAROPE *Phalaropus lobatus* p. 118
Fall/winter: Note dark eye patch, long black needlelike bill.
Flight call (both Red-necked and Red Phalaropes) a sharp *kit* or *whit.*

LEAST SANDPIPER *Calidris minutilla* p. 108
Very small, brown with short wings and tail; faint wing stripe.
Flight call a thin *krreet, krr-eet.*

SEMIPALMATED SANDPIPER *Calidris pusilla* p. 108
Western Sandpiper (p. 108; not shown) similar but bill longer.
Flight call a soft *chirt;* call of Western a strident *jeet.*

BAIRD'S SANDPIPER *Calidris bairdii* p. 110
Larger and longer winged than above two. Size and shape of White-rumped Sandpiper, but rump dark.
Flight call a low, raspy *kreep* or *kree.*

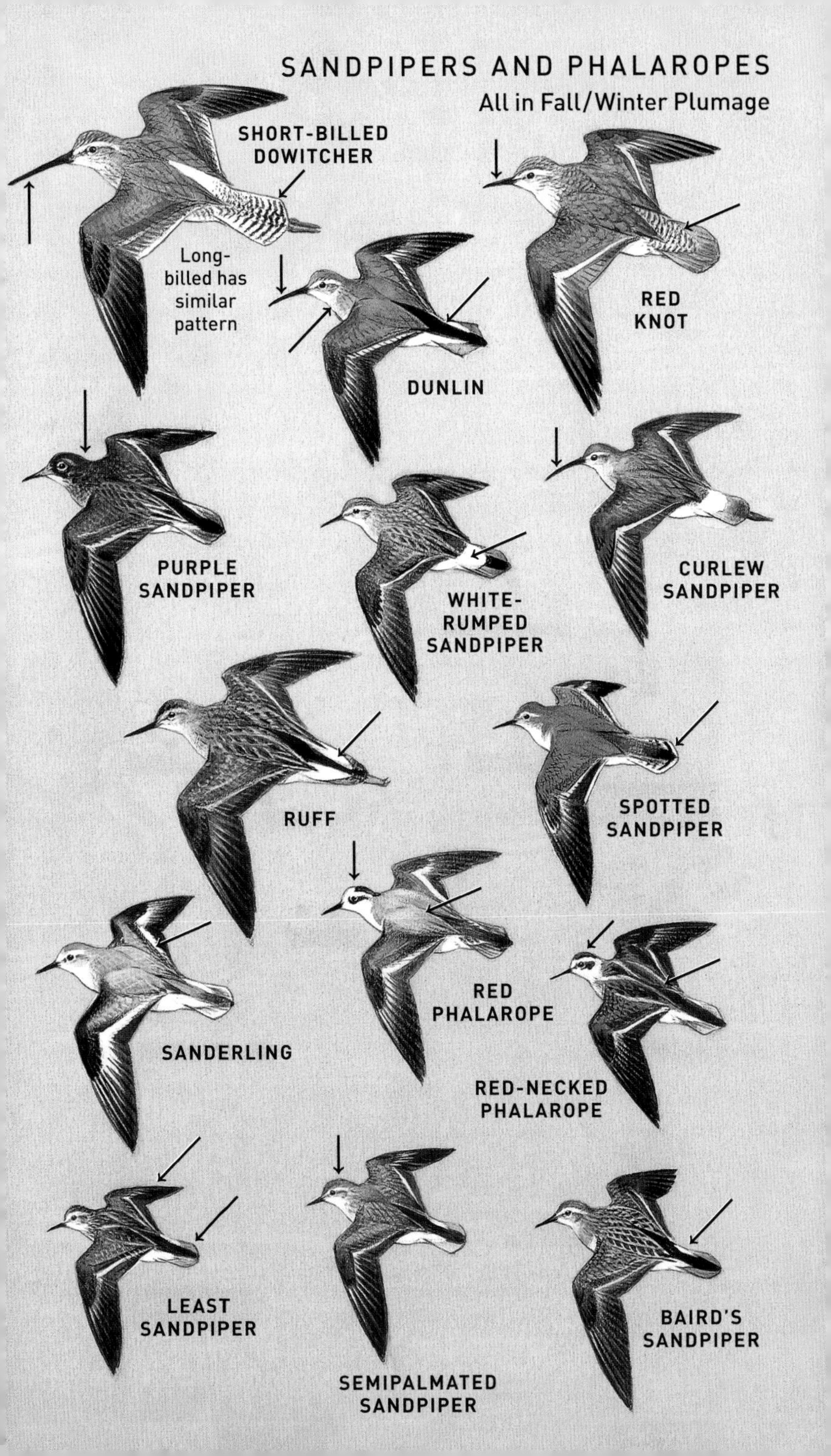

SANDPIPERS AND PHALAROPES
All in Fall/Winter Plumage
SHORT-BILLED DOWITCHER
Long-billed has similar pattern
DUNLIN
RED KNOT
PURPLE SANDPIPER
WHITE-RUMPED SANDPIPER
CURLEW SANDPIPER
RUFF
SPOTTED SANDPIPER
RED PHALAROPE
SANDERLING
RED-NECKED PHALAROPE
LEAST SANDPIPER
SEMIPALMATED SANDPIPER
BAIRD'S SANDPIPER

RARE SHOREBIRDS from EURASIA

PACIFIC GOLDEN-PLOVER *Pluvialis fulva* **Casual vagrant**
10–10¼ in. (25–26 cm). Very similar to American Golden-Plover. Note *wingtips do not extend as far beyond tail tip;* in all plumages, golden spangles on back brighter. In breeding adult, white neck stripe thinner, *extends down to flanks.* **VOICE:** A whistled *chu-wee* or *chu-wee-dle.* **RANGE:** Accidental to casual vagrant from Asia and Alaska.

EUROPEAN GOLDEN-PLOVER *Pluvialis apricaria* **Very rare vagrant**
11 in. (28 cm). Similar to American and Pacific Golden-Plovers but shows *white* underwings. Spring/summer adult like Pacific Golden-Plover but is larger bodied, smaller billed. **VOICE:** A melodic, drawn-out whistle. **RANGE:** Very rare spring vagrant to NL, accidental elsewhere.

NORTHERN LAPWING *Vanellus vanellus* **Casual vagrant**
12–12½ in. (30–32 cm). A distinctive round-winged plover with unique long wispy crest. **RANGE:** Casual European vagrant, mostly in late fall and early winter, from Atlantic Canada south to Mid-Atlantic states; accidental elsewhere. **HABITAT:** Farmland, marshes, mudflats.

SPOTTED REDSHANK *Tringa erythropus* **Casual vagrant**
12½ in. (32 cm). A slender, long-legged, long-billed shorebird. *Spring/summer:* Sooty black with white speckles. Legs *dark red;* bill *reddish basally. Fall/winter and juvenile:* Legs *orange-red,* bill *orange-red* basally. In flight shows *long white wedge* on back. **VOICE:** A sharp, rising whistled *tcheet.* **RANGE:** Casual spring and fall vagrant; records widely scattered.

ESKIMO CURLEW *Numenius borealis* **Almost certainly extinct**
14 in. (36 cm). Last documented record in 1960s; almost assuredly extinct. Much smaller than Whimbrel; bill shorter, thinner, only slightly curved. Wing linings cinnamon-buff. Legs slate gray.

"EURASIAN" WHIMBREL **Casual vagrant**
Numenius phaeopus phaeopus
17–18 in. (43–46 cm). This European subspecies, a casual vagrant along Atlantic Coast, with mostly *white rump,* white wedge up back, paler underwing. **VOICE:** Similar to N. American Whimbrel.

BAR-TAILED GODWIT *Limosa lapponica* **Very rare vagrant**
16–17 in. (41–44 cm). A smaller godwit; bill straighter; legs shorter; underwing plumage distinctive. European birds (*L. l. lapponica*) have white, boldly barred rump. *Spring/summer adult:* Male rich *reddish orange;* female duller. *Fall/winter and first-year:* Both sexes grayish above, white below, underwing whitish with few markings. *Juvenile:* Underparts washed buffy, back with neat buff-and-black pattern. **VOICE:** Flight call a harsh *kirrick;* alarm a shrill *krick.* **RANGE:** Very rare on East Coast; accidental inland. **HABITAT:** Mudflats, shores, tundra.

BLACK-TAILED GODWIT *Limosa limosa* **Accidental vagrant**
16½ in. (42 cm). Resembles Hudsonian Godwit (white rump, white wing stripe, black tail), but bill is straighter. Best field distinction in all plumages is *white* underwing linings in Black-tailed, *black* in Hudsonian. **VOICE:** Flight call a clear *reeka-reeka-reeka.* **RANGE:** Accidental to casual vagrant to East Coast. **HABITAT:** Large lakes with muddy shores.

RARE SHOREBIRDS
PACIFIC GOLDEN-PLOVER
veniles
spring/ summer adult
EUROPEAN GOLDEN-PLOVER
spring/ summer adult
NORTHERN LAPWING
fall/ winter
SPOTTED REDSHANK
spring/ summer
ESKIMO CURLEW
underwing
WHIMBREL
"Eurasian"
Alaskan (accidental in East)
uropean
ll/winter
fall/ winter
fall/ winter
venile
BAR-TAILED GODWIT
spring/ summer male
spring/ summer
BLACK-TAILED GODWIT
fall/ winter

RARE SHOREBIRDS

WOOD SANDPIPER *Tringa glareola* **Accidental vagrant**

8 in. (20 cm). Similar to Solitary Sandpiper but has pale (not dark) underwings, upperparts more *heavily spotted* with pale buff, rump patch *white* (Solitary has dark rump). Legs dull yellow. **VOICE:** A distinctive, sharp, high *chew-chew-chew* or *chiff-chiff-chiff.* **RANGE:** Accidental vagrant from Eurasia in East, but records increasing.

SHARP-TAILED SANDPIPER *Calidris acuminata* **Casual vagrant**

8½ in. (22 cm). Similar to Pectoral Sandpiper, but whitish supercilium bolder and crown brighter rusty. Juveniles have rich *orangey buff breast,* finely streaked on sides only, rather than across breast as in Pectoral. Adults have *dark chevrons* extending to flanks and lack sharp demarcation to breast as in Pectoral. **VOICE:** Trilled *prreeet* or *trrit-trrit,* sometimes twittered. **RANGE:** Asian species; accidental to casual vagrant. **HABITAT:** Borders of wetlands, muddy shores, wet pastures; in summer, tundra.

LITTLE STINT *Calidris minuta* **Casual vagrant**

6 in. (15 cm). Slightly smaller than Semipalmated Sandpiper; bill finer, longer wingtip projection. *Spring/summer:* Rusty orange above and on breast. Similar to Red-necked Stint, but body less elongated, legs longer, and *head with less orange. Juvenile:* Bold white V on mantle, black-centered and rufous-fringed wing coverts and tertials. **VOICE:** A Sanderling-like *tit.* **RANGE:** Widespread vagrant, mostly to coast.

RED-NECKED STINT *Calidris ruficollis* **Accidental vagrant**

6¼ in. (16 cm). Like Semipalmated Sandpiper, but in spring/summer shows *bright rusty head and neck, bordered below by dark streaks. Juvenile:* Has long wingtip projection like Little Stint but plumper body, shorter legs; rusty-fringed scapulars contrast with brown-fringed wing coverts. Bill straight and fine at tip. **RANGE:** Accidental vagrant from Asia.

CURLEW SANDPIPER **Very rare vagrant**

Calidris ferruginea (see also p. 126)

8½–8¾ in. (21–22 cm). A Eurasian species with slim downcurved bill, blackish legs, and white rump in flight. *Spring/summer:* Male variably rich rufous red; female duller with thin pale barring. *Fall/winter and juvenile:* Resemble Dunlin but slightly longer legged, bolder pale supercilium; bill curved more evenly throughout rather than drooping at tip; *white rump.* **VOICE:** A liquid *chirrip.* **RANGE:** Very rare migrant or vagrant along East Coast; casual to accidental inland. **HABITAT:** Marshy pools, mudflats; in summer, tundra.

RUFF *Calidris pugnax* (see also p. 126) **Rare**

Male (Ruff) 12–13 in. (30–32 cm); female (known informally as Reeve) 9 in. (23 cm). Note rather unusual, small-headed, thicker-necked, and *erect stance, oval white patches* on sides of tail in flight. *Spring/summer male:* Unique, with erectile *ruffs* and *ear tufts* that may be black, brown, rufous, buff, white, or barred in various combinations. Legs greenish, yellow, or orange. *Spring/summer female:* Smaller than male; breast *heavily blotched* with dark. *Fall/winter:* Plain with short bill, small head, thick neck. *Juvenile:* Rich buffy head and breast, very scaly on back. **VOICE:** A low *too-i* or *tu-whit.* **RANGE:** Rare but regular vagrant from Eurasia. **HABITAT:** Mudflats, marshes, coastal pools, wet fields; in summer, tundra.

WOOD SANDPIPER
fall/winter
spring/
summer
SHARP-TAILED
SANDPIPER
spring/
summer
juvenile
LITTLE
STINT
spring/
summer
spring/
summer
first-
summer
RED-
NECKED
STINT
juvenile
spring/
summer
male
CURLEW
SANDPIPER
juvenile
female
juvenile
male
male
RUFF
spring/summer
males
spring/summer
female

BITTERNS, HERONS, and ALLIES Family Ardeidae

Medium to large wading birds with long legs and necks, spearlike bills. Hunt with neck erect and roost with head back on shoulders. In flight, neck is folded in an S; legs trail. Plumes develop in winter/spring. Sexes similar. Nest colonially in mangroves or large trees, usually near water. **FOOD:** Fish, frogs, crawfish, other aquatic life; mice, gophers, small birds, insects. **RANGE:** Worldwide except colder regions.

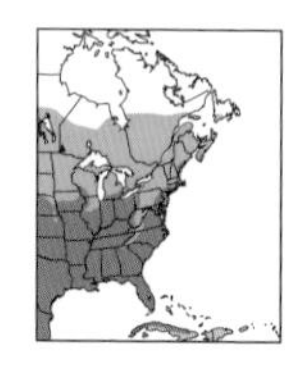

GREAT BLUE HERON *Ardea herodias* — Common

45–47 in. (115–120 cm). A lean gray bird standing 4 ft. (122 cm) tall. Long legs, long neck, daggerlike bill, great size, and blue-gray color mark this species. *Adult:* Crown white with long head, back, and breast plumes in winter through summer. *Juvenile and first-year:* Duller, crown black or mottled white; plumes absent or shorter. White subspecies of s. FL known as "Great White" Heron (p. 134). Presumed intergrades ("Würdemann's" Heron) in FL Keys have white head, including plumes. **VOICE:** Deep, harsh croaks: *frahnk, frahnk, frahnk.* **SIMILAR SPECIES:** Sandhill Crane, Reddish Egret. **HABITAT:** Marshes, swamps, shores, tidal flats, moist fields.

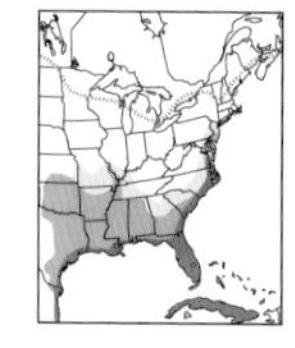

LITTLE BLUE HERON *Egretta caerulea* — Fairly common

24 in. (61 cm). A small, slender heron. *Adult:* Bluish slate with deep maroon-brown neck; legs dark, bill pale blue with dark tip. *First-year* (see p. 134): All white, often with *grayish wingtips* and sometimes blue tinge to crown. Legs *dull olive;* base of bill pale *blue-gray;* lores dull gray-green. Molting one-year-old birds boldly pied (p. 135). **VOICE:** A loud, nasal *scaaah.* **SIMILAR SPECIES:** First-year Reddish Egret slightly larger, longer billed, more brownish, with paler eye and pink-based bill. Juvenile and first-year Little Blue like Snowy Egret except bill slightly thicker and grayer based, lores duller, outer primary tips dusky. **RANGE:** Casual vagrant well north of range. **HABITAT:** Marshes, ponds, mudflats, swamps, rice fields.

TRICOLORED HERON *Egretta tricolor* — Uncommon

26 in. (66 cm). A very slender, dark heron with contrasting *white belly* and white rump. *Long* slender bill. *Adult:* Mostly bluish above, with white crown plumes and pale back plumes in spring/summer. *Juvenile:* Neck dull rusty brown; wing coverts tipped rufous. **VOICE:** A series of drawn-out nasal *quacks.* **SIMILAR SPECIES:** Great Blue and Little Blue Herons. **RANGE:** Casual vagrant inland and well north of range. **HABITAT:** Marshes, swamps, shores.

REDDISH EGRET *Egretta rufescens* — Uncommon

30–31 in. (76–79 cm). Note pinkish, black-tipped bill of adult; habitat almost strictly coastal. *Adult:* Neck and back feathers shaggy; eye pale. Two color morphs: (1) dark morph is neutral gray with bright rusty head and neck (first-year, not shown, duller grayish brown, with short or no plumes to neck, and all-dark bill); (2) white morph is all white with blue-gray legs and feet (see p. 134). When feeding, races about with spread wings. **VOICE:** Sometimes a harsh *kraaak!* **SIMILAR SPECIES:** Habitat and feeding behavior differ from other herons and egrets. Dark first-year can resemble adult Little Blue Heron; white morph suggests Great or Snowy Egret, but bill pinkish-based with black tip, legs and feet blue-gray. **RANGE:** Scarce vagrant to Northeast coast; accidental inland. **HABITAT:** Salt marshes, tidal flats, beaches.

DARK HERONS
AND EGRET
juvenile
herons fly with
neck pulled in
herons' necks may
stretch or be looped
in when they are
standing
WÜRDEMANN'S"
HERON
(intergrade)
adult
adult
GREAT BLUE HERON
(white subspecies
on p. 135)
adult
juvenile
LITTLE BLUE
HERON
(juvenile
and
first-year
on p. 135)
adult
TRICOLORED
HERON
adult
REDDISH
EGRET
dark
morph
(white morph
on p. 135)
Reddish Egret "dancing"
while feeding

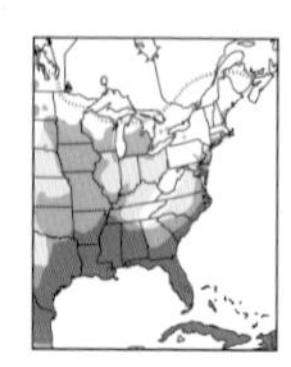

GREAT EGRET *Ardea alba* **Common**

38–39 in. (97–100 cm). A stately, slender white heron with largely *yellow bill.* Legs and feet *black.* In winter through summer, *straight plumes* on back can extend beyond tail; lores greenish. When feeding, leans forward with neck extended. *First-year:* Similar but legs dusky greenish in juvenile; plumes absent or shorter. **VOICE:** A low, hoarse croak. Also *cuk, cuk, cuk.* **SIMILAR SPECIES:** Snowy Egret smaller and with all-black bill, yellow feet. Cattle Egret much smaller. **RANGE:** Casual vagrant well north of range. **HABITAT:** Marshes, ponds, shores, mudflats, moist fields.

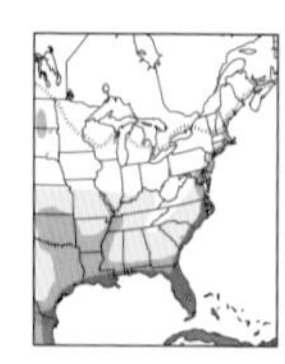

SNOWY EGRET *Egretta thula* **Common**

24 in. (61 cm). Note the *"golden slippers."* A medium-sized heron, with *slender black bill,* yellow lores, black legs, and distinct *yellow feet. Recurved back plumes* and filamentous head plumes during winter through summer. Adults in fall and first-year birds have yellowish or greenish on rear sides of legs; plumes absent or short. **VOICE:** A low croak; in colony, a bubbling *wulla-wulla-wulla.* **SIMILAR SPECIES:** Great Egret. Cattle Egret smaller, squatter, with yellow bill. White first-year Little Blue Heron has blue-gray base to thicker bill, grayer lores, dusky tips to primaries. **RANGE:** Casual vagrant well north of range. **HABITAT:** Marshes, swamps, ponds, shores, tidal flats.

LITTLE EGRET *Egretta garzetta* (not shown) **Casual vagrant**

25 in. (64 cm). A vagrant from Eurasia to East Coast, very similar to Snowy Egret but slightly larger, larger billed, and with duller lores and feet. Develops *two long head plumes* in winter/spring. Young birds very difficult to distinguish. **VOICE AND HABITAT:** Similar to Snowy Egret.

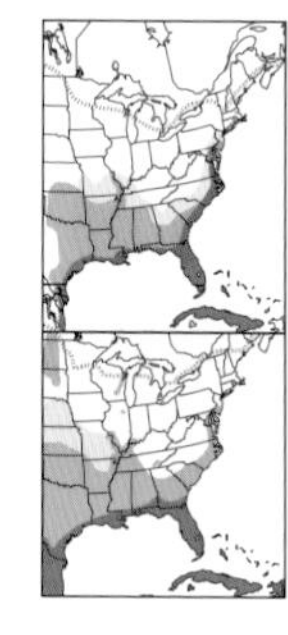

LITTLE BLUE HERON *Egretta caerulea* (adult on p. 132)

First-year: White with dusky wingtips, sometimes bluish tinge to crown. Base of bill blue-gray, lores grayish, legs dull olive. Molting one-year-olds have contrasting gray-and-white feathering.

CATTLE EGRET *Bubulcus ibis* **Uncommon to common**

19–20 in. (48–51 cm). Smaller, squatter, and thicker necked than Snowy Egret. Bill relatively short; bill and legs yellow. *Spring/summer adult:* Has variable (topically applied) *buff-orange* plumes on crown, breast, and back; bill and legs may be pinkish. *Fall/winter adult and first-year:* Have little or no buff. *Juvenile:* May have dusky legs; plumes absent or shorter. **VOICE:** In colony, a series of nasal grunts. **SIMILAR SPECIES:** Snowy Egret larger and more slender, has black bill and legs, yellow feet. See first-year Little Blue Heron and Great Egret. **RANGE:** Casual vagrant well north of range. **HABITAT:** Farms, marshes, fields, highway edges. Often associates with cattle.

REDDISH EGRET *Egretta rufescens* (dark morph on p. 132)

White morph: Note size, feeding behavior, entirely blue-gray legs and feet. Bill blackish in juvenile, pink with black tip in adult. Strictly coastal.

"GREAT WHITE" HERON **Uncommon, local**
Ardea herodias occidentalis

47 in. (120 cm). Our largest white heron, found in s. FL. All white with yellow bill and dull horn-colored legs; slightly smaller Great Egret has blackish legs. Currently regarded as a subspecies (or possibly a morph) of Great Blue Heron, p. 132. **HABITAT:** Mangrove keys, salt bays, marsh banks, open mudflats.

WHITE HERONS AND EGRETS

GREAT EGRET

SNOWY EGRET

molting one-year-old

juvenile

LITTLE BLUE HERON

(adult on p. 133)

fall/winter

spring/summer

CATTLE EGRET

REDDISH EGRET

white morph

(dark morph on p. 133)

"GREAT WHITE" HERON

(see also Great Blue Heron on p. 133)

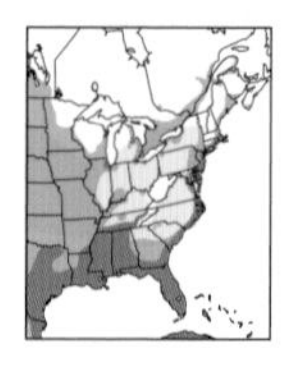

BLACK-CROWNED NIGHT-HERON *Nycticorax nycticorax* Uncommon

25 in. (64 cm). This stocky, thick-billed, short-legged heron is usually hunched and inactive; flies to feed at dusk. *Adult:* Black back and cap contrast with pale gray or whitish underparts, two long white head plumes. Eyes red; legs yellowish or greenish (pinkish in high breeding condition). *Juvenile and first-year:* Brown, streaked and spotted with buff and white. Bill with greenish base; eyes small, reddish. *Second-year:* Has adultlike plumage but paler, washed brown. **VOICE:** A flat *quok!* or *quark!* Most often heard at dusk. **SIMILAR SPECIES:** Juvenile and first-year may be confused with American Bittern and similar-aged Yellow-crowned Night-Heron. **HABITAT:** Marshes, shores, mangroves, marinas; roosts in trees.

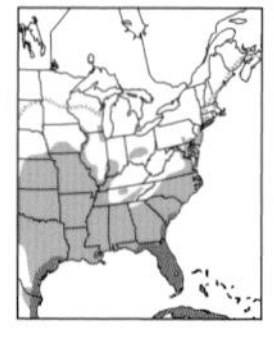

YELLOW-CROWNED NIGHT-HERON *Nyctanassa violacea* Uncommon

24 in. (61 cm). A chunky heron with longer neck and legs than Black-crowned Night-Heron. *Adult:* Gray overall; head black with buffy-white cheek patch and yellowish crown. *Juvenile and first-year:* Similar to Black-crowned Night-Heron but grayer, underparts more finely streaked; back spotting smaller; wing coverts have pale edges. Bill thicker and lacks greenish-yellow base. *Second-year:* Grayer overall, has indistinct adultlike plumage. In flight, entire feet and some of lower legs extend beyond tail. **VOICE:** *Quark,* higher pitched than call of Black-crowned. **RANGE:** Casual vagrant well north of range. **HABITAT:** Swamps, mangroves, bayous, marshes, streams.

GREEN HERON *Butorides virescens* Fairly common

17–18 in. (43–46 cm). A small dark heron that looks crowlike in flight (but flies with bowed wingbeats). When alarmed, stretches neck, elevates shaggy crest, and jerks tail. *Adult:* Comparatively *short* legs are *greenish yellow* or *orange* (when breeding). Back has blue-green gloss; neck deep chestnut. *Juvenile and first-year:* Streaked neck and breast, browner above. **VOICE:** A loud *skyow* or *skewk;* series of *kuck* notes. **HABITAT:** Lakes, ponds, marshes, streams.

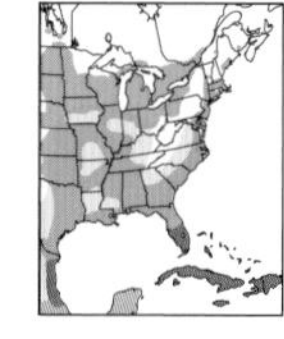

LEAST BITTERN *Ixobrychus exilis* Uncommon, secretive

12–13 in. (31–33 cm). Very small, thin, furtive; straddles reeds. Note large *buff wing patch* (lacking in rails). Back black in adult male, rusty brown in female and juvenile. The dark reddish and blackish "Cory's" morph is extremely rare, seen most often around Great Lakes. **VOICE:** Song a low, muted *coo-coo-coo;* also gives a raspy, rail-like *khak-khak-khak* series. **SIMILAR SPECIES:** Green Heron. **RANGE:** Casual vagrant well north of range. **HABITAT:** Freshwater marshes, reedy ponds.

AMERICAN BITTERN *Botaurus lentiginosus* Uncommon

28 in. (71 cm). A stocky brown heron; size of a young night-heron but warmer brown with longer yellowish bill. In flight, *primaries and secondaries blackish to black* and bill held horizontally (slightly downward in night-herons). At rest or when approached, often stands rigid, bill pointing up. *Black stripe shows on side of neck.* Ages similar. **VOICE:** A "pumping" sound, a low, deep, resonant *oong-ka´ choonk,* etc. Flushing call *kok-kok-kok.* **SIMILAR SPECIES:** First-year night-herons, Green Heron, and (much smaller) Least Bittern. **HABITAT:** Marshes, reedy lakes. Unlike night-herons, seldom sits in trees.

adult

BLACK-CROWNED NIGHT-HERON

juvenile

juvenile

adult

YELLOW-CROWNED NIGHT-HERON

typical

"Cory's"

LEAST BITTERN

adult

juvenile

GREEN HERON

AMERICAN BITTERN

LIMPKINS Family Aramidae

A monotypic family, related to rails and cranes. **FOOD:** Mostly large freshwater snails; a few insects, frogs. **RANGE:** Se. U.S., W. Indies, s. Mex. to Argentina.

LIMPKIN *Aramus guarauna* — Uncommon, local

26 in. (66 cm). A large, spotted wader, long legs and drooping bill give it an ibislike aspect, but no ibis is completely brown with bold white spots and streaks. Flight cranelike, with smart upward flaps. Ages and sexes similar. **VOICE:** A piercing, repeated wail, *kree-ow, kra-ow,* etc., especially at night. **SIMILAR SPECIES:** First-year ibises and night-herons, American Bittern. **RANGE:** Rare vagrant well north of range. **HABITAT:** Fresh swamps, marshes with large snails.

IBISES and SPOONBILLS Family Threskiornithidae

Ibises are long-legged, heronlike waders with slender, decurved bills. Spoonbills have spatulate bills. Both fly in Vs or lines. **FOOD:** Small crustaceans, small fish, insects, etc. **RANGE:** Tropical and temperate regions.

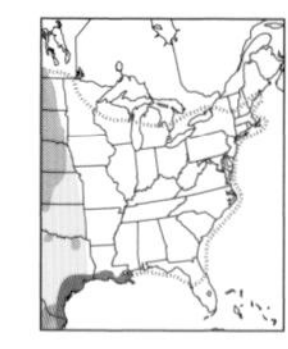

WHITE-FACED IBIS *Plegadis chihi* — Uncommon

23–24 in. (58–62 cm). Similar to Glossy Ibis but adult has *white border of feathers* around face meeting behind eye; pinkish to red facial skin; *red eye.* Juveniles may be impossible to identify, but by mid-fall, iris and facial skin colors develop. **VOICE:** A deep gooselike quacking. **SIMILAR SPECIES:** Glossy Ibis. **RANGE:** Rare east of Mississippi Valley; casual vagrant to East Coast. **HABITAT:** Freshwater marshes, irrigated land.

GLOSSY IBIS *Plegadis falcinellus* — Fairly common

23–24 in. (58–62 cm). A long-legged wader with *long decurved bill.* Flies in lines with neck outstretched, alternately flapping and gliding. *Spring/summer adult:* Dark, with chestnut and bronzy sheen and maroon patch in wing. Dark bluish facial skin with thin cobalt blue borders, and without white feathers around eye of White-faced Ibis; iris brown, without red. *Fall/winter adult:* Head and neck brown, streaked white; facial features dull. *First-year:* Similar to fall/winter adult, but facial skin gray, with indistinct whitish facial stripes; wings flat olive green, without maroon; juveniles sometimes show white feathering in head. **VOICE:** A guttural *ka-onk,* repeated; a low *kruk, kruk.* **SIMILAR SPECIES:** White-faced Ibis. Hybrids with White-faced known. **RANGE:** Casual vagrant inland and well north of range in East. **HABITAT:** Freshwater marshes, irrigated land.

WHITE IBIS *Eudocimus albus* — Common

24–25 in. (62–64 cm). *Adult:* White, with *restricted black in wingtips.* Note *red face,* long *decurved red bill.* Flies with neck outstretched; flocks fly in "roller-coasting" strings, flapping and gliding; may soar in circles. *Juvenile:* Dark brownish with *white belly, white rump,* decurved *orangey pink bill.* Slowly develops white in plumage through first year; head and neck white by second spring. **VOICE:** A low and nasal *uuhhnn!* or *quaahh!* **SIMILAR SPECIES:** Wood Stork larger, with much more black in wing. First-year Glossy Ibis has uniformly dark appearance. **RANGE:** Casual vagrant inland and well north of range in East. **HABITAT:** Salt, brackish, and fresh marshes, rice fields, mangroves.

LIMPKIN AND IBISES
WHITE-FACED IBIS
spring/ summer adults
facial comparison in breeding season
Glossy Ibis
LIMPKIN
GLOSSY IBIS
first-year
adult
juvenile
adult
WHITE IBIS

ROSEATE SPOONBILL *Platalea ajaja* **Uncommon**

32 in. (81 cm). A wading bird with long, flat, spoonlike bill. When feeding, sweeps bill from side to side. In flight, extends neck and often glides between series of wing strokes. *Adult: Bright, shell pink,* with blood red feathers on shoulders in spring through fall; tail orange. Crown and face naked, greenish gray. *Juvenile:* Spatulate bill smooth, yellowish; head feathered white; remainder of plumage whitish, slowly mixed with pale pink feathers through first year, outer primary brown. Full adult appearance assumed in second or third years. **VOICE:** At colony, a low grunting croak. **SIMILAR SPECIES:** Greater Flamingo. **RANGE:** Casual vagrant inland and well north of range. **HABITAT:** Coastal marshes, lagoons, mudflats, mangroves.

STORKS Family Ciconiidae

Large, long-legged, and heronlike, with very large, recurved or decurved bills. Some have naked heads. Sexes alike. Walk is sedate; flight deliberate, with neck and legs extended. **FOOD:** Frogs, crustaceans, lizards, rodents. **RANGE:** Southern U.S. to S. America; Old World.

WOOD STORK *Mycteria americana* **Uncommon, threatened**

39–41 in. (100–105 cm). Very large; wingspan 5½ ft. (168 cm). *Adult:* White, with *dark naked head* and *much black in wing;* black tail. Bill long, thick, slightly decurved. *First-year:* Bill yellowish, head with downy white feathers slowly lost during first one to two years. When feeding, walks on stiff legs. In flight, alternately flaps and glides. Often soars very high on thermals. **VOICE:** A hoarse croak; usually silent. **SIMILAR SPECIES:** In flight, American White Pelican, Whooping Crane. **RANGE:** Casual vagrant inland and well north of range in East. **HABITAT:** Marshes, ponds, lagoons, swamps. East Coast populations threatened.

FLAMINGOS Family Phoenicopteridae

Pinkish white to vermilion wading birds with extremely long neck and legs. Thick bill is bent sharply down and lined with numerous lamellae for straining food. **FOOD:** Small mollusks, crustaceans, cyanobacteria, diatoms. **RANGE:** W. Indies, Yucatán Peninsula in Mex., Galápagos, S. America, Old World.

AMERICAN FLAMINGO *Phoenicopterus ruber* **Rare**

46–47 in. (115–118 cm). W. Indian adults (subspecies *P. r. ruber*) are extremely slim, rose pink wading birds as tall as or taller than a Great Blue Heron but much more slender. Note thick, sharply bent bill. Feeds with bill or head immersed. In flight, shows black secondaries and primaries; extremely long neck and legs. Pale, washed-out birds may be escapees from zoos. First- and second-year flamingos are also much paler pink than wild adults; primary coverts and underwing coverts brownish; bill dull yellowish with indistinct dusky tip. **VOICE:** Gooselike calls, gabbling: *ar-honk,* etc. **SIMILAR SPECIES:** Roseate Spoonbill. Escapees of all five other flamingo species have been recorded in N. America; Chilean Flamingo (*P. chilensis*), most common in captivity, is paler pink and has more black on bill tip (includes angle). **RANGE:** Rare vagrant to FL bays; accidental elsewhere. **HABITAT:** Salt flats, saline lagoons.

VERY LARGE
WADERS
adult
juvenile
ROSEATE
SPOONBILL
adult
WOOD
STORK
first-
year
White Ibis
(p. 139) for
comparison
first-year
adult
adult
AMERICAN
FLAMINGO

CRANES Family Gruidae

Stately birds, more robust than herons, often with red facial skin. Note arching tufted feathering over rump. In flight, neck extended. Migrate in Vs or lines like geese. Large herons are sometimes wrongly referred to as cranes. **FOOD:** Omnivorous. **RANGE:** Nearly worldwide except Cen. and S. America and Oceania.

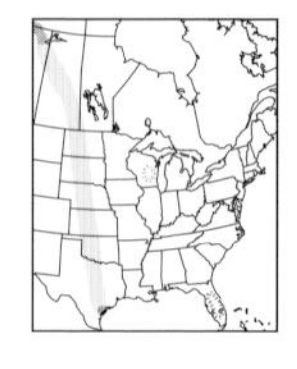

WHOOPING CRANE *Grus americana* **Rare, local, endangered**

51–52 in. (130–132 cm); wingspan 7½ ft. (229 cm). The tallest N. American bird and one of the rarest. *Adult:* Large *white* crane with *bare red forehead and lower face.* Primaries *black. Juvenile:* Plumage washed with rust, especially on head, which is feathered; bill dusky. About three years required to develop full adult plumage and head condition. **VOICE:** A shrill, buglelike trumpeting, *ker-loo! ker-lee-oo!* **SIMILAR SPECIES:** Wood Stork has dark head, more black in wing. Egrets and swans lack black in wings. See also American White Pelican and Snow Goose. **RANGE:** Casual migrant or vagrant in Mississippi Valley. **HABITAT:** Prairies, fields and pastures, coastal marshes; in summer, muskeg.

COMMON CRANE *Grus grus* **Accidental vagrant**

44–50 in. (112–127 cm). Eurasian. *Adult:* Note black neck, white cheek stripe. Feathers arching over rump are blacker than those of Sandhill Crane. *Juvenile:* Entirely gray with yellow bill; develops indistinct adult-like pattern in first year. **RANGE:** This vagrant (probably migrating with Sandhill Cranes from Asia) has been recorded primarily in w. N. America but is accidental in Midwest, most frequently among flocks of Sandhills. Some escapees or presumed escapees have also occurred in e. N. America.

SANDHILL CRANE *Antigone canadensis* **Uncommon, local**

36–48 in. (90–122 cm); wingspan 6–7 ft. (183–213 cm). *Adult:* Note *bare red crown,* bustlelike rear. A long-legged, long-necked gray bird, often stained with rust in spring and summer. *Juvenile:* Browner, with feathered head, yellowish bill; about three years required to develop full adult plumage and head condition. In flight, neck extended and wings flap with an upward flick. **VOICE:** A rolling, bugled *garoo-a-a-a,* repeated. Younger birds also give a very different, cricketlike call. **SIMILAR SPECIES:** Great Blue Heron is sometimes wrongly called a crane. **RANGE AND HABITAT:** Prairies, fields, marshes, tundra. Smaller "Lesser" Sandhill Crane (subspecies *A. c. canadensis*) nests in tundra and winters primarily in w. N. America. Larger "Greater" (*A. c. tabida*) nests in grasslands and bogs and winters across N. America, uncommonly to Midwest and as scarce to casual vagrant to East Coast north of FL. Populations of FL (*A. c. pratensis*) and MS (*A. c. pulla*) are resident, the latter endangered.

storks, ibises, and cranes fly
with neck outstretched
adult
WHOOPING
CRANE
juvenile
COMMON
CRANE
adult
Sandhill
Crane
adult
juvenile
SANDHILL CRANE

COOTS, GALLINULES, and RAILS
Family Rallidae

Rails are rather hen-shaped marsh birds, many of secretive habits and distinctive voices, more often heard than seen. Flight from marshes is brief and reluctant, with legs dangling, although they can also undertake remarkable long-distance migrations at night. Gallinules and coots are much easier to see; they swim and might be confused with small ducks or grebes. They spend most of their time swimming but may also feed on shores. Except in juveniles, ages and sexes generally alike or differ only slightly. **FOOD:** Aquatic plants, seeds, insects, frogs, crustaceans, mollusks. **RANGE:** Nearly worldwide.

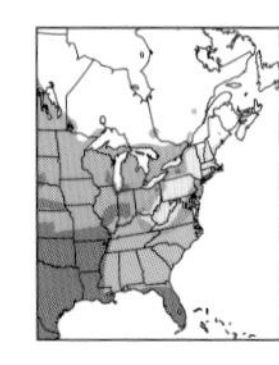

AMERICAN COOT *Fulica americana* **Uncommon to common**

15–15½ in. (38–39 cm). *Adult:* A slaty, ducklike bird with blackish head and neck, slate gray body, *white bill,* and divided white patch under tail. No side striping. Its big feet are lobed ("scallops" on toes). *Juvenile and first-fall:* Paler, throat whiter, bill duller grayish, without shield; plumage becomes grayer and bill-shield develops in first year. Downy chick has bushy *orange-red* feathers on head, a bald crown, and red bill. Gregarious. When swimming, pumps head back and forth. Taking off, it skitters, flight labored, big feet trailing beyond short tail. **VOICE:** A grating *kuk-kuk-kuk-kuk; kakakakakaka;* etc.; also a measured *ka-ha, ha-ha;* various cackles, croaks. **SIMILAR SPECIES:** Juvenile Common Gallinule browner above, has thin white stripe on flanks and warmer colored bill; more solitary than coots. **HABITAT:** Ponds, lakes, marshes; in winter, also fields, park ponds, golf courses, lawns, salt bays.

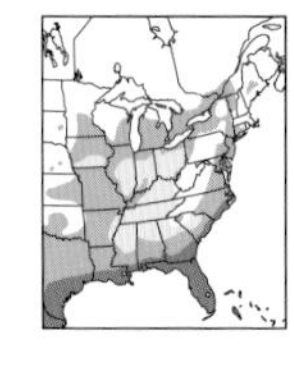

COMMON GALLINULE *Gallinula galeata* **Uncommon**

14 in. (36 cm). Formerly known as Common Moorhen prior to this Eurasian species being split. *Adult:* Rather chickenlike *red bill with yellow tip, red forehead shield,* and white stripe on flanks. When walking, flicks white undertail coverts; while swimming, pumps head like a coot. *Juvenile and first-fall:* Duller, throat whiter, bill duller brownish, without shield; plumage becomes slatier and bill-shield develops in first year. Downy chick has black feathers, a bald crown, and red bill. **VOICE:** A croaking *kr-r-ruk,* repeated; a froglike *kup* and loud, complaining, henlike *kek, kek, kek* (higher than coot's call). **SIMILAR SPECIES:** American Coot, juvenile and first-year Purple Gallinule. **RANGE:** Scarce vagrant well north and west of range. **HABITAT:** Freshwater marshes, reedy ponds.

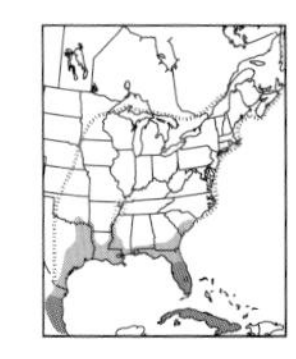

PURPLE GALLINULE *Porphyrio martinicus* **Uncommon**

13 in. (33 cm). *Adult:* Head and underparts *deep violet purple,* back bronzy green. Shield on forehead *pale blue;* bill red with yellow tip. Legs *yellow,* conspicuous in flight. *Juvenile and first-fall:* Buffy brown below, dark above tinged greenish; bill dark; sides unstriped; acquires mixed purple feathering and develops bill during first year. **VOICE:** A henlike cackling, *kek, kek, kek;* also guttural notes, sharp reedy cries. **SIMILAR SPECIES:** Common Gallinule lacks greenish plumage, has white side stripe in all plumages. Also Purple Swamphen (p. 146). **RANGE:** Widespread vagrant north and west of range. **HABITAT:** Freshwater swamps, marshes, ponds. Swims, wades, and climbs bushes.

COOT AND GALLINULES

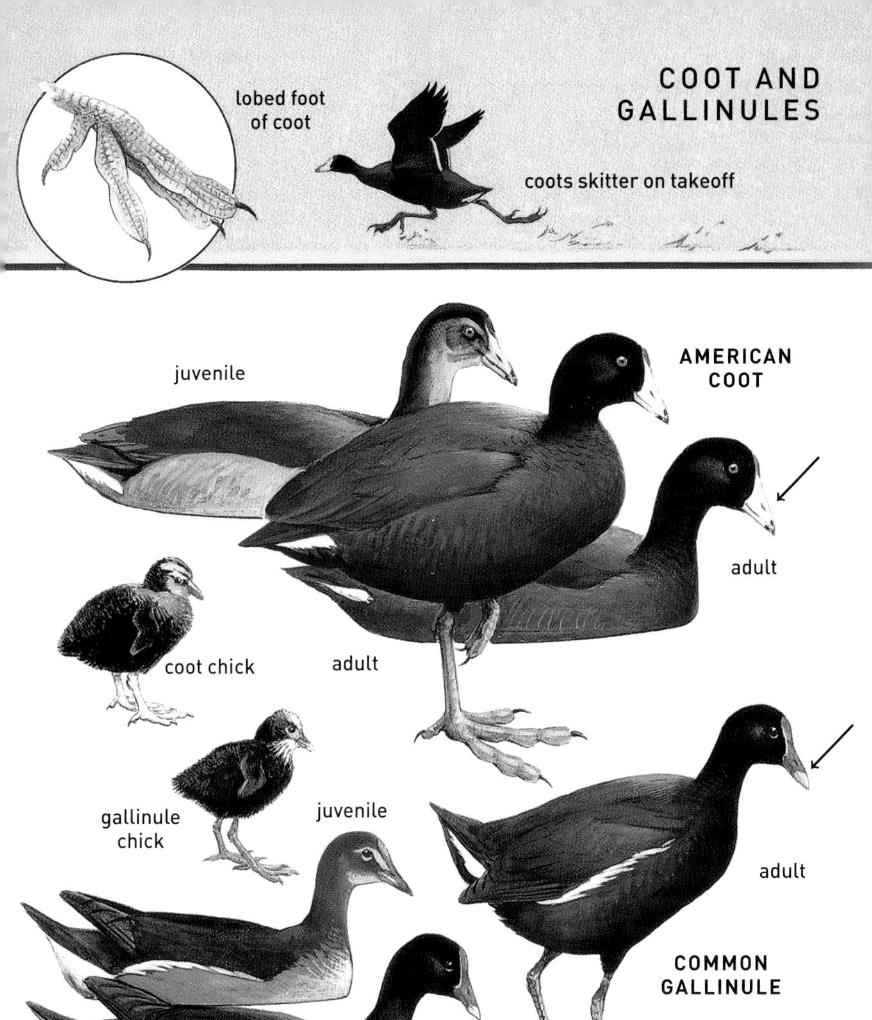

PURPLE SWAMPHEN *Porphyrio porphyrio* **Fairly common, local, introduced**

16–19 in. (40–48 cm). Introduced to se. FL (subspecies *P. p. poliocephalus,* sometimes split as Gray-headed Swamphen), where it's increasing in numbers and range despite control efforts. *Adult:* Purple and blue with turquoise wings, large fearsome red bill, red shield, and bright red legs and feet with dusky joints. *Juvenile and first-fall:* Plumage grayish; bill blackish. **VOICE:** A loud, sharp *ee-erk ee-erk.* **SIMILAR SPECIES:** Purple Gallinule (p. 144) smaller and less blocky, has greener back, pale blue shield, yellow legs. **HABITAT:** Vegetated freshwater wetlands, including lakeshores, ponds, marshes, sloughs.

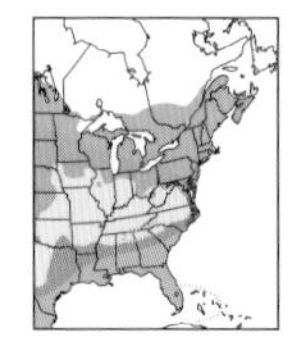

VIRGINIA RAIL *Rallus limicola* **Fairly common**

9½ in. (24 cm). A small rusty rail with gray cheeks, black bars on flanks, and long, slightly decurved, reddish bill with dark tip. Near size of meadowlark; only small rail with *long slender* bill. *Juvenile* (summer only): Shows much black; otherwise ages similar. **VOICE:** A descending grunt, *wuk-wuk-wuk-wuk,* etc.; also *kidick, kidick,* etc.; various "kicking" and grunting sounds. **SIMILAR SPECIES:** Sora has small stubby bill, unbarred undertail coverts. Clapper and King Rails much larger. **HABITAT:** Fresh and brackish marshes; in winter, also salt marshes.

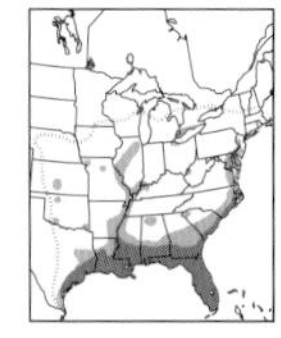

KING RAIL *Rallus elegans* **Uncommon, secretive**

15–16 in. (38–41 cm). A large rusty rail with long slender bill; twice the size of Virginia Rail, or about that of a small chicken. Brighter than Clapper Rail, but note rusty/chestnut cheeks and more contrasting black-and-white flanks, rustier overall with bolder back pattern (blacker feathers with buffier edges); prefers fresh marshes. Juvenile in summer (not shown) has black mottling, similar to juvenile Virginia Rail; otherwise ages similar. **VOICE:** A low, slow, grunting *bup-bup, bup-bup-bup,* etc., or evenly spaced chuck-chuck-chuck (deeper than Virginia Rail). **SIMILAR SPECIES:** Clapper Rail. Virginia Rail half the size, has slaty gray cheeks. *Note:* Hybrids between Clapper and King occur. **RANGE:** Rare to casual vagrant north and west of range. **HABITAT:** Fresh and brackish marshes, rice fields, ditches, swamps. In winter, also salt marshes.

CLAPPER RAIL *Rallus crepitans* **Fairly common**

14–15 in. (35–38 cm). The large "marsh hen" of Atlantic and Gulf Coast marshes. Sometimes swims. Note henlike appearance; strong legs; long, slightly decurved bill; barred flanks; and white patch under short cocked tail, which it flicks nervously. Cheeks gray. Gulf Coast birds (subspecies *R. c. saturatus*) brighter than subspecies found along Atlantic Coast. Juveniles (summer only) duller grayish with blackish mottled flanks; otherwise ages similar. **VOICE:** A clattering *kek-kek-kek-kek,* etc., or *cha-cha-cha,* etc. **SIMILAR SPECIES:** King Rail prefers fresh (sometimes brackish) marshes, has bolder pattern on back and flanks, rusty brown on wings. Its breast is brighter cinnamon, although Clappers along Gulf Coast have warmer tawny tones, approaching those of King. Adult Clapper has grayer cheeks; juvenile not as blackish on head and breast and has duller edging on secondaries. These two rails occasionally co-occur and hybridize in adjacent brackish marshes. **RANGE:** Accidental inland. **HABITAT:** Coastal salt marshes.

PURPLE
SWAMPHEN
adult
VIRGINIA RAIL
juvenile
adult
KING RAIL
Atlantic Coast
Gulf Coast
CLAPPER RAIL
chick

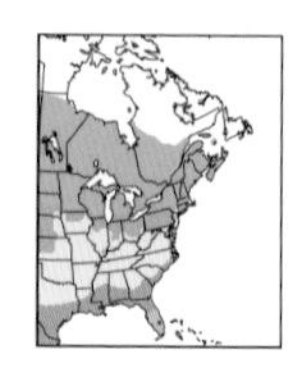

SORA *Porzana carolina* — Fairly common

8½ in. (22 cm). Note *short yellow* bill. *Adult:* A small, plump, gray-brown rail with *black patch* on face and throat, more extensive in male than in female. Short, cocked tail reveals white or buff undertail coverts. *Juvenile and first-winter:* Lack dark throat patch and are browner; acquire duller adult plumage by first spring. **VOICE:** A descending whinny, *whee-ee-ee-ee-ee-ee-e-e-e.* Also a plaintive whistled *keu-wee?* Clapping one's hands can cause startled birds to utter a sharp *keek.* **SIMILAR SPECIES:** Juvenile and first-winter Soras may be confused with smaller and rarer Yellow Rail, which has large white wing patches and blacker-centered feathers above. Virginia Rail has slender bill. **HABITAT:** Freshwater marshes; in migration, also wet meadows; in winter, also salt marshes.

YELLOW RAIL *Coturnicops noveboracensis* — Scarce, secretive

7¼ in. (18 cm). Note *white wing patch* (in flight). A small buffy-and-black rail, suggesting a week-old chick. Bill very short, greenish or yellowish. Back dark, striped, barred, and checkered with buff, white, and black. *Mouselike; very difficult to see.* Ages similar. **VOICE:** Nocturnal ticking notes, often in long series: *tic-tic, tic-tic-tic, tic-tic, tic-tic-tic,* etc., in alternating groups of two and three. Compared with hitting two small stones together. **SIMILAR SPECIES:** Young Sora somewhat larger, buffier overall, lacks dark barring and checkering above, has thin pale trailing edge but no white patch in wing. **RANGE:** Rare to casual migrant throughout interior East. **HABITAT:** Grassy marshes, wet meadows; winters mostly in salt marshes and grain fields.

BLACK RAIL *Laterallus jamaicensis* — Scarce, local, secretive

6 in. (15 cm). A tiny blackish rail with small *black* bill; about the size of a young sparrow. Nape deep chestnut. *Very difficult to glimpse.* Ages of full-grown birds similar. *Caution:* All young rails in downy plumage are black. **VOICE:** Male (mostly at night), *kiki-doo* or *kiki-krrr* (or *kitty go*). Also a growl. **RANGE:** Rare to casual vagrant to Midwest; accidental elsewhere away from range. **HABITAT:** Salt marshes, freshwater marshes, grassy meadows.

JACANAS Family Jacanidae

Jacanas are related to shorebirds but look like gallinules and walk like rails. Adults are dark with very long toes, perfect for walking over lily pads and other floating aquatic vegetation. Sexes alike. **FOOD:** Aquatic insects, seeds, and vegetation. **RANGE:** Pantropical.

NORTHERN JACANA *Jacana spinosa* — Casual vagrant

9½ in. (24 cm). This vagrant has spectacularly long toes for walking on lily pads. *Adult:* Chestnut body with dark head. Yellow bill and forehead frontal shield. Striking yellow primaries and secondaries in flight. Holds wings over head when it lands. *Juvenile and first-winter:* Have white underparts, distinct line behind eye, slightly less yellow in wings; gradually gain incomplete adultlike body plumage by first spring. **VOICE:** A rapid series of high, nasal notes: *jeek-jeek-jeek-jeek.* **RANGE:** Casual vagrant from Mex. to TX. **HABITAT:** Frequents ponds with emergent vegetation, especially lily pads.

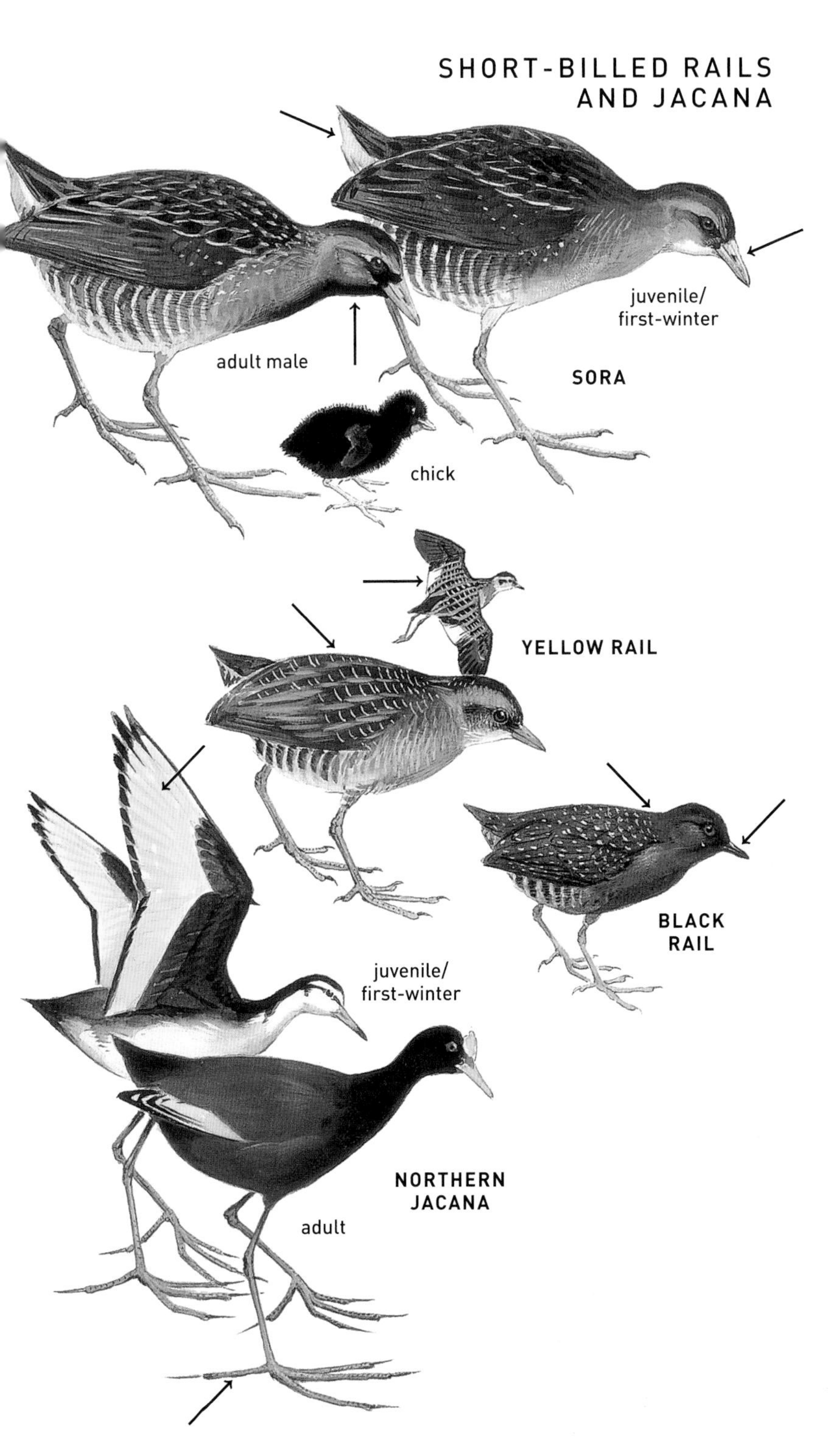
SHORT-BILLED RAILS
AND JACANA
juvenile/
first-winter
adult male
SORA
chick
YELLOW RAIL
BLACK
RAIL
juvenile/
first-winter
NORTHERN
JACANA
adult

CURASSOWS and GUANS Family Cracidae

Tropical forest birds with long tails. Only one species reaches extreme s. U.S. Ages and sexes similar. **FOOD:** Insects, fruit, leaves, seeds. **RANGE:** New World Tropics.

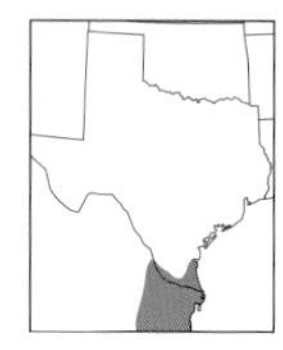

PLAIN CHACHALACA *Ortalis vetula* Fairly common, local

22 in. (56 cm). A large olive-brown bird, shaped somewhat like a half-grown turkey with a small head. Long, rounded, pale-tipped tail, bare red throat. Difficult to observe away from feeding stations; best found in morning when calling raucously from treetops. **VOICE:** Alarm a harsh chickenlike cackle. Characteristic call a raucous three-syllabled *cha-ca-lac,* repeated in chorus from treetops, especially in morning and evening. **SIMILAR SPECIES:** Greater Roadrunner. **HABITAT:** Woodlands, tall brush, well-vegetated residential areas.

GALLINACEOUS BIRDS (TURKEYS, PHEASANTS, GROUSE, PARTRIDGES, and OLD WORLD QUAIL) Family Phasianidae

Often called "upland game birds." Turkeys are very large, with wattles and fan-like tail. Pheasants (introduced) have long pointed tail. Grouse are plump, chickenlike birds, without long tail. Partridges (of Old World origin) are intermediate in size between grouse and quail. Quail are the smallest. Ages generally similar, sexes usually differ. **FOOD:** Insects, seeds, buds, berries. **RANGE:** Nearly worldwide.

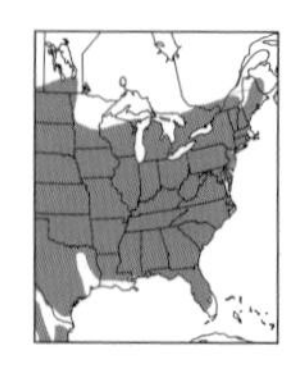

WILD TURKEY *Meleagris gallopavo* Fairly common

Male 46–47 in. (117–120 cm); female 36–37 in. (91–94 cm). A streamlined version of barnyard turkey, with dark (not white) plumage and rusty instead of white tail tips. *Adult male:* Head naked; bluish with red wattles, intensified in display. Tail erected like a fan in display. Bronzy iridescent body; barred primaries and secondaries; prominent "beard" on breast. *Female and first-year male:* Smaller, with smaller and duller head; less iridescent; less likely to have a beard. **VOICE:** "Gobbling" of male like domestic turkey's. Alarm *pit!* or *put-put!* Flock call *keow-keow.* Hen clucks to her chicks. **HABITAT:** Woods, mountain forests, wooded swamps, field edges, clearings. Reintroduced in many areas, and such birds are adapting well to being near people.

RING-NECKED PHEASANT Uncommon, introduced
Phasianus colchicus

Male 31–33 in. (79–84 cm); female 21–23 in. (53–59 cm). A large chickenlike bird introduced from Eurasia. Note long pointed tail. Runs swiftly; flight strong, takeoff noisy. *Male:* Highly colored and *iridescent,* with *scarlet wattles* on face and *white neck ring* (not always present). *Female:* Mottled brown, with *long pointed tail.* **VOICE:** Crowing male gives loud double squawk, *kork-kok,* followed by brief whir of wings. When flushed, harsh croaks. Roosting call a two-syllable *kutuck-kutuck,* etc. **SIMILAR SPECIES:** Female Sharp-tailed Grouse and prairie-chickens have shorter tails, white (Sharp-tailed) or black (prairie-chickens) outer tail feathers, barred upperparts. **HABITAT:** Farms, fields, marsh edges, brush, grassy roadsides. Periodic local releases for hunting.

MISCELLANEOUS CHICKENLIKE BIRDS

WILLOW PTARMIGAN *Lagopus lagopus* **Fairly common**

15 in. (39 cm). Willow and Rock Ptarmigans are fairly similar. In breeding season, Willows are variable, but most males are chestnut brown, redder than any plumage or subspecies of Rock Ptarmigan; females are a warm buffy brown that can overlap brown of Rock. White of wings retained all year and, in flight, contrast with summer body plumage. In winter, white overall with black tail, the latter retained year-round. There is much variation among sexes and between various molts; longest uppertail coverts in summer plumages are barred in females but not males. **VOICE:** Deep raucous calls. Male, a staccato crow, *kwow, kwow, tobacco, tobacco,* etc., or *go-back, go-back.* **SIMILAR SPECIES:** Rock Ptarmigan always has smaller and more slender bill that lacks strong curve on ridge shown by Willow. In winter, male Rock has *black lores* between eye and bill, lacking in both sexes of Willow. Habitats overlap, but Rock tends to prefer higher, more barren hills. **HABITAT:** Tundra, willow scrub, muskeg; in winter, sheltered valleys at slightly lower elevations.

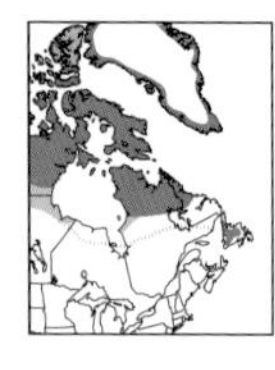

ROCK PTARMIGAN *Lagopus muta* **Uncommon**

14 in. (36 cm). Male in summer and fall is browner or grayer than Willow Ptarmigan, lacking rich chestnut around head and neck. Female can be similar to female Willow Ptarmigan, but Rock has smaller bill. In winter, white male Rock has *black lores* between eyes and bill, reduced or absent in female. **VOICE:** Croaks, growls, cackles; usually silent. **SIMILAR SPECIES:** Willow Ptarmigan. **HABITAT:** Tundra, above timberline in mountains (to lower levels in winter); also near sea level in bleak tundra of northern coasts.

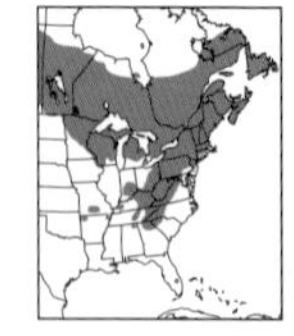

RUFFED GROUSE *Bonasa umbellus* **Uncommon**

17 in. (43 cm). Note short crest, bold flank bars, and fan-shaped tail with broad black band near tip. A large chickenlike bird of brushy woodlands, usually not seen until it flushes with a startling whir. Two color morphs: "rusty" with rufous tail and "gray" with gray tail. Rusty birds more common in southern parts of range, gray birds more common northward. Female slightly smaller and duller than male, and usually has broken black subterminal tail band. **VOICE:** Sound of drumming male suggests a distant motor starting up. Low muffled thumping starts slowly, accelerating into a whir: *Bup . . . bup . . . bup . . . bup . . . bup bup up r-rrrrr.* **SIMILAR SPECIES:** Sharp-tailed and Spruce Grouse. **HABITAT:** Ground and understory of deciduous and mixed woodlands.

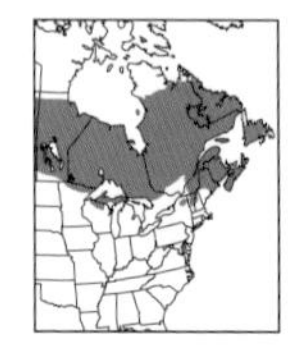

SPRUCE GROUSE *Falcipennis canadensis* **Scarce**

16–17 in. (41–43 cm). Look for this *tame,* dark grouse in deep coniferous North forests. *Male:* Sharply defined *black breast,* with some white spots or bars on sides and *chestnut band* on tip of tail. Erectile red comb above eye is visible at close range. *Female:* Dark rusty or grayish brown, thickly barred, and with black-and-white spotting below; tail short and dark. **VOICE:** Female call an accelerating, then slowing, series of *wock* notes; also cluck notes. Wing flutter from male's courtship display may sound like distant rumble of thunder. **SIMILAR SPECIES:** Ruffed Grouse. **HABITAT:** Coniferous forests, jack pines, muskeg, blueberry patches.

PTARMIGANS AND GROUSE
winter
WILLOW PTARMIGAN
female
spring/summer male
winter
spring/ summer female
spring male
winter males
winter male
spring/ summer female
spring/ summer female
male
ROCK PTARMIGAN
gray morph
RUFFED GROUSE
male display
rusty morph
female
male display
SPRUCE GROUSE
male

SHARP-TAILED GROUSE *Tympanuchus phasianellus* Uncommon

17 in. (43 cm). A pale, speckled-brown grouse of prairies and brushy draws. Note *short pointed tail,* which in display and flight shows *white* at sides. Slight crested look. Marked below by dark bars, spots, and chevrons. Displaying male has yellow eye combs and inflates *purplish* neck sacs; female slightly smaller and duller, has barred crown. **VOICE:** A cackling *cac-cac-cac,* etc. Courting note a single low *coo-oo,* accompanied by quill-rattling, foot-shuffling. **SIMILAR SPECIES:** Prairie-chickens have *rounded, dark* tail and are more barred, rather than spotted, below. Female Ring-necked Pheasant has *long pointed* tail. Ruffed Grouse has banded, *fan-shaped* tail and black neck ruff. **HABITAT:** Prairies, agricultural fields, forest edges, clearings, gullies, open burns and clear-cuts in coniferous and mixed forests.

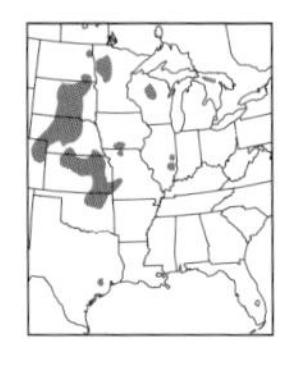

GREATER PRAIRIE-CHICKEN *Tympanuchus cupido* Uncommon, local

17 in. (43 cm). A henlike bird of prairies. Brown, heavily barred. Note *rounded dark tail* (black in male, barred in female). Courting males in communal "dance" inflate orange neck sacs, show off orangey yellow eye combs, and erect black hornlike neck feathers; female slightly smaller and duller, has less elongated neck plumes and barred crown. **VOICE:** "Booming" male in dance makes a hollow *oo-loo-woo,* suggesting sound made by blowing across a bottle mouth. **SIMILAR SPECIES:** Lesser Prairie-Chicken. Sharp-tailed Grouse, slightly paler overall, has more spots or chevrons on underparts, and more pointed, white-edged tail. Female Ring-necked Pheasant slightly larger, has long pointed tail. **HABITAT:** Native tallgrass prairie, now very localized; agricultural land. Populations of coastal TX (subspecies *T. c. attwateri*) endangered.

LESSER PRAIRIE-CHICKEN Scarce, local, threatened
Tympanuchus pallidicinctus

16 in. (41 cm). A small, pale brown prairie-chicken; best identified by range. Male's neck sacs dull *purplish* or *plum colored* (not yellow-orange as in Greater Prairie-Chicken). Breast barring usually paler and thinner than Greater's. **VOICE:** Male's courtship "booming" not as rolling or loud as Greater's. Both sexes give clucking, cackling notes. **SIMILAR SPECIES:** Greater Prairie-Chicken, Sharp-tailed Grouse. **HABITAT:** Sandhill country (sage and bluestem grass, oak shrublands).

male
display
SHARP-TAILED GROUSE
female

GREATER
PRAIRIE-CHICKEN
female
male
display
male
display
LESSER
PRAIRIE-CHICKEN
female

GRAY PARTRIDGE *Perdix perdix* **Uncommon, introduced**

12½–13 in. (32–34 cm). Introduced from Europe. A rotund gray-brown partridge, smaller than grouse but larger than quail; note short *rufous* tail, *rusty face,* chestnut bars on sides; male also has dark U-shaped splotch on belly; female slightly browner (less gray) above and with buffier lores and eye line. **VOICE:** A loud, hoarse *kar-wit, kar-wit.* **SIMILAR SPECIES:** Chukar (which also has rufous tail) prefers rockier habitat, has red bill and legs, black "necklace." **HABITAT:** Cultivated land, hedgerows, bushy pastures, meadows.

CHUKAR *Alectoris chukar* **Exotic**

13½–14 in. (34–36 cm). A popular cage bird, introduced from Asia; established in w. N. America, and escapees occasionally observed in East. Like a large quail; gray-brown with *bright red legs and bill;* light throat bordered by clean-cut black "necklace." Sides *boldly barred;* tail *rufous.* Sexes similar. **VOICE:** A series of raspy *chucks*; a sharp *wheet-u.* **SIMILAR SPECIES:** Gray Partridge. Red-legged Partridge (*Alectoris rufa*), an occasional escapee, is similar but has streaked breast.

NEW WORLD QUAIL Family Odontophoridae

Quail are smaller than grouse. Sexes can be alike or unlike. **FOOD:** Insects, seeds, buds, berries. **RANGE:** Nearly worldwide.

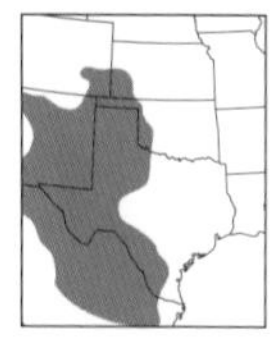

SCALED QUAIL *Callipepla squamata* **Fairly common**

10 in. (25 cm). A pale grayish quail (sometimes called "Blue Quail") of arid country, with scaly markings on breast and back. *Male:* Note *short bushy white crest,* or "cotton top," a common nickname for this species. Runs; often reluctant to fly. *Female:* Has shorter crest than male, duller and finely streaked throat. **VOICE:** A guinea hen–like *che-kar* (also interpreted as *pay-cos*). **HABITAT:** Shrub-grasslands, brush, arid country.

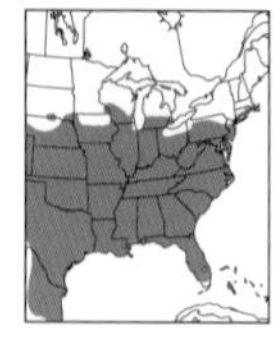

NORTHERN BOBWHITE *Colinus virginianus* **Uncommon**

9½–10 in. (24–26 cm). A small, rotund fowl, near size of a meadowlark. Ruddy, barred and striped, with short dark tail. Male has conspicuous white throat and white eyebrow stripe; in female these are buff. **VOICE:** A clearly whistled *Bob-white!* or *poor, Bob-whoit!* Covey call *ko-loi-kee?* answered by *whoil-kee!* **SIMILAR SPECIES:** No other N. American quail has white throat. Ruffed Grouse larger with fanlike tail. **HABITAT:** Farms, brushy open country, fencerows, roadsides, open woodlands; has been declining.

INTRODUCED GAME BIRDS
AND QUAIL
male
GRAY PARTRIDGE
CHUKAR
Red-legged
Partridge for
comparison
SCALED
QUAIL
female
male
NORTHERN
BOBWHITE

BIRDS of PREY

We tend to call all diurnal raptors with a hooked bill and hooked claws "birds of prey." Actually, they fall into two separate taxonomic groups that recently have been shown to be only distantly related:

1. The hawk group (Pandionidae and Accipitridae)—ospreys, kites, harriers, accipiters, buteos, and eagles
2. The falcon group (Falconidae)—falcons and caracaras

The many raptors can be sorted out by their basic shapes and flight styles. When not flapping, they may alternate between soaring, with wings fully extended and tail fanned, and gliding, with wings slightly pulled back and tail folded. These two pages show some basic silhouettes.

BUTEOS are stocky, with broad wings and a wide, rounded tail. They often soar and wheel high in the open sky.

ACCIPITERS have a small head, short rounded wings, and a longish tail. They typically fly with several rapid beats and a short glide.

HARRIERS are slim, with long, slim, round-tipped wings and a long tail. They fly in open country and glide low, with a vulturelike dihedral.

KITES (except for Snail Kite and Hook-billed Kite) are falcon-shaped, but unlike falcons, they are buoyant gliders, not power fliers.

FALCONS have long pointed wings and a long tail. Their wing strokes are strong and rapid.

OSPREY Family Pandionidae

A monotypic family comprising a single large diurnal raptor that forages above water and plunges feet-first for fish. Sexes alike. **FOOD:** Fish. **RANGE:** All continents except Antarctica.

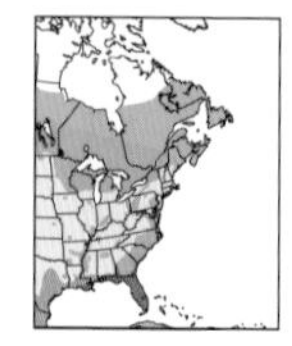

OSPREY *Pandion haliaetus* (see also p. 162) **Fairly common**

23–24½ in. (58–62 cm); wingspan to 6 ft. (183 cm). *Adult:* Blackish brown above, *white below;* head largely white with *broad black mask through eyes.* Flies with distinctive gull-like kink or crook in wings, showing black "wrist" patch to underwing. *Juvenile:* Upperpart feathers fringed whitish or buff, forming scaly pattern. **VOICE:** A series of sharp, annoyed whistles: *cheep, cheep* or *yewk, yewk,* etc. **SIMILAR SPECIES:** First-year Bald Eagle may have dusky "mask." Rough-legged Hawk also has dark wrist mark but lacks wing crook and mask, usually has dark belly patch. **HABITAT:** Rivers, lakes, marshes, bays, coasts.

HAWKS, KITES, EAGLES, and ALLIES
Family Accipitridae

Diurnal birds of prey, with hooked bills and powerful talons. Though formerly persecuted and misunderstood by many, they are very important to the health of ecosystems. **RANGE:** Almost worldwide.

EAGLES

Larger than buteos, with proportionally longer wings. Powerful bills are nearly as long as head. **FOOD:** Bald Eagle eats fish, injured waterfowl, carrion; Golden Eagle eats chiefly rabbits, large rodents, snakes, game birds.

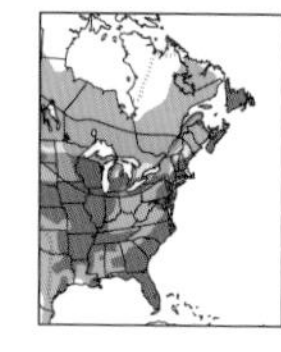

BALD EAGLE *Haliaeetus leucocephalus* (see also p. 162) **Uncommon**

31–37 in. (79–94 cm); wingspan 7–8 ft. (213–244 cm). National bird of U.S. *Adult:* Huge size and dark plumage except *white head* and *white tail* make this bird unmistakable. Bill yellow, massive. Wings held flat when soaring. *Juvenile and first-year:* Mottled dark overall. In second and third year develops variable amounts of *whitish in lower underparts, underwing,* and tail; by fourth and fifth year develops adultlike plumage, some showing a white head with darkish patch through eye, reminiscent of Osprey. **VOICE:** A harsh, high-pitched cackle, *kleek-kik-ik-ik-ik,* or lower *kak-kak-kak.* **SIMILAR SPECIES:** Golden Eagle, Turkey Vulture. **HABITAT:** Coasts, rivers, large lakes; in migration, also mountains, open country.

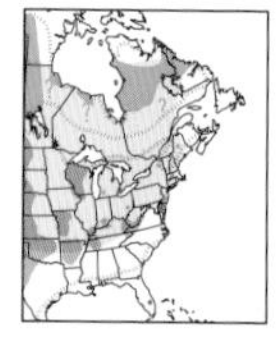

GOLDEN EAGLE *Aquila chrysaetos* (see also p. 162) **Scarce**

30–40 in. (76–102 cm); wingspan 7 ft. (213 cm). Glides and soars flat-winged with occasional shallow wingbeats. *Adult:* Uniformly dark below, or with slight paling at base of obscurely banded tail. On hindneck, a *wash of buffy gold. Juvenile and first-year:* In flight, show *white bases to primaries* and *white tail* with *broad dark terminal band.* Reaches adult plumage by third year. **VOICE:** A yelping bark, *kya;* also whistled notes. **SIMILAR SPECIES:** Younger Bald Eagles have larger head and develop *extensive blotchy white in underwing linings* and lower underparts; tail may be mottled white but is not cleanly banded. Dark-morph buteos smaller, with more rounded wings and different patterns in flight feathers. **HABITAT:** Open mountains, foothills, plains, open country.

OSPREY AND EAGLES

OSPREY, EAGLES, and VULTURES from Below

OSPREY *Pandion haliaetus* p. 160

White body and coverts; black wrist patch; crooked wing.

BALD EAGLE *Haliaeetus leucocephalus* p. 160

Adult: White head and tail.
Juvenile: Some white in underwing linings. Develops more white on belly and elsewhere in second year.

GOLDEN EAGLE *Aquila chrysaetos* p. 160

Adult: Almost uniformly dark; underwing linings dark.
Juvenile: White patch at base of primaries and tail; no white on body.

TURKEY VULTURE *Cathartes aura* p. 180

Mostly brownish black. Two-toned wings held in distinct dihedral. Small head, red in adult, blackish to dark pinkish purple in first- and second-years. Longish tail. Tips and teeters in flight.

BLACK VULTURE *Coragyps atratus* p. 180

Blackish overall. Silver patch in outer primaries. Wings held flat or in slight dihedral. Rapid, shallow wingbeats. Stubby tail. Gray head.

Where Bald Eagle, Turkey Vulture, and Osprey all are found, they can be separated at a great distance by their manner of soaring: Bald Eagle with flat wings; Turkey Vulture with a dihedral; Osprey often with a gull-like kink or crook in its wings.

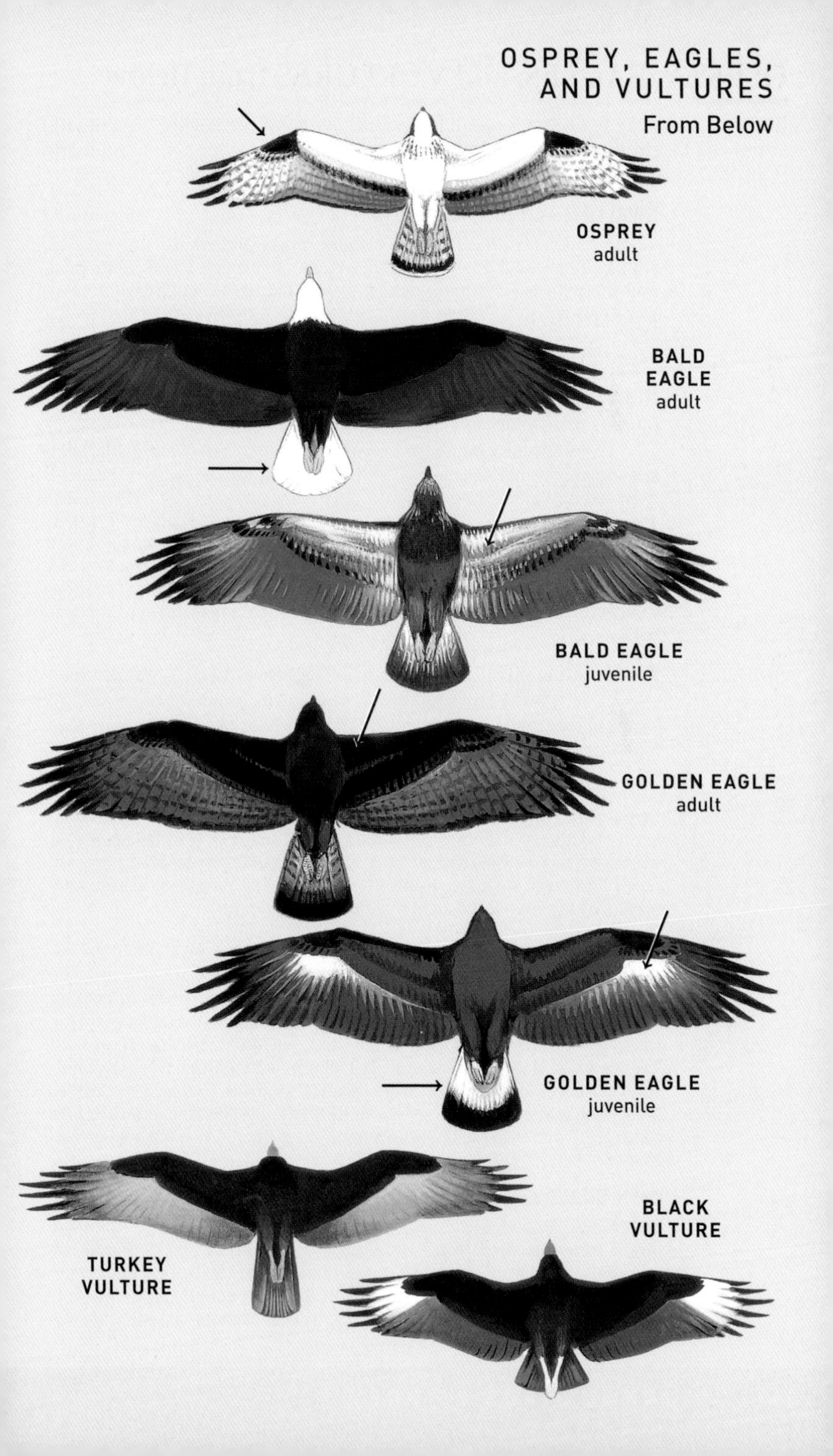
OSPREY, EAGLES, AND VULTURES
From Below
OSPREY
adult
BALD EAGLE
adult
BALD EAGLE
juvenile
GOLDEN EAGLE
adult
GOLDEN EAGLE
juvenile
BLACK VULTURE
TURKEY VULTURE

KITES

Graceful birds of prey, somewhat falconlike, with pointed wings. Ages and sexes can be similar or differ. **FOOD:** Large insects, reptiles, rodents. Snail and Hook-billed Kites specialize in snails.

SWALLOW-TAILED KITE *Elanoides forficatus* **Uncommon**

22–23 in. (55–58 cm). A sleek, elegant, black-and-white kite that flies with incomparable grace. Note blue-black upperparts, clean white head and underparts, and long, deeply forked tail. Ages and sexes similar. **VOICE:** A shrill keen, *ee-ee-ee* or *pee-pee-pee.* **RANGE:** Widespread spring vagrant north of range. **HABITAT:** Wooded river swamps and pine lands, where it feeds mainly on snakes.

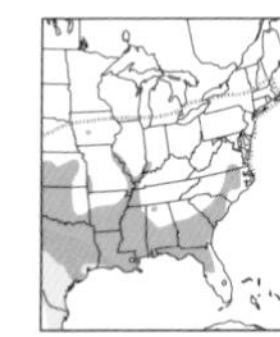

MISSISSIPPI KITE **Fairly common**

Ictinia mississippiensis (see also p. 174)

14–14½ in. (36–37 cm). Falcon-shaped, graceful, gray, and gregarious; spends much time soaring. *Adult:* Dark above, lighter below; head *pale gray;* tail and underwing blackish; note *black unbarred tail. Whitish secondaries* visible from above. *Juvenile:* Heavily streaked, rusty underparts; assumes adultlike plumage by first spring except underwing mottled rusty. **VOICE:** Usually silent; near nest, a two-syllable *phee-phew.* **SIMILAR SPECIES:** Male Northern Harrier, falcons. **RANGE:** Vagrant north of range. **HABITAT:** Nests in riparian woodlands, residential areas, groves, shelterbelts.

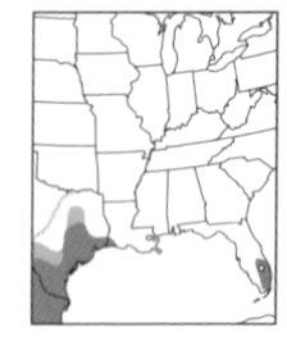

WHITE-TAILED KITE *Elanus leucurus* (see also p. 174) **Uncommon**

15½–16 in. (39–41 cm). Buoyant in flight, with pointed wings and *long white tail* that is slightly notched. Soars and glides like a small gull; *often hovers* and drops to ground with wings up. *Adult:* Pale gray above, with white head (male whiter than female), underparts, and tail. *Large black patch* on fore edge of upperwing. From below, shows oval black patch at carpal joint ("wrist"). *Juvenile:* Like adult but has *rusty mottling on crown, back, and breast;* assumes adultlike body plumage in first fall. **VOICE:** A whistled *kew kew kew,* abrupt or drawn out. **RANGE:** Widespread vagrant north and west of range. **HABITAT:** Open groves, marshes, grasslands. May form communal roosts at night in fall and winter.

SNAIL KITE *Rostrhamus sociabilis* **Scarce, local, endangered**

17 in. (43 cm). Suggests Northern Harrier at a distance but with broader wings and without gliding, tilting flight; flies more floppily on cupped wings, head down, searching for snails. *Adult male:* Slaty black except for broad white band across base of tail; legs, bill, and face red. *Female and juvenile male:* Heavily streaked on buffy body; white stripe over eye; white band across black tail. Male develops adult plumage by third year. **VOICE:** A cackling *kor-ee-ee-a, kor-ee-ee-a.* **HABITAT:** Freshwater marshes and canals with apple snails (*Pomacea* spp.). FL populations (subspecies *R. s. plumbeus*) endangered.

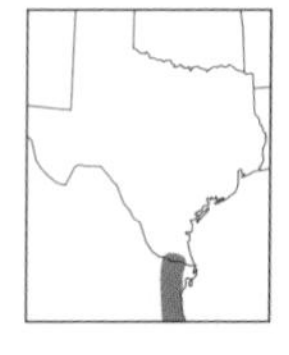

HOOK-BILLED KITE *Chondrohierax uncinatus* **Rare, local**

16½–17½ in. (42–45 cm). Bill has long, hooked tip. Legs yellow. Plumage varies from blackish (rare) or grayish in males to rufous brown in females; juveniles paler below. Adults have horizontally barred underparts. Note *paddle-shaped wings.* **VOICE:** A repeated *kik-kik-kik-kik,* recalling Northern Flicker. **HABITAT:** Subtropical woodlands, soaring only briefly when traveling.

additional overhead
flight patterns
on p. 175
SWALLOW-TAILED
KITE
adult
juvenile
MISSISSIPPI
KITE
adult
juvenile
adult
juvenile
adult
juvenile
WHITE-TAILED KITE
female
adult
male
black-morph
juvenile
adult
male
black-
morph
adult
typical
HOOK-
BILLED
KITE
SNAIL KITE
adult
female
typical

ACCIPITERS

Long-tailed woodland raptors with short, rounded wings, adapted for hunting among trees. Flight mixes quick beats with glides. Adult males have bluer upperparts; females browner and larger. **FOOD:** Chiefly birds, some small mammals. Sometimes stalks backyard bird feeders.

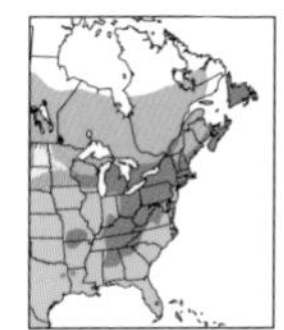

SHARP-SHINNED HAWK Fairly common

Accipiter striatus (see also p. 184)

10–14 in. (25–36 cm). A small, slim hawk, with slim *square-tipped* tail and *short, rounded wings. Adult male:* Dark bluish back, *rusty-barred* breast, red eye. *Adult female:* Browner, with yellower eye. *Juvenile and first-year:* Dark brown above, *thickly streaked* with rusty brown on underparts; yellow eye. **VOICE:** A high *kik, kik, kik* given near nest. **SIMILAR SPECIES:** Cooper's Hawk larger, with *larger head* (protruding farther forward past wings in flight), *rounded* tail with broader white tip, thicker legs. Adult Cooper's has more defined cap; first-year Cooper's *tawnier* headed, whiter breast more *finely streaked.* **HABITAT:** Breeds in extensive forests; in migration and winter, open woodlands, wood edges, parks.

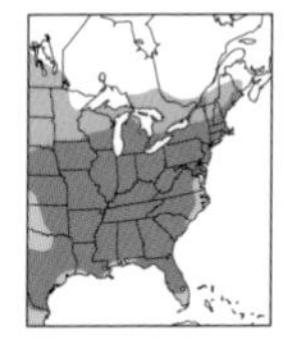

COOPER'S HAWK *Accipiter cooperii* (see also p. 184) Fairly common

14–20 in. (36–51 cm). Similar to Sharp-shinned Hawk but larger, particularly female. See Sharp-shinned Hawk. **VOICE:** Around nest, a rapid *kek, kek, kek;* suggests a flicker. Also a sapsucker-like mewing. **SIMILAR SPECIES:** Sharp-shinned Hawk, Northern Goshawk. **HABITAT:** Like Sharp-shinned but prefers drier and more open areas.

NORTHERN GOSHAWK *Accipiter gentilis* (see also p. 184) Scarce

21–26 in. (53–66 cm). Larger, broader winged, broader tailed, more buteo-like than Cooper's Hawk. *Adult:* Crown and cheek blackish; *broad white stripe over eye.* Underparts *pale gray, finely barred;* paler backed, bluer in male and grayer in female. *Juvenile and first-year:* Buffier overall than young Cooper's, with *bolder eyebrow,* more extensive streaking below. **VOICE:** *Kak, kak, kak* or *kuk, kuk, kuk,* given near nest. **SIMILAR SPECIES:** Cooper's Hawk. A soaring goshawk resembles a Red-shouldered Hawk or other buteo. **HABITAT:** Coniferous and mixed forests, especially in mountains; forest edges; winters also in wooded lowlands.

HARRIERS

Slim raptors with long wings and tail. Flight low, languid, gliding, with wings held in shallow V (dihedral). Sexes not alike. They hunt in open country.

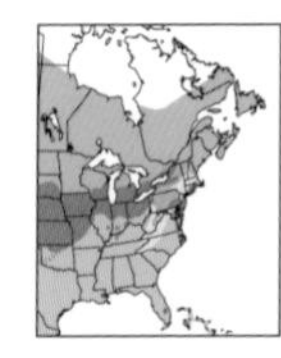

NORTHERN HARRIER Fairly common

Circus hudsonius (see also p. 174)

18–21 in. (46–54 cm). A slim, long-winged, long-tailed raptor of open country. When hunting, flies unsteadily low over ground, with wings held slightly above horizontal. Flaps steadily when migrating. In all plumages, distinct *white rump patch* distinguishes Northern Harrier from most other N. American raptors; note Cooper's Hawks can flare up white flank patches when courting. *Adult male:* Pale gray, whitish beneath, wingtips black as if "dipped in ink." *Adult female:* Brown to grayish brown, streaked below. *Juvenile and first-year:* Russet to warm buff below, fewer or no streaks. **VOICE:** A weak, nasal whistle, *pee, pee, pee.* **SIMILAR SPECIES:** Short-eared Owl. **HABITAT:** Marshes, fields.

ACCIPITERS AND HARRIER

accipiters have small heads, short rounded wings, long tails

juvenile

adult male

SHARP-SHINNED HAWK

additional overhead flight patterns on pp. 175 and 185

adult

juvenile

COOPER'S HAWK

adult male

adult

juvenile

NORTHERN GOSHAWK

adult male

adult

juvenile

adult male

NORTHERN HARRIER

adult female

BUTEOS and BUTEO-LIKE HAWKS

Large, thickset hawks, with broad wings and wide, rounded tails. Many buteos habitually soar high in wide circles. Much variation; sexes similar, females slightly larger. Young birds usually streaked below. Dark morphs often occur. **FOOD:** Small mammals, sometimes small birds, reptiles, grasshoppers. **RANGE:** Widespread in New and Old Worlds.

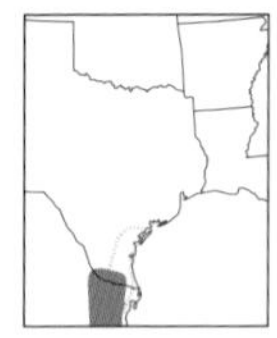

GRAY HAWK *Buteo plagiatus* (see also p. 174) **Scarce, local**

17 in. (43 cm); wingspan 3 ft. (91 cm). A small buteo. *Adult:* Distinguished by its buteo-like proportions, gray back, *thickly barred gray* breast, white rump band, and *banded* tail (similar to Broad-winged Hawk's). *Juvenile and first-year:* Narrowly barred tail, striped buffy breast, bold face pattern, *white U-shaped bar* across rump. **VOICE:** Drawn-out whistles, *ka-lee-oh* or *kleeeeoo.* **SIMILAR SPECIES:** First-year Broad-winged Hawk has weaker face pattern, lacks white U on rump, has shorter tail, more pointed wings. **HABITAT:** Streamside and subtropical woodlands.

WHITE-TAILED HAWK **Uncommon, local**

Geranoaetus albicaudatus (see also p. 174)

21–23 in. (53–58 cm); wingspan 4 ft. (122 cm). A large buteo-like hawk, with long pointed wings. Flies with marked dihedral. *Adult:* White underparts contrasting with dark flight feathers; tail white with black band, shoulders rusty red. *Juvenile and first-year:* Narrower wings and longer tail than adult. Blackish below with white breast patch. Pale U across upper tail. May have dark belly patch like Red-tailed Hawk, but note blacker wing lining. Tail pale gray with weak barring. *Second-year:* Intermediate between juvenile and adult. **VOICE:** A nasal note followed by a high-pitched series of doubled notes: *aaraahh kee-REEK, kee-REEK kee-REEK.* **SIMILAR SPECIES:** Juvenile and first-year Red-tailed Hawk. Adult Swainson's Hawk smaller, has dark chest. **HABITAT:** Coastal prairies, brushlands.

HARRIS'S HAWK **Fairly common**

Parabuteo unicinctus (see also p. 178)

20–21 in. (50–53 cm); wingspan 3½ ft. (107 cm). A blackish-brown, buteo-like hawk, with flashing *white rump* and *white band* at tip of tail. Often hunts cooperatively in small groups. *Adult: Chestnut areas* on thighs, shoulders, and underwing; *rusty shoulders;* conspicuous *white* at base of tail. *Juvenile and first-year:* Underparts streaked pale. **VOICE:** A low-pitched, harsh *raaaah!* **SIMILAR SPECIES:** Dark morphs of Ferruginous and Red-tailed Hawks larger, lack bold rusty patches and white tail base. **RANGE:** Casual vagrant north and east of range. **HABITAT:** Mesquite forest, dry open areas.

ZONE-TAILED HAWK *Buteo albonotatus* (see also p. 178) **Rare, local**

20 in. (51 cm); wingspan 4 ft. (122 cm). Dull *black,* with more *slender* wings than most other buteos. Often mistaken for Turkey Vulture because of proportions, two-toned underwing, and up-tilted wings, but Zone-tailed has larger feathered head, square-tipped tail, barred underwing, yellow cere and legs. *Adult: White tail bands* (pale gray on top side). *Juvenile and first-year:* Narrower tail bands, *small white spots* on breast. **VOICE:** A nasal, drawn-out *keeeeah.* **SIMILAR SPECIES:** Turkey Vulture, other dark-morph buteos. **RANGE:** Accidental vagrant north and east of range. **HABITAT:** Riparian woodlands, mountains, canyons.

BUTEOS AND BUTEO-LIKE HAWKS
GRAY HAWK
adult
juvenile
WHITE-TAILED HAWK
juvenile
adult
adult
juvenile
adult
HARRIS'S HAWK
adults
juvenile
adult
ZONE-TAILED HAWK

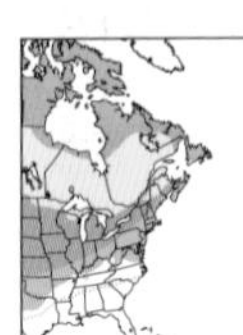

ROUGH-LEGGED HAWK

Uncommon

Buteo lagopus (see also pp. 176 and 178)

21–22 in. (53–55 cm). This open-country hawk often *hovers on beating wings* and has *smaller bill* and feet than other buteos. Legs feathered. Many birds have *solid or blotched dark belly* and *black patch* at "wrist" (carpal joint) of underwing. Some adult males have dark bib but lack blackish belly band. Tail *white*, with *broad black band or bands* toward tip. White flash on upperwing. Juvenile and first-year similar but tail with less distinct band. Dark morph may lack extensive white on tail but shows broad terminal band and extensive white on underwing. **VOICE:** A high-pitched squeal, mostly near nest site. **SIMILAR SPECIES:** Red-tailed Hawk, dark-morph Ferruginous Hawk. **HABITAT:** Nests on tundra escarpments; in winter, open fields, marshes.

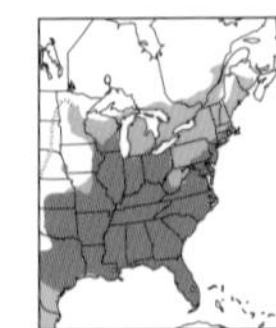

RED-SHOULDERED HAWK

Uncommon to fairly common

Buteo lineatus (see also p. 176)

16–20 in. (40–50 cm). In flight, note *translucent patch* or "window" at base of primaries, longish tail. *Adult:* Heavy black-and-white bands on wings and tail, dark *rufous shoulders* and underwing linings, rufous red underparts. *Juvenile and first-year:* Variably streaked and/or barred below; recognized by tail bands and wing "windows." FL birds (subspecies *B. l. extimus*) smaller and paler; can have whitish heads. **VOICE:** A two-syllable scream, *kee-yer* (dropping inflection), repeated in series. **SIMILAR SPECIES:** Light-morph Broad-winged Hawk has paler underwing linings, more pointed wing, broader bands on tail, lacks wing "windows." Juvenile Cooper's Hawk has longer tail. See also Red-tailed Hawk. **HABITAT:** Woodlands in valleys, along rivers, swamp edges, residential areas.

BROAD-WINGED HAWK

Common

Buteo platypterus (see also pp. 176 and 178)

15–16 in. (38–41 cm). A small, chunky buteo, often seen migrating in fall in spiraling "kettles." *Pale-morph adult:* Note one obvious thick white band visible from below. Underwing linings whitish, the edge trimmed with black. *Juvenile and first-year:* Heavily streaked along sides of neck, breast, and belly; chest often unmarked. Terminal tail band distinct, twice as wide as other bands. Rare dark morph (p. 178), which breeds primarily in Prairie Provinces, similar to dark Short-tailed Hawk but browner (not as black); tail pattern as above; secondaries paler and less barred underneath. **VOICE:** A high-pitched, downward *pwe-eeeeee.* **SIMILAR SPECIES:** Juvenile/first-year Red-shouldered Hawk. See also juvenile Gray and Short-tailed Hawks, accipiters. **HABITAT:** Woods, groves.

SHORT-TAILED HAWK

Uncommon, local

Buteo brachyurus (see also p. 178)

15–16 in. (38–41 cm). A small black or black-and-white buteo. Dark morph has blackish brown body and black underwing linings; light-morph adult blackish above, white below, with dark cheeks, *two-toned* underwing pattern, white underwing linings. *Light-morph juvenile and first-year:* Similar to juvenile Broad-winged Hawk but less streaked below; secondaries darker underneath, with more distinct barring. **VOICE:** A descending, high-pitched scream: *kleeear!* **SIMILAR SPECIES:** Broad-winged Hawk. See also Swainson's Hawk. **HABITAT:** Cypress swamps, mangroves, hardwood hammocks.

ROUGH-LEGGED
HAWK
dark
morph
light-morph
adults
additional overhead
flight patterns on
pp. 177 and 179
pale FL
subspecies
adults
RED-SHOULDERED
HAWK
juvenile
juvenile
adult
juvenile
adult
dark
morph
light-
morph
juvenile
BROAD-WINGED
HAWK
dark
morph
dark-morph
adult
light-morph
adult
light-morph
adult
dark-morph
adult
SHORT-TAILED HAWK

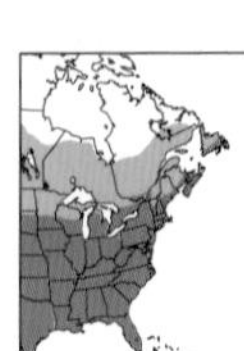

RED-TAILED HAWK
Common

Buteo jamaicensis (see also pp. 176 and 178)

19–22 in. (48–56 cm). The common conspicuous hawk of roadsides and woodland edges. *Adult:* When soaring, has diagnostic *rufous* on top side of tail, paler reddish below. On light-morph birds, note mottled *white patches* on scapulars; rather diagnostic *dark patagial bar* on fore edge of wing from below. Otherwise body plumage quite variable. *Juvenile and first-year:* Tail brownish with narrow, dark banding. Underparts typically "zoned" (light breast, dark *belly band*). Dark-morph birds (rare to uncommon in East) variably dark brown to blackish; red tail of adults diagnostic; broad wing shape and tail pattern help identify juvenile and first-year birds. On Great Plains and wintering to se. U.S., whitish "Krider's" morph has whitish tail that may be tinged with pale rufous. Dark-morph "Harlan's" Red-tailed Hawk (*B. j. harlani*), an uncommon breeder in AK and wintering to lower Mississippi Valley, is sootier and tail is usually dirty white, with *longitudinal* mottling and freckling of gray, black, sometimes with red, merging into dark subterminal band. **VOICE:** An asthmatic squeal, *keeer-r-r* (slurring downward). **SIMILAR SPECIES:** Rough-legged, Ferruginous, Swainson's, Red-shouldered, and Broad-winged Hawks. **HABITAT:** Open country, woodlands, prairie groves, mountains, plains, roadsides.

SWAINSON'S HAWK
Fairly common

Buteo swainsoni (see also pp. 176 and 178)

19–21 in. (48–53 cm). A buteo of plains, quite variable in body plumage. Slimmer than Red-tailed Hawk, with narrower, more pointed wings at tips. When gliding, holds wings slightly above horizontal. When perched, *wingtips extend to tail tip.* In light and intermediate morphs, *pale underwing linings contrast with dark flight feathers* from below. *Adult:* Light morph has dark breast-band and often a dark-hooded look; tail gray-brown above, often pale toward base. Dark- and rufous-morph birds best identified by shape, shaded flight feathers, and tail pattern. *Juvenile and first-year:* Light morph variably streaked below, usually *more heavily marked on breast than belly;* white band across rump; often best identified by shape and wing pattern. **VOICE:** A shrill, plaintive whistle, *kreeeeeeer.* **SIMILAR SPECIES:** Swainson's wing shape distinctive for a buteo. Lacks white scapular patches and dark patagial marks of bulkier Red-tailed. In TX, see White-tailed Hawk. **RANGE:** Uncommon in eastern Plains; casual fall visitor or vagrant to East Coast. **HABITAT:** Plains, grasslands with sparse trees, agricultural land.

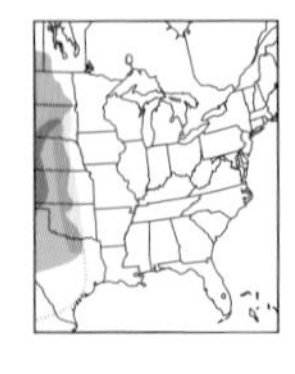

FERRUGINOUS HAWK *Buteo regalis* (see also pp. 174 and 178)
Scarce

23–24 in. (58–61 cm). A large buteo of plains. Note large bill, long gape line, *long tapered wings* with *pale panel* on upper surface of primaries, *mostly white tail. Adult:* Rufous above, especially shoulder; pale tail, with rufous thighs forming *dark V* on birds from below. Rare dark morphs show whitish flight feathers and whitish tail. *Juvenile and first-year:* Similar to adult but duller, tail with indistinct bars or marks. **SIMILAR SPECIES:** Red-tailed Hawk, dark-morph Rough-legged Hawk. **RANGE:** Casual vagrant east of range, accidentally to coast. **HABITAT:** Plains, grasslands, agricultural fields.

BUTEOS
rufous adult
Western juvenile
Eastern juvenile
dark adult
Western adult
Eastern adult
RED-TAILED HAWK
"Harlan's" adult
"Harlan's" adult
juvenile
adult
dark morph
light morph
light morph
adult
SWAINSON'S HAWK
light morph
adult
light morph
dark morph
FERRUGINOUS HAWK

KITES, HARRIERS, and PALE BUTEOS from Below

WHITE-TAILED KITE *Elanus leucurus* p. 164
Adult: White body; whitish tail; dark underside to primaries, black mark at primary coverts.

MISSISSIPPI KITE *Ictinia mississippiensis* p. 164
Falcon-shaped. *Adult:* Pale gray head, black tail, dark gray and blackish wings, gray body.
Juvenile: Streaked breast; banded square-tipped or notched tail. First-spring and summer adultlike but with underwing mottled brownish.

NORTHERN HARRIER *Circus hudsonius* p. 166
Male: Whitish wings with black tips and dark trailing edge. Gray hood.
Female: Brown, heavily streaked; note long, slim wings and tail.
Juvenile and first-year (not shown): Warmer brown than adult female, unstreaked body, dark head. From above, all plumages have characteristic white rump.

FERRUGINOUS HAWK *Buteo regalis* (light morph) p. 172
Whitish underparts, with dark V formed by reddish thighs in adult. Wings and tail long for a buteo. A bird of western plains and open range.

GRAY HAWK *Buteo plagiatus* p. 168
Stocky. Broadly banded tail (suggestive of Broad-winged Hawk); adults have gray-barred underparts. Scarce resident of Rio Grande Valley.

WHITE-TAILED HAWK *Geranoaetus albicaudatus* p. 168
Adult: Whitish underparts, gray head. White tail with black band near tip. Soars with marked dihedral. Resident of coastal prairie of TX.

Kites (except Snail and Hook-billed Kites) are falcon-shaped but, unlike falcons, are buoyant gliders, not power fliers. All are southern.

KITES
From Below

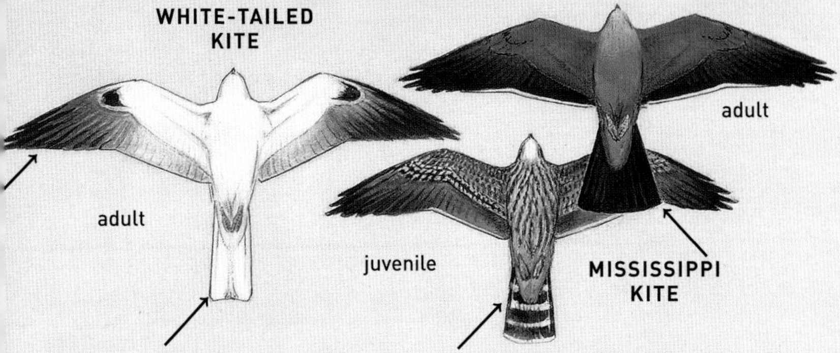

WHITE-TAILED KITE

adult

adult

juvenile

MISSISSIPPI KITE

HARRIER
From Below

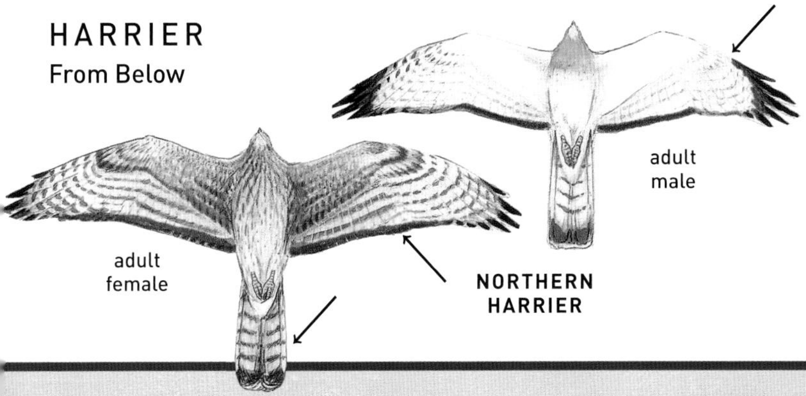

adult male

adult female

NORTHERN HARRIER

PALE BUTEOS
From Below

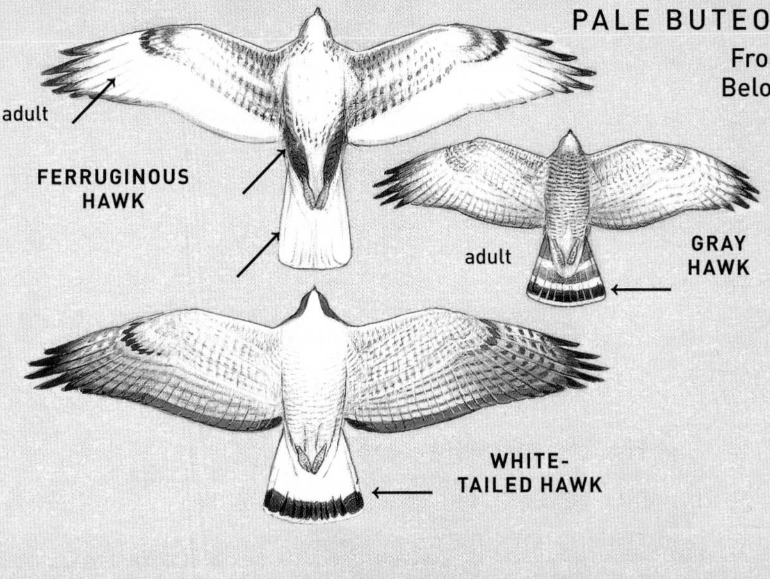

adult

FERRUGINOUS HAWK

adult

GRAY HAWK

WHITE-TAILED HAWK

PALE BUTEOS from Below

RED-TAILED HAWK *Buteo jamaicensis* (light morph) p. 172

Reddish tail and dark patagial bar at fore edge of wing are best mark from below. *Adult:* Light chest, streaked belly (often forming belly band); tail plain with little or no banding.
Juvenile and first-year: Streaked below, tail without red and with light banding.

SWAINSON'S HAWK *Buteo swainsoni* (light morph) p. 172

Adult: Dark breast-band. Long, pointed, two-toned wings.
Juvenile and first-year: Similar but with streaks on underbody.

RED-SHOULDERED HAWK *Buteo lineatus* p. 170

Adult: Tail strongly banded (white bands narrower than black ones). Body and underwing coverts barred or mottled reddish.
Juvenile and first-year: Chest and belly heavily streaked brown. Both first-year and adult have light crescent "window" on outer wings, longish tail.

BROAD-WINGED HAWK *Buteo platypterus* (light morph) p. 170

Smaller and chunkier than Red-shouldered Hawk, with shorter tail, more pointed wings. *Adult:* Widely banded tail (white bands wider); underwing pale with dark rear margin and tip.
Juvenile and first-year: Body usually streaked, tail narrowly banded, the outermost dark band widest and most distinct.

ROUGH-LEGGED HAWK *Buteo lagopus* (light morph) p. 170

Note black primary covert patch at "wrist" contrasting with white primaries and secondaries. Broad, blackish band ("cummerbund") across belly (blacker and more solid than in Red-tailed Hawk) is distinctive. Tail light, with broad, dark subterminal band.

Buteos are chunky, with broad wings and a broad, rounded tail. They often soar and wheel high in the air.

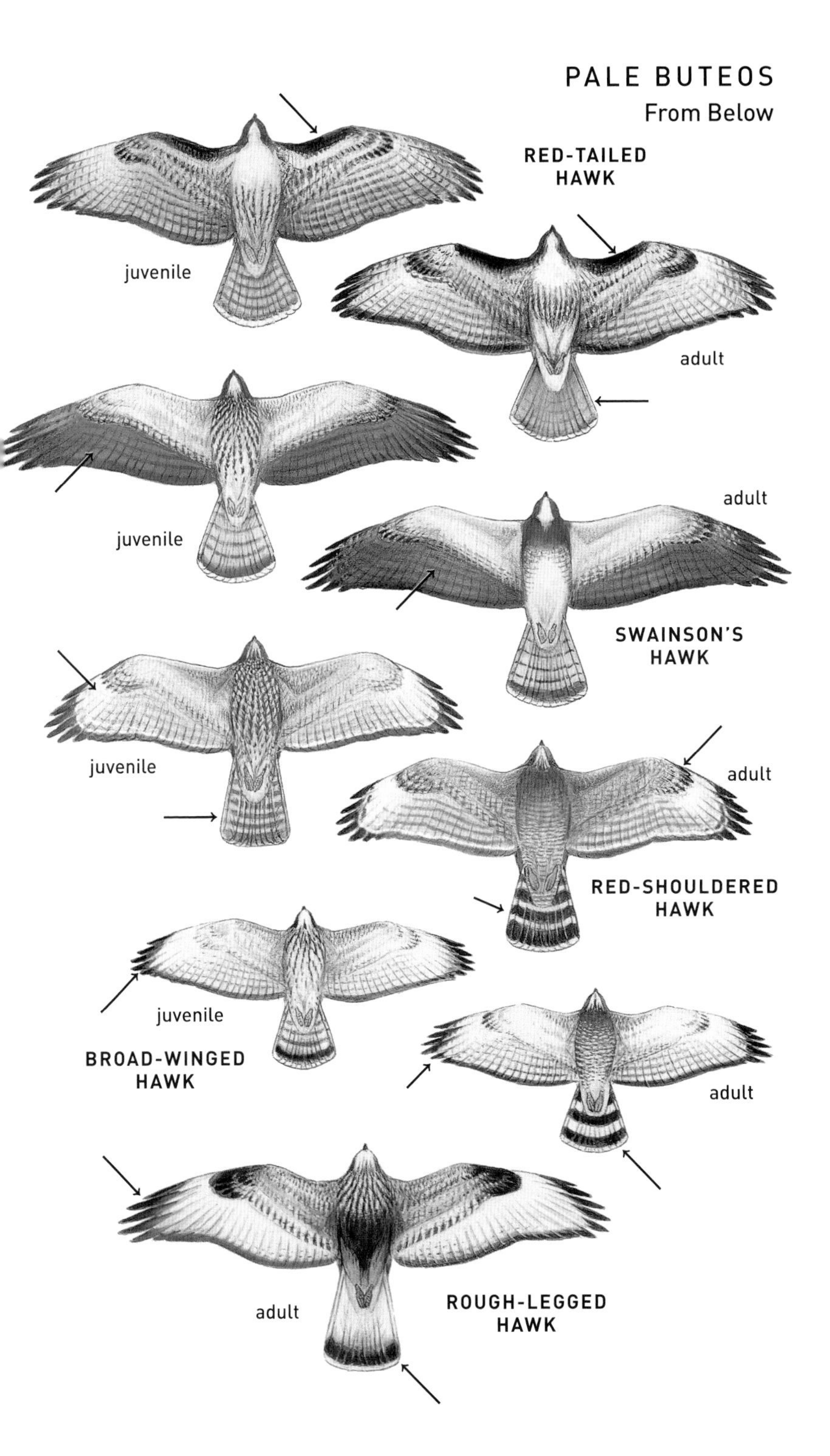
PALE BUTEOS
From Below
RED-TAILED HAWK
juvenile
adult
juvenile
adult
SWAINSON'S HAWK
juvenile
adult
RED-SHOULDERED HAWK
juvenile
BROAD-WINGED HAWK
adult
adult
ROUGH-LEGGED HAWK

DARK BIRDS of PREY from Below

CRESTED CARACARA *Caracara cheriway* p. 180

Whitish chest, black belly, large *pale patches* in primaries, white tail with black band. Elongated neck, stiff-winged flight.

ROUGH-LEGGED HAWK *Buteo lagopus* (dark morph) p. 170

Dark body and underwing linings; *whitish flight feathers;* tail light from below, with one broad, *black terminal band* in female; additional bands in male.

FERRUGINOUS HAWK *Buteo regalis* (dark morph) p. 172

Similar to dark-morph Rough-legged Hawk, but tail whitish, without dark banding. Note also white wrist marks, or "commas."

SWAINSON'S HAWK *Buteo swainsoni* (dark morph) p. 172

Pointed wings are usually dark throughout, *including flight feathers;* tail narrowly banded, whitish undertail coverts. Rufous morph (not shown) may be rustier, with lighter rufous underwing linings.

RED-TAILED HAWK *Buteo jamaicensis* (dark morph) p. 172

Typical chunky shape of Red-tailed; tail reddish, brighter above than below; variable. Dark patagial bar on leading edge of wing obscured.

"HARLAN'S" RED-TAILED HAWK *Buteo jamaicensis harlani* p. 172

Similar to dark-morph Red-tailed Hawk. Breast mottled white; tail tends to be mottled with gray and whitish and with dusky subterminal band, usually lacking obvious red; primary tips barred dark and light.

BROAD-WINGED HAWK *Buteo platypterus* (dark morph) p. 170

Typical small size and broad-winged shape. Tail pattern and flight feathers as in light morph, but body and underwing linings dark brownish to brownish black. Note whiter flight feathers than Short-tailed Hawk.

ZONE-TAILED HAWK *Buteo albonotatus* (first-year) p. 168

Slim and longish, *two-toned wings* (suggesting Turkey Vulture) with barred flight feathers. Several white bands on slim tail (only one visible on folded tail). Yellow legs. Wings held in slight dihedral.

SHORT-TAILED HAWK *Buteo brachyurus* (dark morph) p. 170

Jet-black body and underwing linings. Lightly banded tail; flight feathers more shaded and often more distinctly barred than in dark Broad-winged Hawk.

COMMON BLACK HAWK *Buteogallus anthracinus* (sitting bird not shown)

Very broad black wings; faint light patches near wingtips. Short, broad tail with broad white band at *mid-tail* and very broad black subterminal band.

HARRIS'S HAWK *Parabuteo unicinctus* p. 168

Chocolate brown body, chestnut underwing linings. Very broad white band at base of black tail, narrow white terminal band. Flies more buoyantly than buteos.

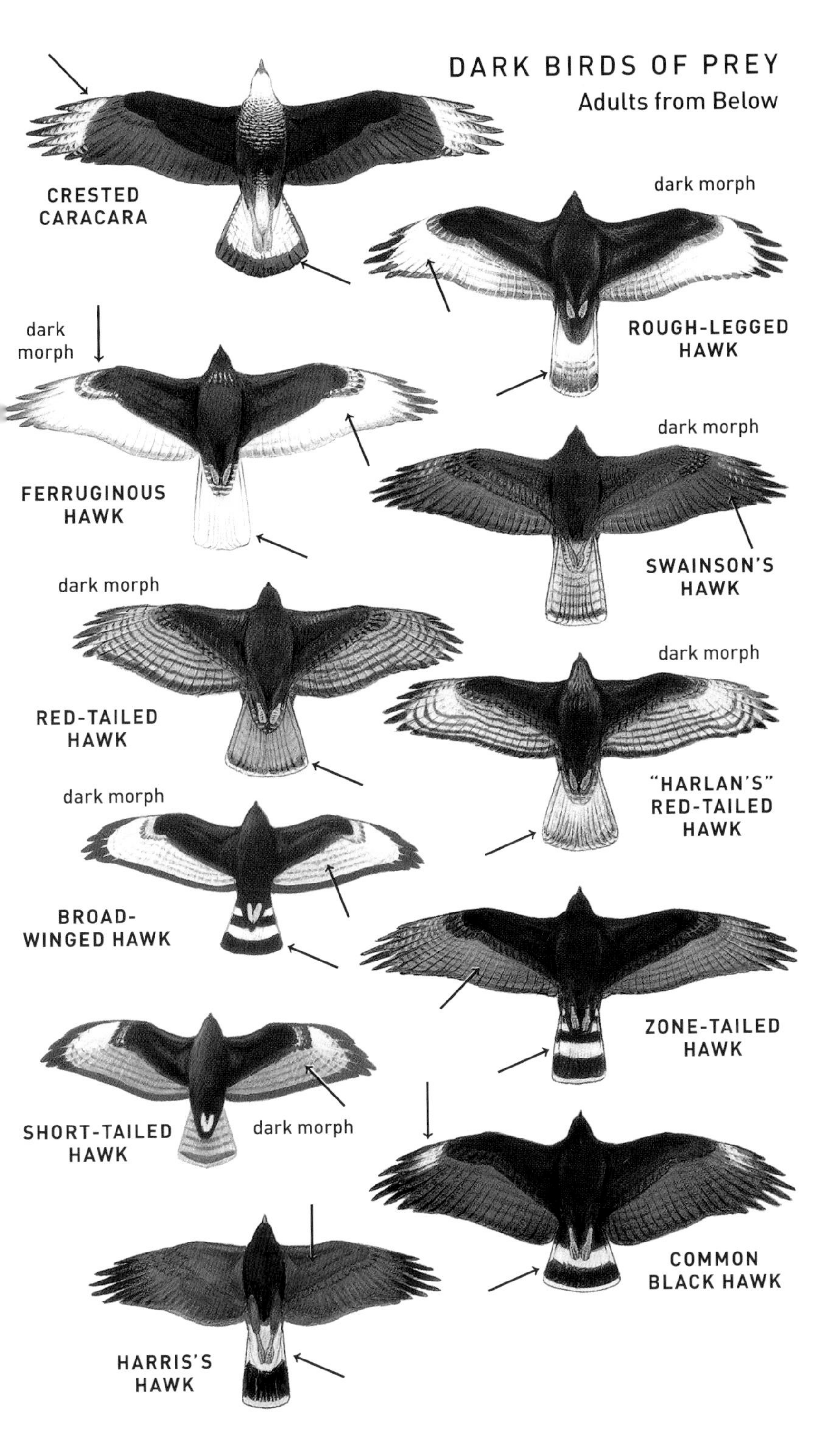
DARK BIRDS OF PREY
Adults from Below
CRESTED CARACARA
dark morph
ROUGH-LEGGED HAWK
dark morph
FERRUGINOUS HAWK
dark morph
SWAINSON'S HAWK
dark morph
RED-TAILED HAWK
dark morph
"HARLAN'S" RED-TAILED HAWK
dark morph
BROAD-WINGED HAWK
ZONE-TAILED HAWK
SHORT-TAILED HAWK
dark morph
COMMON BLACK HAWK
HARRIS'S HAWK

NEW WORLD VULTURES Family Cathartidae

Blackish; often seen soaring high in wide circles. Their naked heads are relatively smaller than those of hawks and eagles. Vultures are often locally called "buzzards." Silent away from nest site. Ages vary in plumage and head features; sexes alike. **FOOD:** Carrion. **RANGE:** S. Canada through S. America.

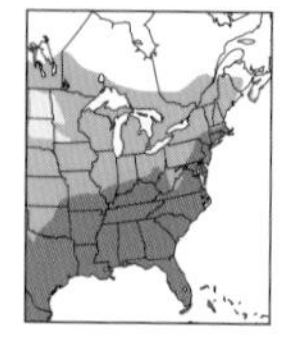

TURKEY VULTURE *Cathartes aura* (see also p. 162) **Common**

26–27 in. (66–69 cm); wingspan 6 ft. (183 cm). Nearly eagle-sized. From below, note dark color with *two-toned wings* (flight feathers paler). Soars with *wings in dihedral* (shallow V); rocks and tilts unsteadily. At close range, small, naked *red head* of adult is evident; juvenile has dark bill and grayish head with black mask and bristlelike feathers, head becoming purplish in first year and not fully naked and red until third year. **SIMILAR SPECIES:** Black Vulture. Eagles and Zone-tailed Hawk, the latter of which "mimics" Turkey Vulture in flight profile, all have larger, feathered heads, shorter tails; eagles also soar with wings held in steady flat plane. **HABITAT:** Seen soaring in sky, on ground feeding, or perched on dead trees or posts; suns with wings outstretched. Ubiquitous through much of range.

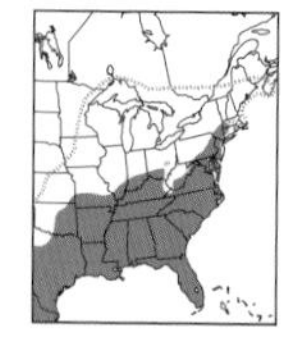

BLACK VULTURE *Coragyps atratus* (see also p. 162) **Common**

25 in. (64 cm); wingspan less than 5 ft. (152 cm). This dark scavenger is readily identified by its short, square tail that barely projects beyond rear edge of wings and by *whitish patch* toward wingtip. Legs longer and whiter than in Turkey Vulture; in flight, feet visible beyond tail. Note distinctive *shallow and quick flapping,* alternating with short glides. **SIMILAR SPECIES:** Turkey Vulture has longer, rounded tail; flapping is slower, less frequent; soars with noticeable dihedral. *Beware:* Juvenile and first-year Turkey Vultures have dark heads but show paler bills and structural and flight-style differences noted above. **RANGE:** Widespread vagrant well north of breeding range. **HABITAT:** Similar to Turkey Vulture but prefers wetter lowland areas, sometimes scavenges in dumps.

CARACARAS and FALCONS Family Falconidae

Recently found to be more closely related to parrots than to other diurnal raptors. Caracaras are large, long-legged birds of prey, some with naked faces. Sexes alike. **FOOD:** Our one U.S. species feeds mostly on carrion. **RANGE:** Southern U.S. to Tierra del Fuego, Falklands. Falcons are streamlined birds of prey with pointed wings, longish tails. Ages and sexes similar or vary in different combinations. **FOOD:** Birds, rodents, reptiles, insects. **RANGE:** Worldwide.

CRESTED CARACARA **Uncommon, local, threatened**

Caracara cheriway (see also p. 178)

23 in. (58 cm). A large, long-legged, big-headed, long-necked bird of prey, often seen feeding on carrion or roadkill with vultures. *Adult: Black crest* and *red face* distinctive. In flight, underbody presents alternating areas of light and dark: white chest, black belly, and whitish, dark-tipped tail. Note combination of *pale wing patches, pale chest, and pale tail panel,* giving impression of "white at all four corners." *Juvenile and first-year:* Browner, streaked on breast. Second-year intermediate. Sexes similar. **VOICE:** A weird, guttural series of croaks and rattles. **RANGE:** Casual vagrant well north of range. **HABITAT:** Prairies, rangeland, deserts; in FL, central plains, where subspecies *C. c. audubonii* considered threatened.

VULTURES AND CARACARA
adult
juvenile
TURKEY
VULTURE
juvenile
adult
BLACK
VULTURE
juvenile
adults
CRESTED
CARACARA

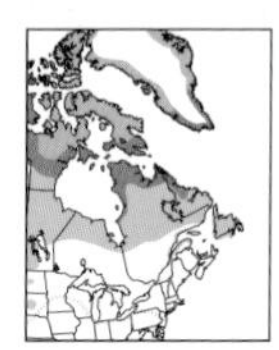

GYRFALCON *Falco rusticolus* (see also p. 184) **Scarce**

20–25 in. (51–64 cm). A very large Arctic falcon; *tail broader and longer* than Peregrine Falcon's, *wingtips not reaching near tail tip when perched.* Has brown, gray, and white color morphs. Ages and sexes similar. **VOICE:** A harsh *kak-kak-kak* series. **SIMILAR SPECIES:** Peregrine and Prairie Falcons. **RANGE:** Vagrant south of normal winter range. **HABITAT:** Arctic barrens, seacoasts; in winter, open country, coastlines.

PRAIRIE FALCON *Falco mexicanus* (see also p. 184) **Scarce**

16–19 in. (41–50 cm). Like a sandy-colored Peregrine Falcon, with *white eyebrow stripe* and *narrower mustache.* In flight from below, shows *blackish patches* in axillars ("wingpits") and inner coverts. Ages and sexes similar. **SIMILAR SPECIES:** Other falcons darker and lack black underwing patches. **RANGE:** Vagrant east of range, accidentally to coast. **HABITAT:** Open country, tundra grasslands, agricultural land.

PEREGRINE FALCON *Falco peregrinus* (see also p. 184) **Fairly common**

16–20 in. (41–51 cm). A robust, medium-large falcon with pointed wings, narrow tail, and quick, powerful wingbeats. Note *wide black mustache. Adult:* In widespread "N. American" subspecies (*F. p. anatum*), upperparts slaty blue, breast washed rose, barred and spotted black below breast. *Juvenile and first-year:* Brown, heavily streaked below. Smaller migratory "Tundra" Peregrine (subspecies *F. p. tundrius*) slimmer, adult paler, and juvenile has pale crown. **VOICE:** A rapid *kek kek kek kek,* a repeated *we'chew.* **SIMILAR SPECIES:** Merlin, Gyrfalcon. **HABITAT:** Nests on cliffs and ledges; open country, from mountains to coasts. Breeds on building ledges and bridges in major cities.

APLOMADO FALCON *Falco femoralis* (see also p. 184) **Rare, local**

15–16½ in. (38–42 cm). A medium-sized falcon, a little smaller than Peregrine Falcon. *Long wings and tail.* Note *dark underwing* and *black belly,* contrasting with white or pale cinnamon breast. Thighs and undertail coverts orange-brown. **VOICE:** A high-pitched *klee-klee-klee-klee!* **RANGE:** Formerly a very rare visitor from Mex. Reintroduced in s. TX.

AMERICAN KESTREL *Falco sparverius* (see also p. 184) **Fairly common**

9½–10½ in. (24–27 cm). Note small size, *rufous back and tail.* Male has blue-gray wings. Both sexes have black-and-white face with double mustache. *Hovers* for prey; sits fairly erect, occasionally lifting tail. Ages similar. **VOICE:** A rapid, high *klee klee klee* or *killy killy killy.* **SIMILAR SPECIES:** Merlin lacks rufous and is more compact, flies much faster, never hovers. **HABITAT:** Open country, farmland, wood edges, dead trees, wires, roadsides.

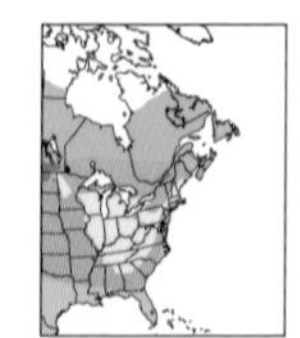

MERLIN *Falco columbarius* (see also p. 184) **Uncommon**

11–12 in. (28–31 cm). Suggests a miniature Peregrine Falcon but with less distinct mustache. *Adult male:* In common N. American subspecies (*F. c. columbarius,* "Taiga" Merlin), upperparts darkish blue-gray, tail *gray* with broad black bands, underparts streaked reddish brown. *Female, juvenile, and first-year:* Dark brown above, with banded tail; boldly streaked below. "Prairie" subspecies (*F. c. richardsoni*) paler gray or brown, lacks or has indistinct mustache. **VOICE:** A high, rapid *kee-kee-kee-kee.* **SIMILAR SPECIES:** American Kestrel, Peregrine and Prairie Falcons. **HABITAT:** Cliffs, grasslands, tundra; in migration and winter, open country, marshes, beaches, locally in cities and neighborhoods.

gray
morph
GYRFALCON
gray
morph
white morph
brown
morph
brown
morph
PRAIRIE
FALCON
PEREGRINE
FALCON
"Tundra"
juvenile
"Tundra"
adult
"North
American"
adults
APLOMADO
FALCON
adult
male
female
male
female
male
AMERICAN
KESTREL
"Taiga"
female
"Taiga"
male
"Prairie"
male
MERLIN

ACCIPITERS and FALCONS from Below

COOPER'S HAWK *Accipiter cooperii* p. 166

Underparts rusty (adult). Tail rounded and tipped with broad white terminal band. Note head and neck projecting noticeably beyond leading edge of wing.

NORTHERN GOSHAWK *Accipiter gentilis* p. 166

Adult with bold facial pattern, underbody heavily barred with pale gray. Tail and wings broad.

SHARP-SHINNED HAWK *Accipiter striatus* p. 166

Small. When folded, tail square or notched, with narrow pale tip. Fanned tail slightly rounded. Note small head and short neck barely projecting beyond wing.

PEREGRINE FALCON *Falco peregrinus* p. 182

Falcon shape; large; bold face pattern; longer wings than Merlin or Kestrel.

AMERICAN KESTREL *Falco sparverius* p. 182

Small; banded rufous tail. Paler underwing and less heavily marked underparts than Merlin.

MERLIN *Falco columbarius* p. 182

Small; heavily marked underparts and dark underwing; heavily banded tail.

GYRFALCON *Falco rusticolus* p. 182

Larger than Peregrine Falcon, without that bird's contrasting facial pattern, and with broader wings and tail. Varies in color from brown to gray to white.

APLOMADO FALCON *Falco femoralis* p. 182

Black belly band or vest, light chest, orange undertail. Tail barred with black.

PRAIRIE FALCON *Falco mexicanus* p. 182

Size of Peregrine Falcon. *Dark axillars* ("wingpits") and inner coverts.

Accipiters (hawks) have short, rounded wings and a long tail. They fly with several rapid beats and a short glide. They are better adapted to hunting in woodlands than most other hawks. Females are larger than males. Juveniles (not shown) have a streaked breast.

Falcons have long, pointed wings and a relatively long tail. Wing strokes are typically rapid and continuous.

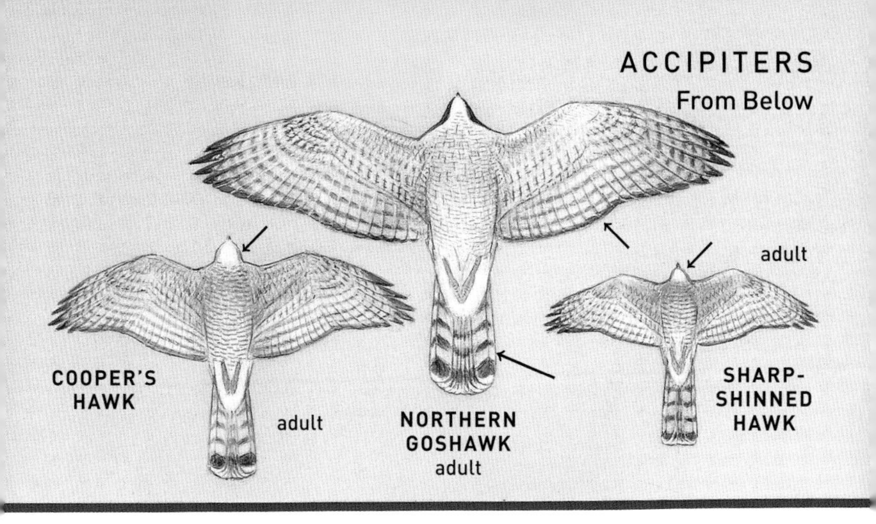
ACCIPITERS
From Below
adult
COOPER'S HAWK
adult
NORTHERN GOSHAWK
adult
SHARP-SHINNED HAWK

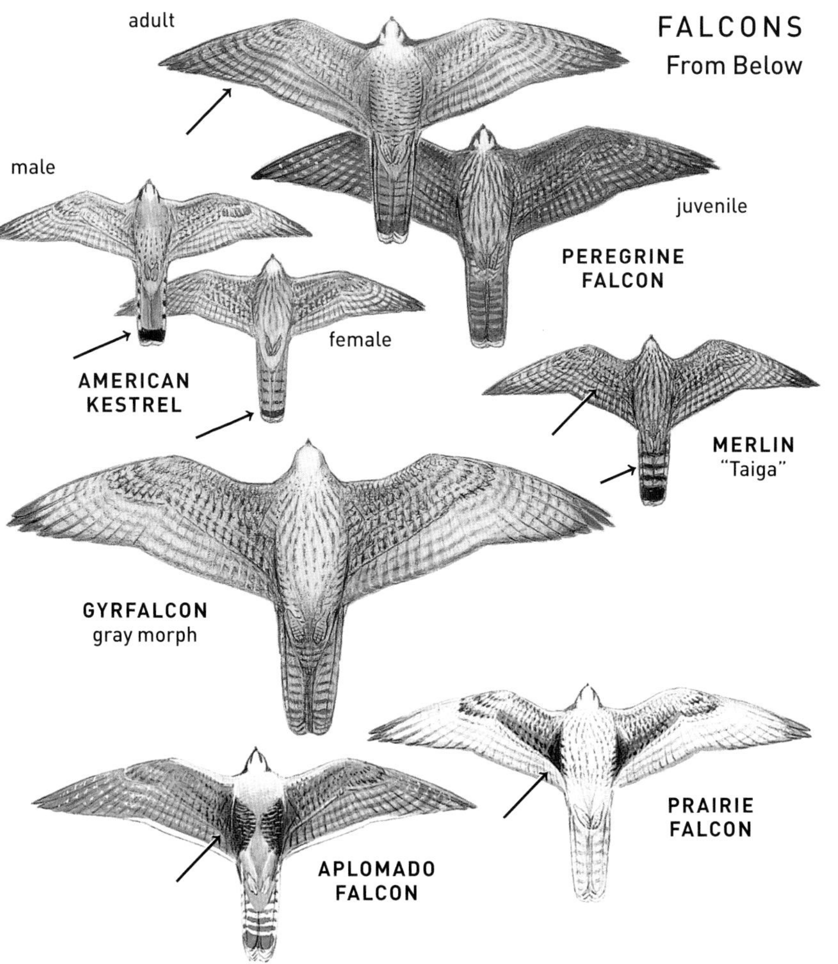
adult
FALCONS
From Below
male
juvenile
PEREGRINE FALCON
female
AMERICAN KESTREL
MERLIN
"Taiga"
GYRFALCON
gray morph
PRAIRIE FALCON
APLOMADO FALCON

OWLS Families Tytonidae (Barn Owls) and Strigidae (Typical Owls)

Chiefly nocturnal birds of prey, with large heads and flattened faces forming facial disk; large, forward-facing eyes; hooked bill and claws; usually feathered feet. Flight noiseless, mothlike. Some species have "horns," or ear tufts. Ages and sexes largely similar; females larger. **FOOD:** Rodents, birds, reptiles, fish, large insects. **RANGE:** Nearly worldwide.

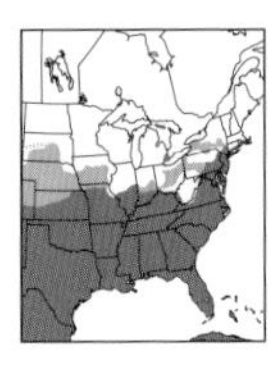

BARN OWL *Tyto alba* Uncommon

16 in. (41 cm). A long-legged, pale, monkey-faced owl. *White heart-shaped face and dark eyes;* no ear tufts. Distinguished in flight as an owl by large head and quiet, mothlike flight; as this species by unstreaked whitish, buff, or pale cinnamon underparts (ghostly at night) and warm tawny brown back. **VOICE:** A shrill, rasping hiss or screech: *kschh* or *shiiish.* **SIMILAR SPECIES:** Short-eared Owl streaked, has darker face and underparts, *yellow* eyes. **HABITAT:** Open country, groves, farms, barns, towns, cliffs, marshes.

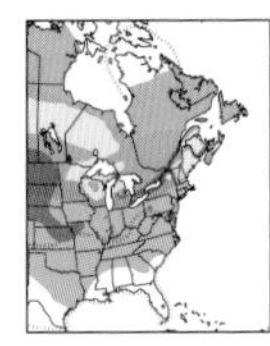

SHORT-EARED OWL *Asio flammeus* Uncommon

15 in. (38 cm). An owl of open country; often foraging at dawn and dusk or during cloudy days. Sometimes tussles with Northern Harrier, with which it shares habitats and prey resources. Streaked, tawny brown color and diagnostic, irregular flopping flight identify it. Large buffy wing patches with *black carpal ("wrist") marks* and pale trailing edge on secondaries. Females average darker than males. **VOICE:** An emphatic, sneezy bark: *kee-yow!, wow!,* or *waow!* **SIMILAR SPECIES:** Long-eared Owl similar in flight but with jerkier wing action, more orangey eyes, darker feathering without white trailing edge on secondaries. **RANGE:** Winter range and numbers vary from year to year. **HABITAT:** Grasslands, fresh and salt marshes, dunes, tundra. Roosts on ground, rarely in trees.

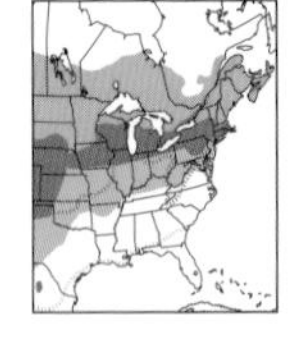

LONG-EARED OWL *Asio otus* Uncommon

15 in. (38 cm). A slender, crow-sized owl with long ear tufts. Usually seen "frozen" close to trunk of a tree. Much smaller and slimmer than Great Horned Owl; underparts streaked *lengthwise,* not barred crosswise. Ears *closer together, erectile;* much black around eyes. **VOICE:** One or two long *hooos*; usually silent. Also a catlike whine and doglike bark. **SIMILAR SPECIES:** Short-eared Owl in flight, Great Horned Owl. **HABITAT:** Coniferous and deciduous woodlands, groves. Often roosts in groups in fall and winter. Hunts over open country.

GREAT HORNED OWL *Bubo virginianus* Common

21–22 in. (54–56 cm). A *very large* owl with ear tufts, or "horns." Heavily *barred* beneath; conspicuous *white throat bib.* In flight, large, looks neckless, large-headed, broad-winged. Varies geographically and individually from very dark to rather pale. Often active just before dark. Subarctic birds frostier than others. **VOICE:** Male usually utters five or six resonant hoots: *hoo hu-hu-hu, hoo! hoo!* Female's hoots slightly higher pitched than male's, one note fewer. Young birds make catlike screams, especially when begging in late summer and fall. **SIMILAR SPECIES:** Long-eared Owl smaller (crow-sized in flight), with lengthwise streaking rather than crosswise barring beneath; ears closer together; lacks white bib. **HABITAT:** Forests, woodlots, deserts, residential areas, open country.

BARN OWL
female
(male is whiter below)
SHORT-EARED OWL
female
female
male
fly on awkward stiff wings, often at dusk
LONG-EARED OWL
flies like Short-eared but not as often seen at dusk
typical
subarctic
GREAT HORNED OWL

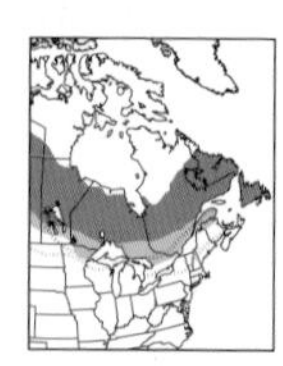

NORTHERN HAWK OWL *Surnia ulula* **Scarce**

16 in. (41 cm). A medium-sized, slender, day-flying owl, with *long, rounded tail* and *barred underparts.* Often *perches at tip of tree* and jerks tail like a kestrel. **VOICE:** A falconlike chattering *kikikiki* and kestrel-like *illy-illy-illy-illy.* Also a harsh scream. **RANGE:** Sporadically appears well south of normal range. **HABITAT:** Open coniferous forests, birch scrub, tamarack bogs, muskeg, field edges.

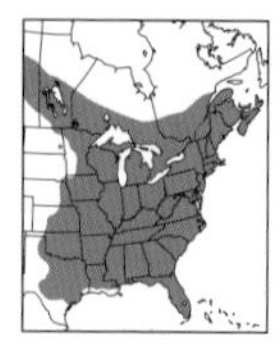

BARRED OWL *Strix varia* **Fairly common**

20–21 in. (51–53 cm). A large, brown, puffy-headed woodland owl with large, moist *brown* eyes. Barred *across* chest and streaked *lengthwise* on belly. **VOICE:** Usually eight accented hoots, in two groups of four: *hoohoo-hoohoo, hoohoohooHOOaaw.* Sometimes rendered as *who cooks for you, who cooks for you-all.* The *all* at end is characteristic and sometimes uttered singly or as *hoo-aww.* **SIMILAR SPECIES:** Other large owls except Barn Owl have ear tufts and/or yellow eyes. **HABITAT:** Woodlands, wooded river bottoms, wooded swamps.

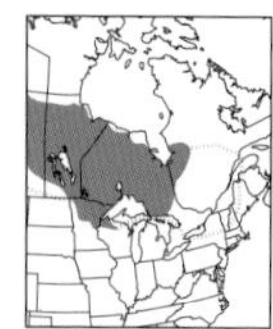

GREAT GRAY OWL *Strix nebulosa* **Scarce**

26–28 in. (67–73 cm). Our largest N. American owl; very tame. Plumage soft, dusky gray, heavily striped *lengthwise* on underparts. Round-headed, without ear tufts; large, *strongly lined facial disk* dwarfs *yellow* eyes. Note *black chin spot* bordered by two broad *white mustaches.* Tail long for an owl. **VOICE:** A deep *whoo-hoo-hoo.* Also deep single *whoos.* **SIMILAR SPECIES:** Barred Owl much smaller, browner. **RANGE:** An irruptive species. Invades well to south one year, then may be rare for several years. **HABITAT:** Coniferous forests, adjacent meadows, bogs. Often hunts by day, particularly in winter.

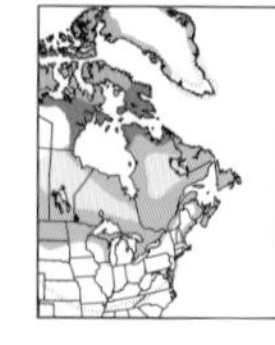

SNOWY OWL *Bubo scandiacus* **Scarce**

22–24 in. (56–61 cm). An irruptive, large, mostly *white,* Arctic, day-flying owl. Round head, *yellow eyes.* Variably flecked or barred with black to dusky, adult males less so than females and young birds. **VOICE:** Usually silent. Flight call when breeding a loud, repeated *krow-ow;* also a repeated *rick.* **SIMILAR SPECIES:** Barn Owl whitish on underparts only; much smaller and has dark eyes. Many downy young owls are whitish. See Gyrfalcon (white morph). **RANGE:** Has cyclic winter irruptions southward into U.S., with accidental vagrants as far as FL and Bermuda. **HABITAT:** Prairies, fields, marshes, beaches; in summer, Arctic tundra. Perches on dunes, posts, haystacks, ground in open country, sometimes buildings.

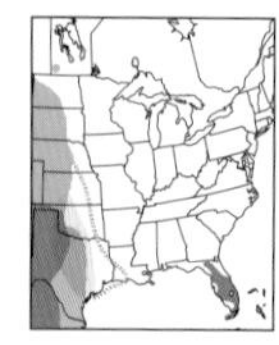

BURROWING OWL *Athene cunicularia* **Uncommon, local**

9½ in. (24 cm). A medium-small owl of open country, often seen by day standing erect on ground or low perches near nesting or wintering burrows, culverts, or drain-pipe entrances. Note *long legs.* Barred and spotted, with white chin stripe, round head. Bobs and bows when agitated. **VOICE:** A rapid, chattering *quick-quick-quick.* At night, a mellow *co-hoo,* higher than Mourning Dove's *coo.* Also a Barn Owl-like screech. Juvenile in burrow rattles like rattlesnake to deter predators. **RANGE:** Uncommon resident in FL; widespread vagrant to north and west. **HABITAT:** Open grasslands, unplowed prairies, farmland, airfields, golf courses.

BARRED OWL
NORTHERN HAWK OWL
GREAT GRAY OWL
first-year female
adult male
BURROWING OWL
SNOWY OWL

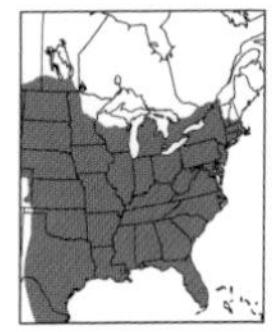

EASTERN SCREECH-OWL *Megascops asio* Common

8½ in. (22 cm). A widespread small owl with conspicuous ear tufts. Yellow eyes. Two color morphs: red and gray. No other owl is bright foxy red. *Juvenile:* Fluffy and may lack conspicuous ear tufts. **VOICE:** A mournful whinny or wail; tremulous, *descending* in pitch. Sometimes a series of notes on one pitch. **SIMILAR SPECIES:** The only small eastern owl with ear tufts. **HABITAT:** Deciduous woodlands, shade trees, residential areas.

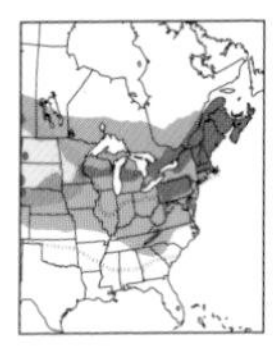

NORTHERN SAW-WHET OWL *Aegolius acadicus* Uncommon

8 in. (20 cm). A very tame little owl at daytime roosts; smaller than a screech-owl, without ear tufts. Underparts have blotchy, reddish brown streaks. Bill black. Forehead streaked white. *Juvenile:* Chocolate brown in summer, with conspicuous white eyebrows; belly *tawny ocher.* **VOICE:** Song a mellow, whistled note repeated in endless succession, often 100–130 times per minute: *too, too, too, too,* etc. Longer and faster than in Northern Pygmy-Owl, which is also more apt to vary tempo. Also raspy, squirrel-like yelps. **SIMILAR SPECIES:** Boreal Owl. **HABITAT:** Coniferous and mixed woods, swamps.

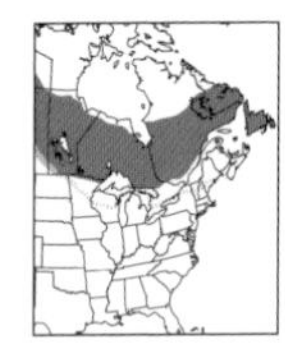

BOREAL OWL *Aegolius funereus* Scarce

10 in. (25 cm). A small, flat-headed, earless owl of northern and high-elevation coniferous forests. Tame. Similar to Northern Saw-whet Owl but a bit larger; facial disk pale grayish white, *framed with black;* bill pale horn color or *yellowish;* forehead *thickly spotted* with white. *Juvenile:* Similar to juvenile Northern Saw-whet but duskier; eyebrows grayish; belly obscurely blotched. **VOICE:** An accelerating series of hoots, similar to a winnowing snipe; call includes a raspy *skew.* **SIMILAR SPECIES:** Northern Saw-whet Owl. **RANGE:** Sporadically appears well south of normal range. **HABITAT:** Spruce, fir, and pine forests; muskeg.

FERRUGINOUS PYGMY-OWL *Glaucidium brasilianum* Scarce

6½–6¾ in. (16–17 cm). Hunts and calls by day, particularly early and late. Often mobbed by birds. Streaking on breast *brownish* rather than black; crown has fine pale streaks (not dots). Tail *rusty, barred with black.* **VOICE:** *Chook* or *puip;* sometimes repeated monotonously two or three times per second. **SIMILAR SPECIES:** Northern Pygmy-Owl (note habitat). **HABITAT:** In s. TX, mesquite and subtropical woods.

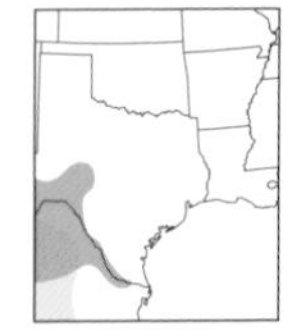

ELF OWL *Micrathene whitneyi* Uncommon

5¾ in. (15 cm). A tiny, short-tailed, earless owl. Underparts softly striped rusty; eyebrows white. Favors woodpecker holes in saguaros, telephone poles, or trees. Found at night by call. **VOICE:** A rapid, high-pitched *whi-whi-whi-whi-whi-whi* or *chewk-chewk-chewk-chewk,* etc., puppylike, with chattering in middle of series. **SIMILAR SPECIES:** Eastern Screech-Owl. **HABITAT:** Saguaro and mesquite woodlands and deserts, wooded canyons.

red
morph
EASTERN
SCREECH-OWL
gray morph
juvenile
NORTHERN
SAW-WHET
OWL
adult
BOREAL
OWL
note "eye
pattern" on
nape
FERRUGINOUS
PYGMY-OWL
ELF OWL

GOATSUCKERS (NIGHTJARS)
Family Caprimulgidae

Nocturnal birds with ample tails, large eyes, tiny bills, large bristled gapes, and very short legs. By day, rest on limbs or on ground, camouflaged by their "dead-leaf" patterns. Ages similar; sexes can differ in wing and tail patterns. Most species best detected and identified at night by voice. **FOOD:** Nocturnal insects. **RANGE:** Nearly worldwide in temperate and tropical land regions.

COMMON NIGHTHAWK *Chordeiles minor* Uncommon

9½ in. (24 cm). A slim-winged, gray-brown bird, often seen high in air; flies with easy strokes, changing gear to quicker erratic strokes. Prefers dusk but may be abroad at midday. Note *broad white bar* across pointed wing. Barred white-and-gray undertail coverts. Male has white bar across notched tail, whiter throat, and larger white bars in wings than female. At rest, *tertials extend well past white wing patch,* and wingtips extend to or beyond tail tip; in flight, white bars occur about halfway out primaries. Some western subspecies tawnier, more similar to Lesser Nighthawk in color. **VOICE:** A nasal *peer* or *pee-ik.* In aerial display, male dives, then zooms up sharply with sudden deep whir of wings. **SIMILAR SPECIES:** Antillean Nighthawk, regular in FL Keys, best distinguished by voice. Lesser Nighthawk's white on wing closer to tip of primaries, wings more bluntly tipped; buffier and less barred than Commons in East. **HABITAT:** Country from mountains to lowlands; open pine woods; sagebrush; often seen in air over cities, towns. Also over ponds. Sits on ground, posts, rails, roofs, limbs.

ANTILLEAN NIGHTHAWK *Chordeiles gundlachii* Scarce, local

8–8½ in. (20–22 cm). This W. Indian species is a regular late-spring and summer visitor to FL Keys and Dry Tortugas. Somewhat tawnier and smaller than Common Nighthawk but readily distinguished from it only by call. **VOICE:** A katydid-like *killy-kadick* or *pity-pit-pit.* **SIMILAR SPECIES:** Common and Lesser Nighthawks. **HABITAT:** Open fields, suburban areas.

LESSER NIGHTHAWK *Chordeiles acutipennis* Uncommon, local

8½–9 in. (21–23 cm). Slightly smaller than Common Nighthawk; white bar (*buffy* in female) *closer to tip of wing;* at rest, this bar even with or slightly beyond tips of tertials. More extensive brown spotting on inner primaries. Undertail coverts browner, less sharply barred. Readily identified by odd calls. Does not power-dive. **VOICE:** A low *chuck chuck* and soft purring or whinnying sound, much like trilling of a toad. **SIMILAR SPECIES:** Common Nighthawk. **RANGE:** A few vagrants wander north as far as Canada and east along Gulf Coast as far as FL. **HABITAT:** Lowlands; arid scrub, dry grasslands, farm fields, dirt roads. Also seen in air over ponds. Sits on branches and ground.

NIGHTHAWKS
female
male
COMMON NIGHTHAWK
male
male
female
ANTILLEAN NIGHTHAWK
male
female
male
LESSER NIGHTHAWK
male

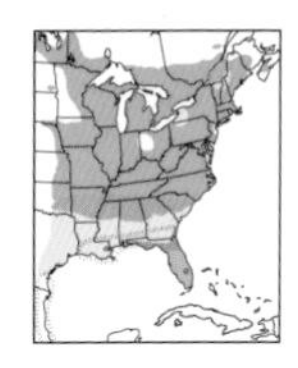

EASTERN WHIP-POOR-WILL *Antrostomus vociferus* **Uncommon**

9½–9¾ in. (24–25 cm). A voice in the night woods, this species is more often heard than seen. When flushed by day, flits away on rounded wings, like a large brown moth. Male has large *white tail patches;* in female these are smaller and buffy. At rest, tail extends beyond wings, unlike nighthawk's. Note *black throat* and *broad black crown stripe.* **VOICE:** At night, a rolling, tiresomely repeated *WHIP poor-WEEL,* or *purple-rib,* etc. **SIMILAR SPECIES:** Chuck-will's-widow. **HABITAT:** Deciduous forests, drier second-growth woodlands.

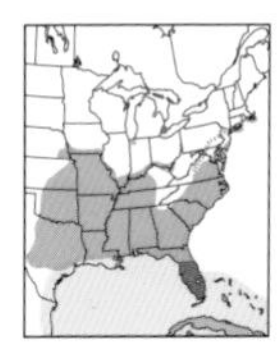

CHUCK-WILL'S-WIDOW *Antrostomus carolinensis* **Uncommon**

12 in. (30 cm). Similar to Eastern Whip-poor-will but larger and with flat, bull-headed appearance, much browner, with *brown* (not blackish) throat and *streaked crown;* white areas in tail of adult male restricted. Tail of first-year males and females has buff tips instead of white. **VOICE:** Call a four-syllable *chuck-will-widow* (less vigorous than effort of Eastern Whip-poor-will); *chuck* often very low and difficult to hear. **SIMILAR SPECIES:** Eastern Whip-poor-will, Common Poorwill. **HABITAT:** Pine and mixed forests, river woodlands, groves.

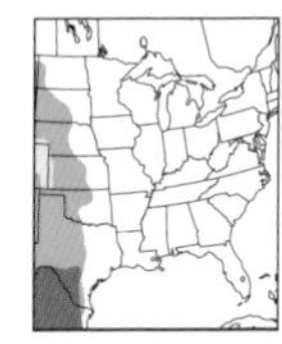

COMMON POORWILL *Phalaenoptilus nuttallii* **Uncommon**

7½–7¾ in. (19–20 cm). Best known by its night cry in arid hills. Appears smaller than a nighthawk, has shorter, more rounded wings *(no white bar),* and the short and rounded tail has *white corners;* these are slightly buffier in female, but sexes otherwise very similar in plumage. Short wings and tail give it a *compact look* at rest. **VOICE:** At night, a loud, repeated *poor-will* or *poor-jill.* **SIMILAR SPECIES:** Eastern Whip-poor-will has white in wing, different tail pattern; does not overlap in breeding range and habitat with Poorwill. **HABITAT:** Dry or rocky hills, including open pine forests, juniper; roadsides.

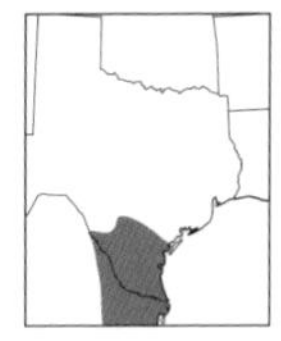

COMMON PAURAQUE *Nyctidromus albicollis* **Uncommon, local**

11 in. (28 cm). Larger than Eastern Whip-poor-will. Dark brown, with long, round wings and tail. Flight floppy with deep wingbeats. Note *broad white band* across pointed wing of male; bar in female buffy. Extensive *white in middle tail feathers* on each side forms obvious double stripe in males; more confined to feather tips in females. At rest, note *pale-edged scapulars.* Recognized by its call. **VOICE:** A hoarse, slurred whistle: *purr-WEE-eeerrr.* **SIMILAR SPECIES:** Other nightjars. Tail pattern and habits differ from nighthawks. **HABITAT:** Dense brushy woodlands, farmlands.

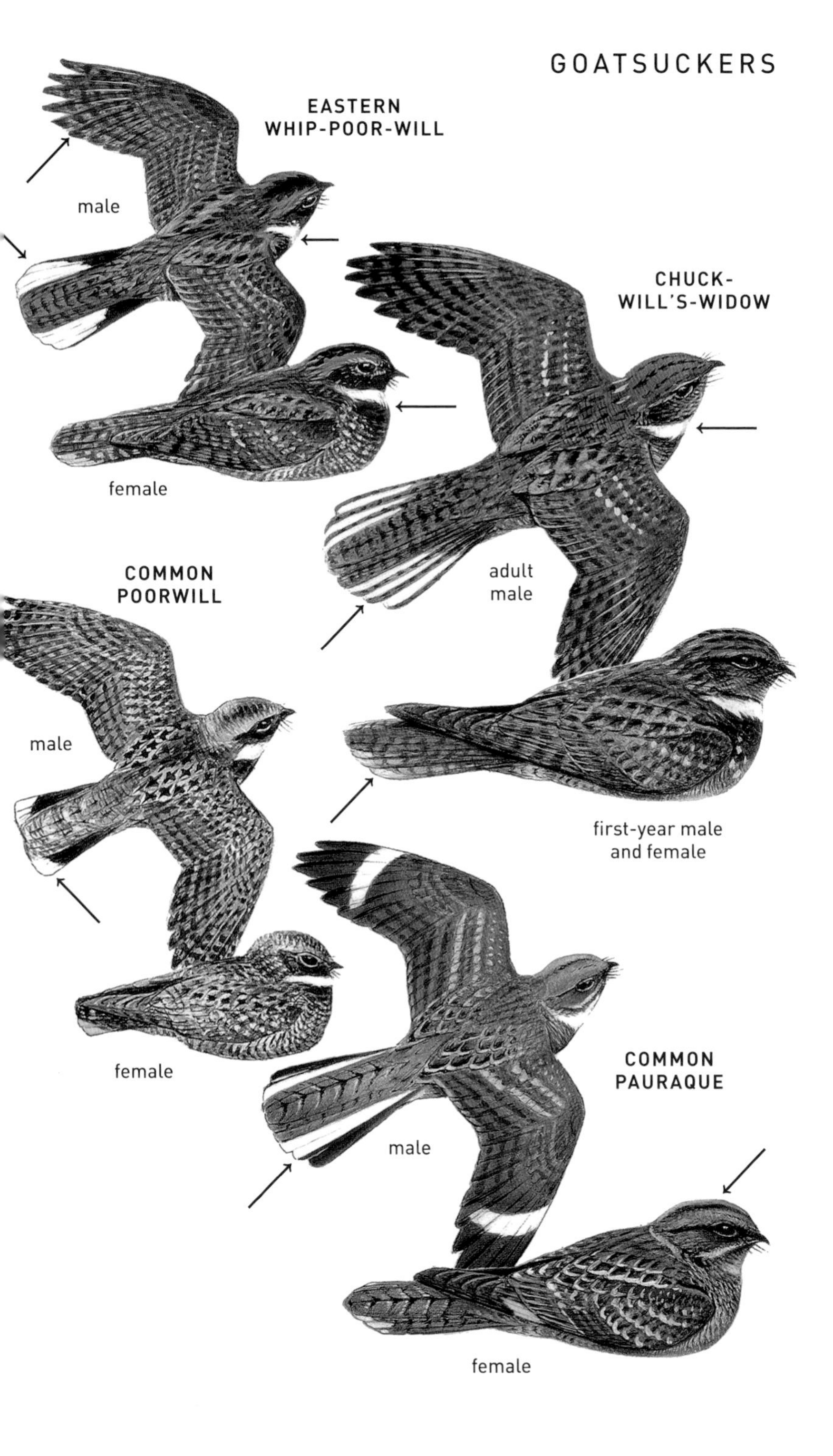
GOATSUCKERS
EASTERN
WHIP-POOR-WILL
male
female
CHUCK-
WILL'S-WIDOW
adult
male
first-year male
and female
COMMON
POORWILL
male
female
COMMON
PAURAQUE
male
female

PIGEONS and DOVES Family Columbidae

Plump, fast-flying birds with small heads and low, cooing voices; nod heads as they walk. Some have fanlike tail (such as Rock Pigeon) and others have pointed tail (such as Mourning Dove). Ages and sexes mostly similar; juveniles scaled above. **FOOD:** Seeds, waste grain, fruit **RANGE:** Nearly worldwide except in arctic regions. Passenger Pigeon (*Ectopistes migratorius*), formerly abundant throughout e. N. America, became extinct in 1914 (p. 199).

BAND-TAILED PIGEON *Patagioenas fasciata* Casual vagrant

14½–15 in. (37–38 cm). Heavily built; might be mistaken for Rock Pigeon except for its woodland habitat and tendency to alight in trees. Note *broad pale band* across end of tail; *white band* on nape. Feet *yellow*. Bill *yellow* with *dark tip*. **VOICE:** A hollow, owl-like *whoo-oo-whoo,* repeated. **SIMILAR SPECIES:** Rock Pigeon. **RANGE:** Casual visitor or vagrant from West to Great Plains and vagrant farther east, accidentally to coast.

RED-BILLED PIGEON *Patagioenas flavirostris* Scarce, local

14–14½ in. (36–37 cm). A large, all-dark pigeon (in good light deep maroon), including underbelly. Bill red with yellowish tip. Shy, mostly arboreal. **VOICE:** *Whoo, whoo, whoooooo.* **SIMILAR SPECIES:** Rock Pigeon. **HABITAT:** Riparian woodlands with tall trees and brush. Recent decline in numbers in lower Rio Grande Valley, TX.

WHITE-CROWNED PIGEON Uncommon, local
Patagioenas leucocephala

13½ in. (34 cm). A stocky, shy pigeon, completely dark except for immaculate white crown. **VOICE:** A low, owl-like *wof, wof, wo, co-woo.* **SIMILAR SPECIES:** Rock Pigeon. **HABITAT:** Mangrove keys, thickets, hardwood hammocks. Perches on power lines, treetops.

AFRICAN COLLARED-DOVE *Streptopelia roseogrisea* Exotic

12 in. (30 cm). An escaped cage bird, also known as Ringed Turtle-Dove. Occurs as escapees, but has declined with arrival of Eurasian Collared-Dove. Paler than Eurasian, especially undertail coverts and underside of flight feathers. Paler (leucistic?) Eurasians, complicate identification. **VOICE:** Two-syllable cooing notes rather than three as in Eurasian.

EURASIAN COLLARED-DOVE *Streptopelia decaocto* Common

12½–13 in. (32–33 cm). A recent colonizer of N. America from Eurasia via the Caribbean; has rapidly increased throughout our area. Slightly chunkier than Mourning Dove, *paler beige,* and with *square-cut tail.* Note *narrow black ring on hindneck. Grayish undertail coverts.* Three-toned wing pattern in flight. **VOICE:** A *three*-noted *coo-COOO-cup.* **SIMILAR SPECIES:** African Collared-Dove. White-winged Dove smaller, lacks neck collar, white in wing obvious. **HABITAT:** Towns, field edges, cultivated land.

ROCK PIGEON (ROCK DOVE, DOMESTIC PIGEON) Common, introduced
Columba livia

12½ in. (32 cm). Typical birds are silvery gray with iridescent purple and green head and breast, *whitish rump, two black wing bars,* and broad, dark tail band. Domestic or feral birds may have many color variants, ranging from blackish to mottled reddish and white, to entirely white. **VOICE:** A soft, gurgling *coo-roo-coo.* **SIMILAR SPECIES:** Band-tailed Pigeon, White-crowned Pigeon. **HABITAT:** Cities, farms, cliffs, bridges.

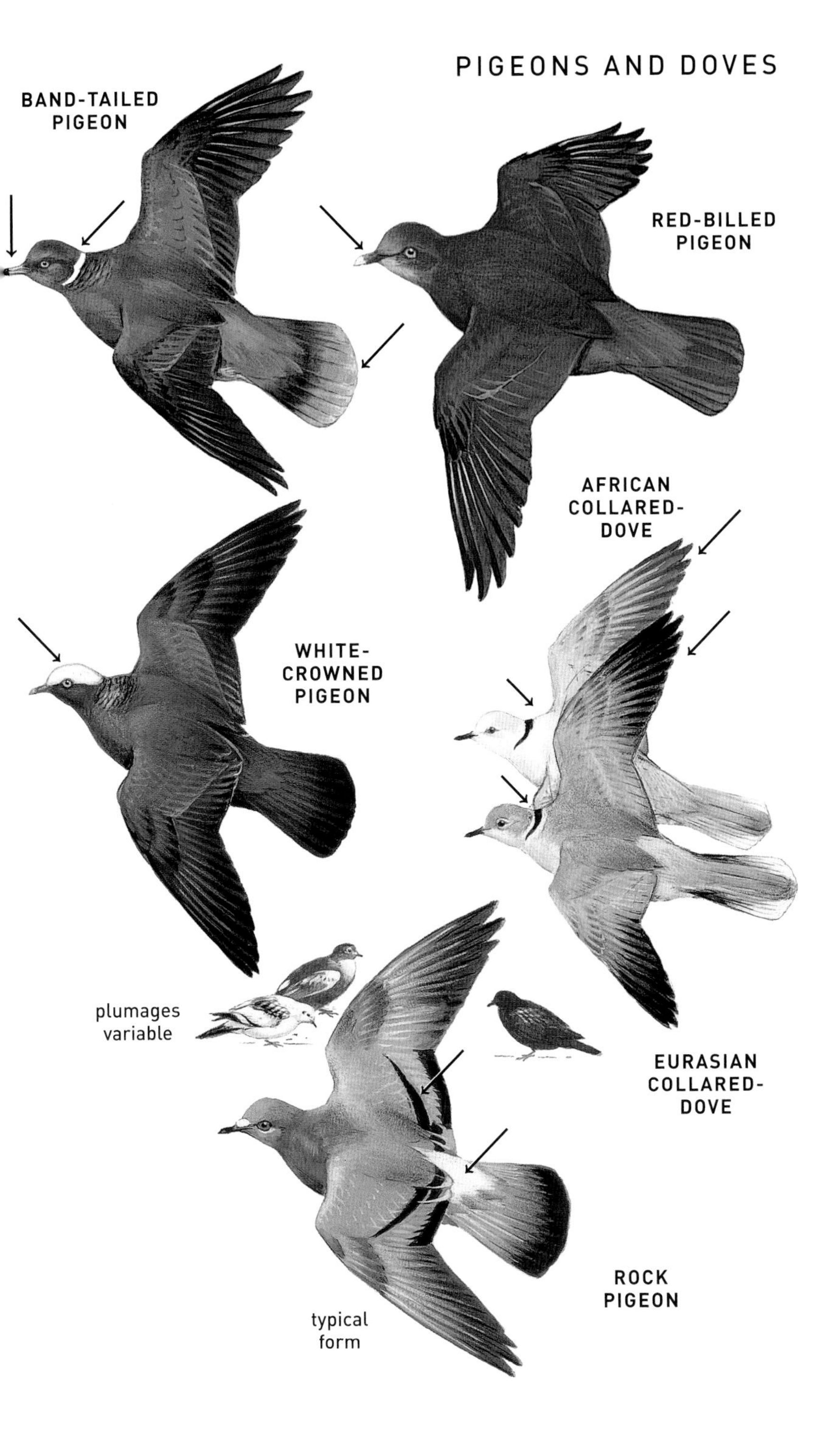
PIGEONS AND DOVES
BAND-TAILED PIGEON
RED-BILLED PIGEON
AFRICAN COLLARED-DOVE
WHITE-CROWNED PIGEON
plumages variable
EURASIAN COLLARED-DOVE
ROCK PIGEON
typical form

WHITE-WINGED DOVE *Zenaida asiatica* **Locally common**

11½–12 in. (29–30 cm). A dove of dry western habitats, readily known by *white wing patches, large when bird is in flight, narrow when at rest.* Otherwise similar to Mourning Dove, but tail *rounded* and tipped with broad white corners, bill slightly longer, eye orangey red. **VOICE:** A harsh cooing, *ooo-uh-CUCK oo* (*who cooks for you?*). **SIMILAR SPECIES:** Mourning Dove, Eurasian Collared-Dove. **RANGE:** Widespread vagrant north of range. **HABITAT:** River woods, groves, towns, feeders.

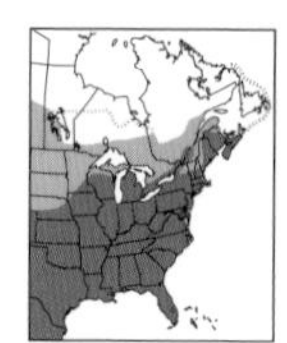

MOURNING DOVE *Zenaida macroura* **Common**

12 in. (30–31 cm). The common widespread wild dove. Brown; smaller and slimmer than Rock Pigeon and Eurasian Collared-Dove. Note *pointed tail* with large white spots. *Male* with slightly bluer crown and rosier breast than *female;* juvenile scaled above. **VOICE:** A hollow, mournful *coah, cooo, coo, cooo.* At a distance, only the three *coos* are audible. **SIMILAR SPECIES:** White-winged Dove, Eurasian Collared-Dove. **HABITAT:** Farms, towns, open woods, fields, scrub, roadsides, grasslands, feeders.

WHITE-TIPPED DOVE *Leptotila verreauxi* **Uncommon, local**

11½ in. (29 cm). A large, stocky dove with broad, dark wings. Short tail has *white corners.* Body pale, underwings cinnamon. **VOICE:** A long, drawn-out, hollow *who—whoooooooooo.* **SIMILAR SPECIES:** White-winged and Mourning Doves. **HABITAT:** Often seen walking in shadows of brushy tangles or dense woods; flies furtively away.

RUDDY GROUND-DOVE *Columbina talpacoti* **Casual vagrant**

6½–6¾ in. (16–17 cm). This rare stray from Mex. to TX is similar to Common Ground-Dove but is slightly larger, longer tailed, and longer billed; has *dark, grayish base* to bill, *lacks all scaliness, and has dark underwing lining* (rufous in Common Ground-Dove). *Male:* Washed rufous, crown pale blue. *Female and juvenile:* Plain brown and gray; wing coverts and *scapulars* have blackish *streaks* rather than spots as in female Common. **VOICE:** A cooing similar to Common Ground-Dove, but faster and more repetitive: *pity-you pity-you pity you.* **SIMILAR SPECIES:** Inca Dove, Common Ground-Dove. **HABITAT:** Farms, livestock pens, fields, brushy areas. Often found with Inca Dove and Common Ground-Dove.

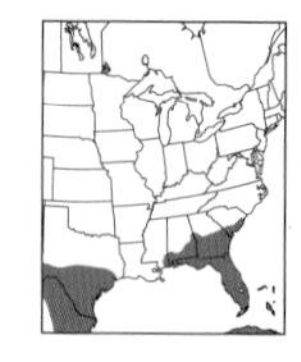

COMMON GROUND-DOVE *Columbina passerina* **Uncommon**

6¼–6½ in. (15–16 cm). A very small dove. Note *stubby black tail,* scaly breast, pinkish or orangey base of bill. Rounded wings flash *rufous* in flight; *bronzy* spots on wing coverts; underwing coverts rufous. Feet yellow or pink. *Male:* Body tinged pinkish. *Female:* Browner. **VOICE:** A soft, monotonously repeated *woo-oo, woo-oo,* etc. May sound monosyllabic—*wooo,* with rising inflection. **SIMILAR SPECIES:** Inca Dove, Ruddy Ground-Dove. **RANGE:** Casual vagrant north of range. **HABITAT:** Farms, orchards, brushy areas, roadsides.

INCA DOVE *Columbina inca* **Fairly common**

8¼–8½ in. (21–22 cm). A very small, slim dove with *scaly* look. *Rufous* in primaries (as in ground-doves) but has *longer tail* with *white sides and corners,* noticeable in flight. **VOICE:** A monotonous *coo-hoo* or *no-hope.* **SIMILAR SPECIES:** Common Ground-Dove has short tail without obvious white, lacks scaling on back. **HABITAT:** Towns, parks, farms.

DOVES
PASSENGER PIGEON
extinct 1914
WHITE-WINGED DOVE
MOURNING DOVE
male
female
WHITE-TIPPED DOVE
male
RUDDY GROUND-DOVE
female
male
COMMON GROUND-DOVE
male
INCA DOVE
female
female

CUCKOOS, ROADRUNNERS, and ANIS
Family Cuculidae

Slender, long-tailed birds; feet zygodactyl (two toes forward, two backward). Sexes alike. **FOOD:** Cuckoos eat caterpillars, other insects; roadrunners eat reptiles, rodents, large insects, small birds; anis eat seeds, fruit. **RANGE:** Warm and temperate regions of world; some cuckoos (but not ours) are parasitic.

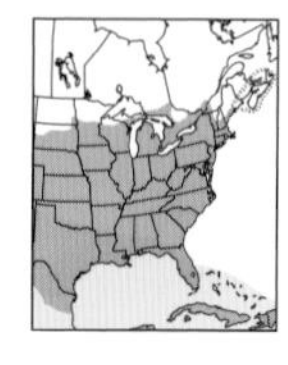

YELLOW-BILLED CUCKOO *Coccyzus americanus* **Fairly common**
12 in. (30–31 cm). Slim with brown back and white underparts. *Rufous* in wings, *large white* spots at tips of dark undertail feathers, *yellow* lower mandible on slightly curved bill, and dusky orbital ring; juvenile and first-fall birds have less distinct tail spots and yellowish orbital ring. **VOICE:** Distinctive throaty *ka-ka-ka-ka-ka-ka-ka-ka-ka-kow-kow-kowlp-kowlp—kowlp—kowlp* (slowing toward end). Often heard during hot afternoons. **SIMILAR SPECIES:** Black-billed Cuckoo. **HABITAT:** Riparian woodlands.

MANGROVE CUCKOO *Coccyzus minor* **Uncommon, local**
12 in. (30–31 cm). Adult and first-fall birds similar to Yellow-billed Cuckoo (both found in s. FL), but belly creamy buff; no rufous in wing. Note black ear patch. **VOICE:** An accelerating series of guttural grunting notes: *unh unh unh unh unh unh aanngg aanngg.* Final two notes longer. **SIMILAR SPECIES:** Other cuckoos, especially Yellow-billed. **RANGE:** Casual vagrant along Gulf Coast. **HABITAT:** Mangroves, hardwood forests.

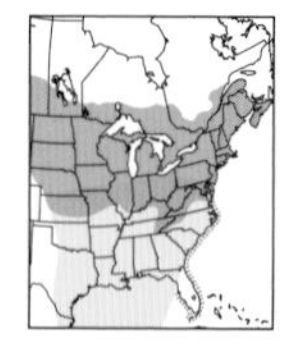

BLACK-BILLED CUCKOO *Coccyzus erythropthalmus* **Uncommon**
11½–12 in. (29–30 cm). *Adult:* Similar to Yellow-billed Cuckoo, but *bill dark gray to blackish;* adult has narrow *red orbital ring. No rufous in wing;* undertail spots small. *Juvenile and first-fall:* Have less distinct undertail spots, greenish orbital ring and slight rufous in wing; thus more like Yellow-billed Cuckoo. **VOICE:** A fast, rhythmic *cucucu, cucucu, cucucu,* etc. The grouped rhythm (three or four) is typical, but often employs irregular cadences. May sing at night. **HABITAT:** Wood edges, groves, thickets.

GROOVE-BILLED ANI *Crotophaga sulcirostris* **Uncommon, local**
13–13½ in. (33–34 cm). A coal black, grackle-sized bird with long and loose-jointed tail, short wings, and *large bill with high curved ridge* and noticeable angle to lower mandible (giving it puffinlike profile), with distinct bill grooves or ridges, more prominent in older adults. Flight weak; alternately flaps and sails. Often moves in groups. **VOICE:** A repeated *whee-o* or *tee-ho,* first note slurring up. **SIMILAR SPECIES:** Grackles, Smooth-billed Ani. **RANGE:** Widespread vagrant north of range. **HABITAT:** Thickets, open woodlands.

SMOOTH-BILLED ANI *Crotophaga ani* **Casual vagrant**
14–14½ in. (35–37 cm). Similar to Groove-billed Ani but larger, bill with higher ridge and lacking grooves. **VOICE:** A whining whistle. A querulous *que-lick.* **RANGE AND HABITAT:** Brushy edges, thickets. Recently extirpated as a breeding species in FL and now a vagrant from Caribbean.

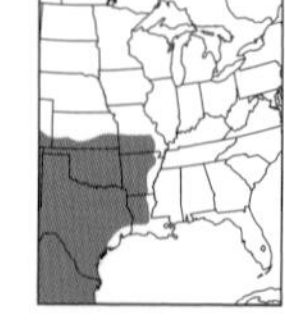

GREATER ROADRUNNER *Geococcyx californianus* **Uncommon**
22–23 in. (56–58 cm). A large, slender, streaked bird with long, white-edged tail; shaggy crest; long legs. **VOICE:** Six to eight low, dovelike *coos,* descending in pitch. **SIMILAR SPECIES:** Thrashers much smaller. **HABITAT:** Open country with scattered cover, brush.

YELLOW-BILLED
CUCKOO
adults
adult
BLACK-
BILLED
CUCKOO
first-
fall
MANGROVE
CUCKOO
adults
GROOVE-
BILLED
ANI
SMOOTH-
BILLED ANI
GREATER
ROADRUNNER

PARAKEETS and PARROTS Family Psittacidae

Noisy, compact birds with stout, hooked bills. Parakeets smaller, with long, pointed tail. **RANGE:** Worldwide in Tropics and subtropics. Carolina Parakeet (*Conuropsis carolinensis*) formerly in e. U.S.; became extinct around 1918.

MONK PARAKEET *Myiopsitta monachus* **Locally fairly common, exotic**
11 in. (28 cm). Native to Argentina. Pale gray face and chest, buff band across belly. Established in spots from CT to FL and west to IL and TX. Massive colonial stick nest. Raucous calls. Comes to feeders.

GREEN PARAKEET **Locally fairly common, exotic**
Psittacara holochlorus
10–12 in. (25–30 cm). This large parakeet has long pointed tail and is green above, yellow-green below. **VOICE:** Sharp, squeaky notes, shrill noisy chatter. **RANGE AND HABITAT:** Tropical ne. Mex. to s. Nicaragua. Exotic population in Miami. Populations established in s. Rio Grande Valley, TX, may include natural strays from Mex.

NANDAY PARAKEET *Aratinga nenday* **Locally fairly common, exotic**
12 in. (30 cm). Native to S. America; locally established on west coast of FL. Also known as Black-hooded Parakeet, in reference to its diagnostic black face and crown. Long tail, remiges black below; undertail black; uppertail green, tipped blue. Loud, shrill, ternlike calls.

WHITE-WINGED PARAKEET **Locally uncommon, exotic**
Brotogeris versicolurus

YELLOW-CHEVRONED PARAKEET **Locally fairly common, exotic**
Brotogeris chiriri
9 in. (23 cm). Native to S. America, these two similar *Brotogeris* parakeets are locally established in Miami area. Green; note wing patterns. White-winged formerly more common but has since been outnumbered by Yellow-chevroned.

RED-CROWNED PARROT **Locally fairly common, exotic**
Amazona viridigenalis
12 in. (30 cm). Large, with red crown (reduced in first-year), blue nape, red wing panels. **VOICE:** Loud, familiar, raucous notes and squeals of Amazon parrots. **RANGE:** Native to ne. Mex. Exotic populations in s. FL and s. TX; the latter may include natural strays from Mex.

LILAC-CROWNED PARROT *Amazona finschi* **Unestablished exotic**
(Mex.) 12½–13½ in. (30–34 cm). Small populations in Miami and s. TX.

BUDGERIGAR *Melopsittacus undulatus* **Unestablished exotic**
(Australia) 7 in. (18 cm). Usually as shown; may also be blue, yellow, or white. Formerly established in FL; now extirpated. Escapees still found.

MITRED PARAKEET *Psittacara mitrata* **Unestablished exotic**
(S. America) 15 in. (38 cm). Moderate populations in s. FL; found occasionally elsewhere in N. America.

YELLOW-HEADED PARROT *Amazona oratrix* **Unestablished exotic**
(Mex. and Belize) 14–15 in. (36–38 cm). Small populations found locally in s. FL and s. TX.

PARAKEETS
AND PARROTS
Established
MONK
PARAKEET
GREEN
PARAKEET
NANDAY
PARAKEET
WHITE-
WINGED
PARAKEET
YELLOW-
CHEVRONED
PARAKEET
LILAC-CROWNED
PARROT
Unestablished Exotics
BUDGERIGAR
RED-CROWNED
PARROT
MITRED
PARAKEET
YELLOW-HEADED
PARROT

WOODPECKERS Family Picidae

Chisel-billed, wood-boring birds with strong zygodactyl feet (two toes front, two rear), remarkably long tongues, and stiff spiny tails that act as props for climbing. Flight usually undulating. **FOOD:** Tree-boring insects and grubs; some species eat ants, flying insects, berries, acorns. **RANGE:** Most wooded parts of world; absent in Australia, Madagascar.

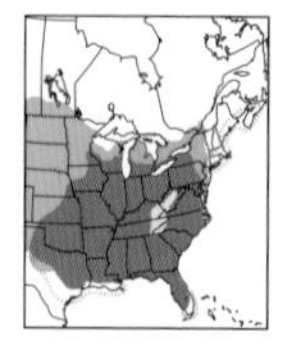

RED-HEADED WOODPECKER *Melanerpes erythrocephalus* **Uncommon**

9¼ in. (24 cm). *Adult:* A black-backed woodpecker with *entirely red* head (other woodpeckers may have patch of red). Back *solid black,* rump white. Large, square *white patches* conspicuous on wing, including when sitting on a tree. Sexes similar. *Juvenile:* Dusky-headed; wing patches mottled with dark through first year. **VOICE:** A loud *queer* or *queeah.* **SIMILAR SPECIES:** Red-bellied Woodpecker has only partially red head. **HABITAT:** Groves, farm country, shade trees in towns, large scattered trees.

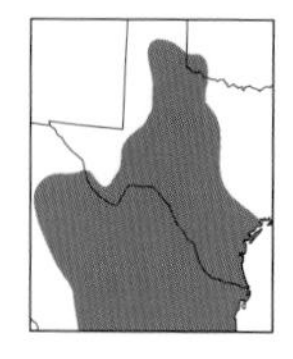

GOLDEN-FRONTED WOODPECKER *Melanerpes aurifrons* **Common**

9½ in. (25 cm). A zebra-backed woodpecker with light underparts and white rump. Has white wing patch in flight. *Male:* Note *multicolored head* (yellow near bill, poppy red on crown, orange nape). *Female:* Lacks red crown patch. *Juvenile:* Has tan head and nape, lacking color. **VOICE:** A tremulous *churrrr;* flickerlike *kek-kek-kek-kek.* **SIMILAR SPECIES:** Note aberrant Red-bellied Woodpecker can have yellow lores. **HABITAT:** Mesquite, woodlands, groves.

RED-BELLIED WOODPECKER *Melanerpes carolinus* **Common**

9¼ in. (24 cm). *Adult:* A *zebra-backed* woodpecker with *red cap, white rump.* Red covers both crown and nape in male, *only nape in female. Juvenile:* Also zebra-backed, but has brown head, devoid of red. **VOICE:** Call *kwirr, churr,* or *chaw;* also *chiv, chiv* and a muffled flickerlike series. **SIMILAR SPECIES:** Golden-fronted and Red-headed Woodpeckers. **HABITAT:** Woodlands, groves, orchards, towns, feeders.

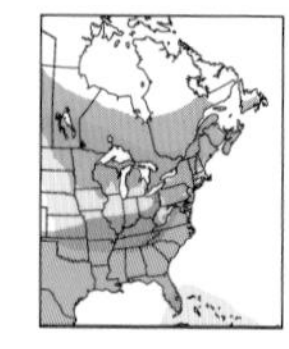

YELLOW-BELLIED SAPSUCKER *Sphyrapicus varius* **Fairly common**

8½ in. (22 cm). Sapsuckers drill orderly rows of small holes in trees for sap and the insects it attracts. Note characteristic *sapsucker white wing patches,* red forehead. *Adult:* Male has all-red throat, female white. *Juvenile:* Head and body mottled brown; unlike other sapsuckers, retains brown plumage through winter. **VOICE:** A nasal mewing note, *cheerrrr;* drum in this is several rapid thumps followed by several slow, rhythmic thumps. **SIMILAR SPECIES:** Red-naped Sapsucker. **HABITAT:** Coniferous, mixed, and deciduous woods, shade trees.

RED-NAPED SAPSUCKER *Sphyrapicus nuchalis* **Casual vagrant**

8½ in. (22 cm). Very similar to Yellow-bellied Sapsucker, but note *red nape.* Black frame around throat *broken* toward rear. Female has white chin. Occasionally hybridizes with Yellow-bellied. **VOICE:** Similar to Yellow-bellied Sapsucker. **SIMILAR SPECIES:** Yellow-bellied Sapsucker. **RANGE AND HABITAT:** Casual vagrant to Gulf Coast, found in similar habitats as Yellow-bellied.

WOODPECKERS
juvenile
RED-HEADED
WOODPECKER
adult
GOLDEN-FRONTED
WOODPECKER
male
female
male
female
RED-BELLIED
WOODPECKER
female
male
female
female
male
juvenile
YELLOW-BELLIED
SAPSUCKER
RED-NAPED
SAPSUCKER

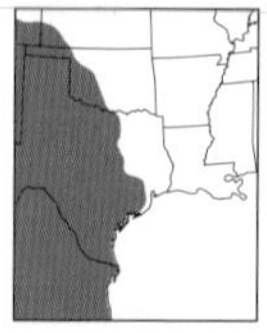

LADDER-BACKED WOODPECKER *Dryobates scalaris* Fairly common

7¼ in. (18 cm). The black-and-white zebra-backed woodpecker found in more arid country. Male has red crown. **VOICE:** A rattling series, *chikikikikikikikikikik,* diminishing. Call a sharp *pick* or *chik* (like Downy Woodpecker). **SIMILAR SPECIES:** Downy Woodpecker. **HABITAT:** Canyons, pinyon-juniper, riparian woodlands, arid brush.

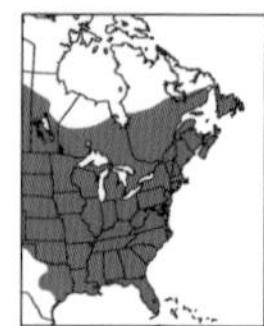

DOWNY WOODPECKER *Dryobates pubescens* Common

6½–6¾ in. (17 cm). Note *white back* and *small bill.* Downy and Hairy Woodpeckers are almost identical in pattern, checkered and spotted with black and white; male has small red patch on back of head, female does not. Outer tail feathers spotted, red nape patch of male in unbroken square. **VOICE:** A rapid whinny of notes, descending in pitch. Call a flat *pick,* not as sharp as Hairy's *peek!* **SIMILAR SPECIES:** Hairy Woodpecker larger, has larger bill and clean white outer tail feathers. Ladder-backed Woodpecker has similar call. **HABITAT:** Woods, riparian thickets, residential areas, suet feeders, even corn and cattail stems.

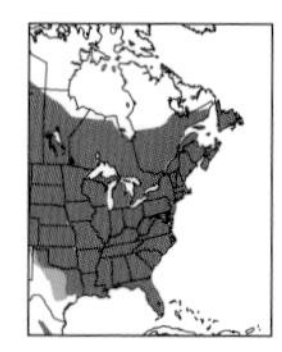

HAIRY WOODPECKER *Dryobates villosus* Fairly common

9–9¼ in. (23–24 cm). Hairy is like an exaggerated Downy Woodpecker, especially its bill. *Juvenile:* May have red to yellowish crown patch, more extensive in male. **VOICE:** A kingfisher-like rattle, quicker than that of Downy. Call a sharp *peek!* (Downy says *pick.*) **SIMILAR SPECIES:** Downy Woodpecker. American Three-toed Woodpecker has some barring on back and barred sides. **HABITAT:** Coniferous forests, deciduous woods, shade trees, suet feeders.

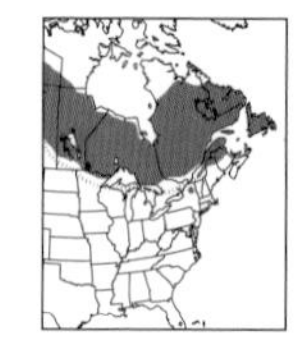

AMERICAN THREE-TOED WOODPECKER *Picoides dorsalis* Scarce

8½–8¾ in. (22 cm). Males of this and the next species are our only woodpeckers that have three toes and, normally, *yellow caps* (note some juvenile Hairy Woodpeckers have sparse yellow in crown). Both have *barred sides.* This species is distinguished by irregular white *bars* on back. Female lacks yellow cap and can suggest Downy or Hairy Woodpecker, but note *barred sides.* **VOICE:** A level-pitched whinny and a flat *pyik.* **SIMILAR SPECIES:** Black-backed Woodpecker, Hairy Woodpecker. **RANGE:** Casual vagrant south of range; birds from the Rockies accidental in Great Plain States. **HABITAT:** Coniferous forests, particularly in burned areas and where deadwood is present.

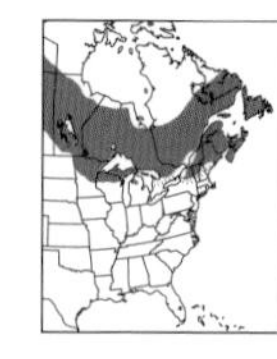

BLACK-BACKED WOODPECKER *Picoides arcticus* Scarce

9½ in. (24 cm). Note combination of *solid black back* and *barred sides.* Male has *yellow cap.* This and preceding species inhabit boreal and montane forests; their presence can be detected by patches of bark scaled from dead conifers. **VOICE:** A low, flat *kuk* or *puk* and a short buzzy call. **SIMILAR SPECIES:** American Three-toed and Hairy Woodpeckers. **HABITAT:** Coniferous forests, particularly in burned areas and where deadwood is present.

WOODPECKERS
LADDER-BACKED
WOODPECKER
female
male
HAIRY
WOODPECKER
male
male
female
female
DOWNY
WOODPECKER
female
male
Rockies
male
northern male
female
AMERICAN THREE-
TOED WOODPECKER
BLACK-BACKED
WOODPECKER

NORTHERN FLICKER *Colaptes auratus* **Common**

12–12½ in. (30–32 cm). Conspicuous *white rump* in flight, barred *brown back*, and *black patch* across chest mark this bird as a flicker. Unlike other woodpeckers, often hops awkwardly on ground, feeding on ants. Two basic subspecies groups: "Yellow-shafted" Flicker (*C. a. auratus*), of North and East, has *golden yellow* underwings and tail. *Red crescent* on nape; *gray crown; tan-brown cheeks;* male has *black* mustache. "Red-shafted" Flicker (*C. a. cafer*) of West has *salmon red* underwing and undertail. Both sexes lack red crescent on nape; have *brownish crown* and *gray cheeks;* male has *red* mustache. Red-shafted Flickers and intergrades are fairly common on eastern edge of Great Plains and, rarely in winter, farther East. **VOICE:** A loud *wick wick wick wick wick,* etc. Also a loud *klee-yer* and a squeaky *flick-a, flick-a,* etc. See also Pileated Woodpecker. **HABITAT:** Open forests, woodlots, towns.

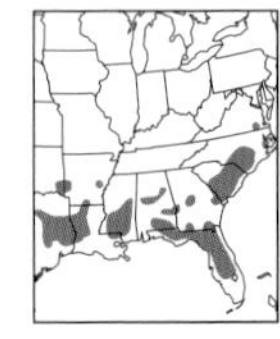

RED-COCKADED WOODPECKER **Rare, local, endangered**
Dryobates borealis

8½ in. (22 cm). Zebra-backed, with black cap. White cheek is obvious field mark. Male's tiny red cockade hard to see. Forms colonial "clans." **VOICE:** A rough, rasping *sripp* or *zhilp* (suggests flocking note of young starling). Sometimes a higher *tsick.* **SIMILAR SPECIES:** Downy and Hairy Woodpeckers. **HABITAT:** Open pine woodlands that have trees with heartwood disease.

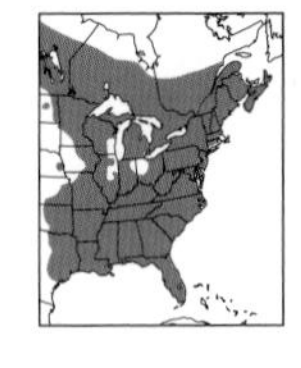

PILEATED WOODPECKER *Dryocopus pileatus* **Uncommon**

16½–17 in. (42–44 cm). A spectacular, *crow-sized* woodpecker, black with flaming red *crest.* Female has blackish forehead, lacks red on mustache. Great size, sweeping wingbeats, and flashing white underwing coverts identify Pileated in flight. Large foraging pits in dead or dying trees—large *oval* or *oblong* holes—indicate its presence. **VOICE:** Call resembles that of a flicker, but louder, deeper, irregular: *kik-kik-kikkik-kik-kik,* etc. Also a more ringing, double-note call that may rise or fall slightly in pitch and volume. **SIMILAR SPECIES:** Ivory-billed Woodpecker (probably extinct). **HABITAT:** Coniferous, mixed, and hardwood forests with large mature trees; woodlots.

IVORY-BILLED WOODPECKER **Almost certainly extinct**
Campephilus principalis

19–21 in. (48–53 cm). Separated from Pileated Woodpecker by its slightly larger size, longer wings, ivory white bill, large white upperwing patch in secondaries (visible at rest), and all-white underwing pattern with black line through it. Female has black crest. **VOICE:** A single loud tooting note constantly uttered as bird forages—a sharp nasal *kent* suggesting to some a big nuthatch. Drum is a quick double knock, unique among N. American woodpeckers. **SIMILAR SPECIES:** Pileated Woodpecker. **RANGE:** Throughout Southeast. Reports persist, but the last documented sightings of this conspicuous species were in 1940s. **HABITAT:** Bottomland hardwood forests, wooded bayous and swamps.

WOODPECKERS
NORTHERN
FLICKER
"Red-
shafted"
"Yellow-
shafted"
"Red-
shafted"
female
"Red-
shafted"
male
males
"Yellow-shafted"
male
male
RED-COCKADED
WOODPECKER
below
(female lacks
red "cockade")
female
below
male
PILEATED
WOODPECKER
male
above
IVORY-BILLED
WOODPECKER

KINGFISHERS Family Alcedinidae

Chiefly solitary birds, with large heads, long pointed bills, and small syndactyl feet (two toes partially joined). Most are fish eaters, perching above water or hovering and plunging headfirst. **FOOD:** Mainly fish; some species eat insects, lizards. **RANGE:** Almost worldwide.

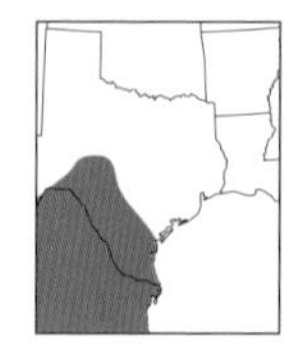

GREEN KINGFISHER *Chloroceryle americana* — Uncommon, local

8½–8¾ in. (22 cm). Kingfisher shape, small size; flight buzzy, direct. Upperparts deep green with white spots; collar and underparts white, sides spotted. *Adult male:* Has *rusty* breast-band. *Adult female:* Has one or two greenish bands. (The reverse is true in Belted Kingfisher, in which female has rusty band.) *First-year:* Birds of both sexes have mixed rufous and green feathers in breast. **VOICE:** A sharp clicking, *tick tick tick;* also a sharp squeak. **SIMILAR SPECIES:** Larger and much larger-billed Amazon Kingfisher (*C. amazona;* not shown) of Mex. is accidental in TX. **RANGE:** Accidental north of range. **HABITAT:** Small rivers and streams.

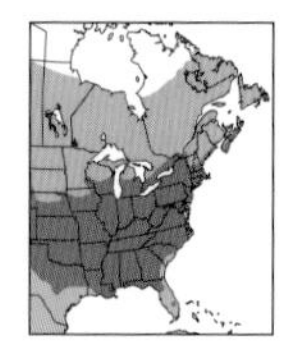

BELTED KINGFISHER *Megaceryle alcyon* — Fairly common

13 in. (33 cm). Our common widespread kingfisher, this big-headed and big-billed fisher hovers on rapidly beating wings in readiness for the plunge, or flies with uneven wingbeats, rattling as it goes: it is easily recognized. *Adult male:* Blue-gray above, with ragged bushy crest and broad gray chest-band. *Adult female:* Has an additional rusty breast-band. *First-year:* Similar in both sexes, except blue chest-band mottled with rusty feathers. **VOICE:** A distinctive loud, dry rattle. **SIMILAR SPECIES:** Ringed Kingfisher in TX. **HABITAT:** Streams, lakes, bays, coasts; nests in banks, perches on wires.

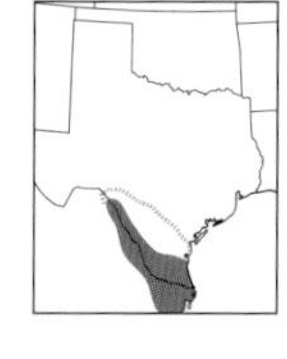

RINGED KINGFISHER *Megaceryle torquata* — Uncommon, local

16 in. (41 cm). Larger than Belted Kingfisher; bill very large. *Male:* Has entirely chestnut breast and belly (ages similar). *Adult female:* Has broad blue-gray band across breast, separated from chestnut belly by narrow white line (chest mixed with rufous feathers in first-year female). **VOICE:** A rusty *cla-ack* or *wa-ak* or rolling rattle after a loud *chack.* **SIMILAR SPECIES:** Belted Kingfisher. **HABITAT:** Slow rivers (particularly Rio Grande in our area), marshes.

SWIFTS Family Apodidae

Swallowlike in habits but structurally distinct, with shorter forewing, flat skull, and feet with all four toes pointing forward. Flight very rapid, "twinkling," sailing between spurts; narrow wings often stiffly bowed. Ages and sexes similar. **FOOD:** Flying insects. **RANGE:** Nearly worldwide.

CHIMNEY SWIFT *Chaetura pelagica* — Common

5¼ in. (13 cm). Like a cigar with wings. A dark swift with long, slightly curved, stiff wings and stubby tail. Rapid, twinkling wingbeats interspersed with bowed-winged glides. **VOICE:** A loud, rapid ticking or twittering notes. **SIMILAR SPECIES:** Longer, slimmer wings than swallows. Flies differently than Vaux's Swift (*C. vauxi;* not shown), a rare vagrant to East (particularly along Gulf Coast); Vaux's smaller, throat and rump paler, flies with shallower, more twinkling flight. **HABITAT:** Open sky, especially over cities, towns; nests and roosts in chimneys (originally in large hollow trees and cliff crevices).

KINGFISHERS AND SWIFT
adult female
adult male
hovering
first-year female
GREEN KINGFISHER
plunging
BELTED KINGFISHER
adult male
RINGED KINGFISHER
male entirely chestnut below
CHIMNEY SWIFT

HUMMINGBIRDS Family Trochilidae

The smallest of birds, with needlelike bills for sipping nectar. Adult males of most species adorned by jewel-like, iridescent gorget throat and sometimes crown feathers; in poor light, feathers can appear dark. Hummingbirds hover when feeding and can fly backward; their wingbeats are so rapid that they appear as a blur. Pugnacious. Vocal differences can be important identification aids. **FOOD:** Nectar (red flowers favored), small insects, spiders. **RANGE:** W. Hemisphere; majority in Tropics.

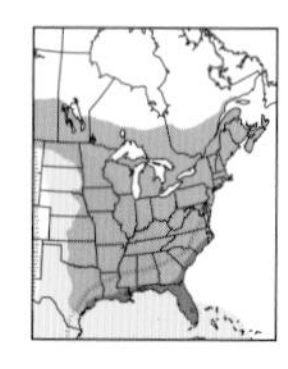

RUBY-THROATED HUMMINGBIRD *Archilochus colubris* Fairly common

3¾ in. (10 cm). *Adult male: Fiery red throat,* iridescent green back, forked tail. *Female:* Lacks red throat; tail blunt, with white spots. *First-fall male:* Like female but tail slightly forked, a few scattered ruby feathers molt in during fall. *The only widespread species in East;* several other hummers may turn up as strays, especially in se. states in fall and winter. **VOICE:** Male's wings hum in courtship display. Chase calls high, squeaky. Other call a soft *chew.* **SIMILAR SPECIES:** Male Broad-tailed Hummingbird lacks forked tail, typically makes wing-trill sound. Female and first-fall male similar to Black-chinned Hummingbird but have *brighter green crown and back,* shorter bill; *outermost primary narrower and straighter at tip (more club-shaped in Black-chinned).* See also Anna's Hummingbird. **HABITAT:** Flowers, gardens, wood edges, over streams.

BUFF-BELLIED HUMMINGBIRD Uncommon, local
Amazilia yucatanensis

4¼ in. (11 cm). Note combination of buff underparts, rufous tail, and green throat. Bill orange-red with dark tip. Ages and sexes similar. **VOICE:** Call a surprisingly loud *smak smak smak.* Aggressive flight call an unmusical buzz: *chr chr chr chr chr.* **RANGE:** Casual vagrant to Gulf Coast. **HABITAT:** Open woodlands, gardens, feeders.

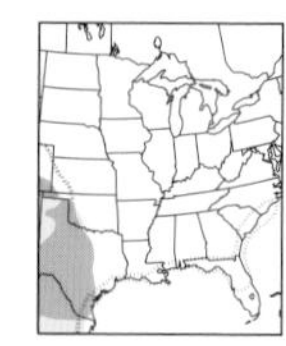

BLACK-CHINNED HUMMINGBIRD Uncommon to rare
Archilochus alexandri

3¾ in. (10 cm). *Adult male:* Note *black throat* and conspicuous white collar; iridescent blue-violet of lower throat. *Female and first-fall male:* Similar to these plumages in Ruby-throated Hummingbird, but crown often grayish, back duller, bill longer. **VOICE:** Like Ruby-throated. **SIMILAR SPECIES:** Ruby-throated and Anna's Hummingbirds. **RANGE:** Casual vagrant to East and Gulf Coasts. **HABITAT:** Riparian woodlands, wooded canyons, feeders.

RUFOUS HUMMINGBIRD *Selasphorus rufus* Scarce to rare

3¾ in. (9–10 cm). *Adult male:* Note bright *rufous or red-brown upperparts,* sometimes mottled green, but rufous predominates; throat flaming orange-red. *Female and first-fall male:* Green-backed; dull *rufous on sides and at base of outer tail feathers* (visible when tail fully spread); some iridescent orange-red feathers on throat. **VOICE:** Aggressive flight call a buzzy *zap* or *zeee chippity chippity.* Male's wings make high trill in flight. **SIMILAR SPECIES:** Allen's, Calliope, and Broad-tailed Hummingbirds. **RANGE:** Casual vagrant to East and Gulf Coasts. **HABITAT:** Wooded or brushy areas, feeders.

RUBY-THROATED HUMMINGBIRD
adult male
adult male
female and first-fall male
sphinx moths resemble hummingbirds
BUFF-BELLIED HUMMINGBIRD
adult male
female and first-fall male
BLACK-CHINNED HUMMINGBIRD
adult male
female and first-fall male
RUFOUS HUMMINGBIRD

RARE HUMMINGBIRDS

BROAD-BILLED HUMMINGBIRD *Cynanthus latirostris* **Casual vagrant**

4 in. (10 cm). *Adult male:* Dark green above and below; *blue throat;* black tail; *reddish* bill with black tip. *Female: Dull orange-red base to bill; dark tail; unmarked, pearly gray* throat; thin white line behind eye. *First-year male:* Like female but mottled blue. **VOICE:** A Ruby-crowned Kinglet–like chattering. **RANGE AND HABITAT:** Casual to accidental vagrant from Southwest; this and the following rare hummingbirds are observed mostly at feeders.

GREEN-BREASTED MANGO *Anthracothorax prevostii* **Casual vagrant**

4¾ in. (12 cm). Large, with long downcurved bill. *Adult male:* Dark emerald green; tail purple. *Female and first-year male:* Green back, light underparts with irregular dark stripe from throat to belly. Dusky tail. **VOICE:** A high-pitched *tzat.* **RANGE:** Stray from Mex., primarily to s. TX but also to NC and WI.

ANNA'S HUMMINGBIRD *Calypte anna* **Accidental vagrant**

4 in. (10 cm). *Adult male:* Note *rose red crown* and throat. *Female and first-summer male:* Slightly larger and with messier underparts than other hummers, grayer below, with *green sides* and more heavily spotted throat. **VOICE:** Chase call a raspy chatter. **RANGE:** Accidental vagrant from w. U.S.

MEXICAN VIOLETEAR *Colibri thalassinus* **Casual vagrant**

4¾ in. (12 cm). Until recently, included with Lesser Violetear (*C. cyanotus;* not shown) of Neotropics as a single species, Green Violetear (*C. thalassinus*). Large, long-billed, dark green hummingbird with violet ear patch, bluish tail. Sexes mostly similar. **VOICE:** A series of dry *chips.* **RANGE:** Vagrant species from Mex.; primarily in summer in TX.

BROAD-TAILED HUMMINGBIRD **Casual vagrant**
Selasphorus platycercus

4 in. (10 cm). *Adult male:* Larger than Ruby-throated Hummingbird, with longer tail; wings trill in flight. *Female and first-fall male:* Larger and larger-tailed than female Rufous Hummingbird. **RANGE:** Casual vagrant from w. U.S., primarily along Gulf Coast.

CALLIOPE HUMMINGBIRD *Selasphorus calliope* **Accidental vagrant**

3¼ in. (8 cm). Our smallest hummingbird, with wedge-shaped tail containing some rufous. *Adult male:* Throat with purple-red rays on white background showing as a dark inverted V when folded. *Female and first-fall male:* Similar to female Broad-tailed and Rufous Hummingbirds but smaller; *wingtips extend beyond tail at rest.* **VOICE:** High-pitched chips and buzzes. **RANGE:** Accidental vagrant from w. U.S.

ALLEN'S HUMMINGBIRD *Selasphorus sasin* **Accidental vagrant**

3¾ in. (9–10 cm). Like Rufous Hummingbird, but adult male with *green or mostly green back;* female and first-fall male distinguished with difficulty, only by Allen's having narrower rectrices within each age and sex class. **RANGE:** Accidental vagrant from w. U.S.

BROAD-BILLED HUMMINGBIRD
adult male
female
adult male
female
GREEN-BREASTED MANGO
adult male
MEXICAN VIOLETEAR
adult male
adult female
ANNA'S HUMMINGBIRD
adult male
female
BROAD-TAILED HUMMINGBIRD
CALLIOPE HUMMINGBIRD
adult male
adult male
female and first-fall male
ALLEN'S HUMMINGBIRD

PASSERINES Order Passeriformes

Passerines, also known as "perching birds" or "songbirds," comprise the rest of the species in this book. They are distinguished from other birds by having one toe back and three forward, ideal for perching.

TYRANT FLYCATCHERS Family Tyrannidae

New World flycatchers, or tyrant flycatchers, make up the largest family of birds in the world, with approximately 425 known species. A large number are very similar and require attention to details to separate them. Most species perch quietly, sitting upright on exposed branches, from which they sally forth to snap up insects. Bills flattened, with bristles at base. Ages and sexes similar in most but not all species. **FOOD:** Mainly flying insects. Some species also eat fruit in winter. **RANGE:** New World; the vast majority in Neotropics.

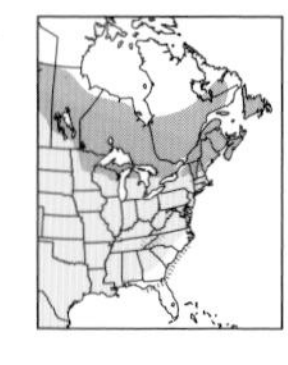

OLIVE-SIDED FLYCATCHER *Contopus cooperi* **Uncommon**

7½ in. (19 cm). A stout, large-headed flycatcher; often perches on dead snags at tops of trees. Note large bill and *dark chest patches* separated by narrow strip of white (like unbuttoned vest). A *cottony tuft* may poke from behind wing. **VOICE:** Call a two- or three-note *pip-pip-pip.* Song a spirited whistle, *I SAY there* or *Quick three beers!,* middle note highest, last one sliding. **SIMILAR SPECIES:** Wood-pewees. **HABITAT:** Coniferous forests, bogs, burns.

WESTERN WOOD-PEWEE *Contopus sordidulus* **Rare**

6¼ in. (16 cm). Very similar to Eastern Wood-Pewee but often appears more "vested" below (with "top button buttoned"); black bill usually has only a small amount of pale at base of lower mandible. Best separated by voice. **VOICE:** A nasal *peeeer* (less commonly, *pee-yee*), more guttural (less clear) than in Eastern Wood-Pewee. **SIMILAR SPECIES:** Eastern Wood-Pewee. Olive-sided Flycatcher larger, more strongly "vested," different voice. **RANGE AND HABITAT:** Very rare migrant to Great Plains from West, and casual vagrant to East Coast; often in pine-oak forests.

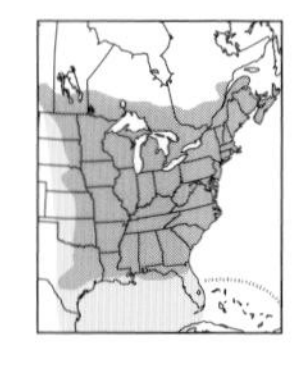

EASTERN WOOD-PEWEE *Contopus virens* **Fairly common**

6¼ in. (16 cm). A dark, medium-small flycatcher found in mid-canopy. Note *two narrow wing bars, no eye-ring,* and variably pale orangish lower mandible. *Slightly larger* than *Empidonax* flycatchers but with no eye-ring; wings extend farther down tail; *does not flick tail.* Very similar to Western Wood-Pewee but slightly greener or paler gray above and clearer below (vest "not buttoned"); best distinguished by voice, range. **VOICE:** A sweet, plaintive whistle, *pee-a-wee,* slurring down then up (less commonly, *pee-ur*), and a *chip.* **SIMILAR SPECIES:** Western Wood-Pewee. Eastern Phoebe lacks wing bars; bobs tail downward. **HABITAT:** Woodlands, groves.

NORTHERN BEARDLESS-TYRANNULET **Uncommon, local**
Camptostoma imberbe

4¼ in. (11 cm). A very small, nondescript flycatcher suggesting a kinglet or Bell's Vireo. Grayish olive, with *slight crested* look. *Dull wing bars* and indistinct pale supercilium. **VOICE:** A thin *peeee-yuk.* A gentle, descending *ee, ee, ee, ee, ee.* **SIMILAR SPECIES:** *Empidonax* flycatchers larger and larger-headed. **HABITAT:** Lowland woods, mesquite, stream thickets. Builds a globular nest with entrance on side.

FLYCATCHERS
OLIVE-SIDED
FLYCATCHER
WESTERN
WOOD-PEWEE
EASTERN
WOOD-PEWEE
NORTHERN
BEARDLESS-TYRANNULET

EMPIDONAX FLYCATCHERS

Flycatchers of this genus show light eye-rings and two pale wing bars. Species are notoriously difficult to separate, especially cross-continental vagrants; breeding habitats and vocalizations are the most useful means of identification. For silent birds, including migrants, note size and shape of bill, color of lower mandible, shape and boldness of eye-ring, color of wings and wing-feather edging, pattern of underparts, primary (wingtip) projection, and tail length.

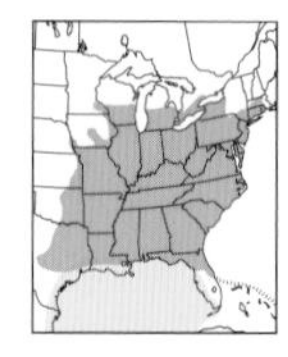

ACADIAN FLYCATCHER *Empidonax virescens* Fairly common

5¾ in. (15 cm). A large and elongated *greenish Empidonax* with *pale* underparts, thin eye-ring, and thin, *long* bill with yellow lower mandible. First-fall birds (p. 220) brighter, yellower below. **VOICE:** "Song" a sharp explosive *pit-see!* or *wee-see!* (sharp upward inflection); also a sharp *peet.* **SIMILAR SPECIES:** Whitish chin and throat separate Acadian from Yellow-bellied Flycatcher. **HABITAT:** Shady deciduous forests, ravines, swampy woods, beech and hemlock groves.

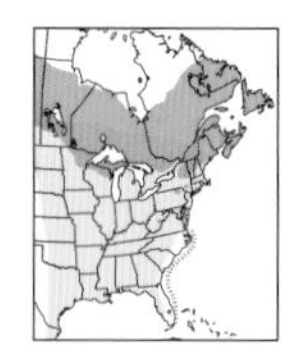

YELLOW-BELLIED FLYCATCHER *Empidonax flaviventris* Uncommon

5½ in. (14 cm). Has green back, rounded yellowish eye-ring, dusky breast-band, and blackish wings with bold whitish edging. First-fall birds (p. 220) much yellower below, including chin and throat. **VOICE:** Song a simple, spiritless *chi-lek;* also a rising *chu-wee;* call an explosive *peeyup,* distinct among *Empidonax.* **SIMILAR SPECIES:** Acadian Flycatcher. See "Western" Flycatchers (Cordilleran and Pacific-slope, p. 220). **HABITAT:** In summer, boreal forests, muskeg, bogs; in migration, wood edges.

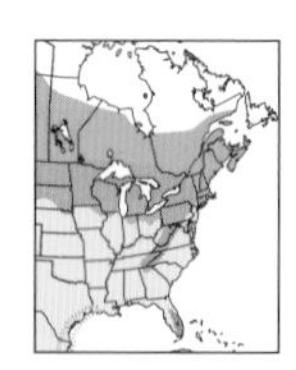

LEAST FLYCATCHER *Empidonax minimus* Fairly common

5¼ in. (13 cm). A small *Empidonax,* plumage variable but usually *grayish* above and *pale* below with *bold white eye-ring,* medium-short wingtip projection, and short, wide-based bill. Whitish wing bars on mostly blackish wing. First-fall birds (p. 220) fresher, greener and yellower. Actively flicks tail. **VOICE:** Song an emphatic, sharply snapped *che-bek!* Call a sharp, dry *whit.* **SIMILAR SPECIES:** Willow and Alder Flycatchers browner above with bigger bill, longer wingtips, weaker eye-ring. See Hammond's, Dusky, and Gray Flycatchers (p. 220). **HABITAT:** Deciduous and mixed woodlands, poplars, aspens.

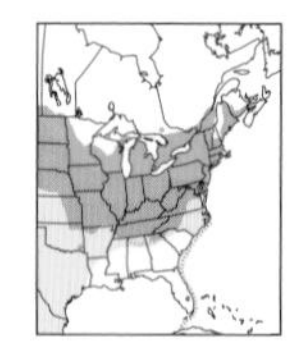

WILLOW FLYCATCHER *Empidonax traillii* Fairly common

5¾ in. (15 cm). Willow and Alder Flycatchers are nearly identical in appearance, a bit larger, longer billed, and often browner than Least Flycatcher. They are separated from each other mainly by voice and breeding habitat. Willow averages paler and browner (less olive) and has grayer head than Alder, has slightly weaker or no eye-ring, and duller wing-feather edging on average. First-fall birds (p. 220) fresher, greener. **VOICE:** Song a sneezy *fitz-bew,* unlike the *fee-BE-o* of Alder. Call a soft *whit.* **HABITAT:** Willow thickets, brushy fields; often in drier habitats than Alder, but can be found in close proximity where ranges overlap.

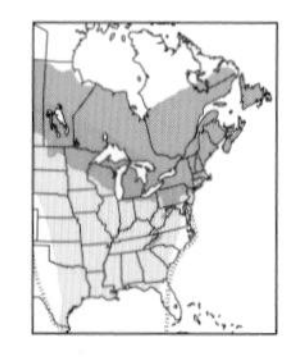

ALDER FLYCATCHER *Empidonax alnorum* Fairly common

5¾ in. (15 cm). The northern counterpart of Willow Flycatcher. Greener, smaller billed, and with brighter wing-feather edging than Willow, but best distinguished by voice. First-fall birds (p. 220) brighter. **VOICE:** Song an accented *fee-BE-o* or *rree-BE-o.* Call *kep* or *pit,* sharper than in Willow. **HABITAT:** Willows, alders, brushy swamps, swales. Usually in moister areas than Willow.

dults on Breeding Grounds
EMPIDONAX
FLYCATCHERS
Empidonax flycatchers are often best identified by voice. Breeding habitat is also a helpful clue.
pit-see!
deciduous woods,
pecially beech trees;
wooded swamps;
s. and cen. U.S.
ACADIAN
FLYCATCHER
greener than
Least, Alder,
or Willow
chi-lek
coniferous woods,
bogs; Canada, n.
edge of U.S.
YELLOW-BELLIED
FLYCATCHER
throat and
breast washed
with yellow
LEAST
FLYCATCHER
che-BEK or
chebek
farms, orchards,
groves, open woods;
n. U.S. and Canada
grayest of eastern
species
fitz-bew
wet and dry thickets,
brushy pastures, old
orchards, willows;
mostly in U.S.
WILLOW FLYCATCHER
fee-bee'-o
alder swamps, wet thickets,
usually near water;
n. U.S., Canada
ALDER FLYCATCHER

FIRST-FALL *EMPIDONAX* FLYCATCHERS

First-fall *Empidonax* flycatchers differ in appearance from worn breeding adults, averaging brighter, having buffier wing bars, and showing more orange on mandible. Vagrants of Western species can occur in our area, often in late fall, making identification challenging.

ACADIAN FLYCATCHER — Fairly common

Empidonax virescens (see also p. 218)

5¾ in. (15 cm). Note long wings and tail. Chin and throat white versus yellow in Yellow-bellied Flycatcher; brighter than "Western" Flycatchers.

YELLOW-BELLIED FLYCATCHER — Uncommon

Empidonax flaviventris (see also p. 218)

5½ in. (14 cm). From "Western" Flycatchers; note brighter green back; blacker wings with brighter edging; eye-ring rounded.

"WESTERN" FLYCATCHERS — Casual vagrants

Empidonax difficilis and *occidentalis* (adults not shown)

5½ in. (14 cm). Pacific-slope Flycatcher (*E. difficilis*) and Cordilleran Flycatcher (*E. occidentalis*) are largely inseparable. Back duller olive, wings browner with duller edging than in Yellow-bellied, eye ring almond.

ALDER FLYCATCHER — Fairly common

Empidonax alnorum (see also p. 218)

5¾ in. (15 cm). Head and upperparts greener (less olive, grayish, or brownish) than in Willow; wing-feather edging brighter; bill smaller.

WILLOW FLYCATCHER — Fairly common

Empidonax traillii (see also p. 218)

5¾ in. (15 cm). See Alder Flycatcher. Eastern Willow is greener and can be more difficult to separate from Alder.

LEAST FLYCATCHER — Fairly common

Empidonax minimus (see also p. 218)

5¼ in. (13 cm). First-fall birds variable, generally grayish, sometimes tinged olive above and washed lemon below. Bill broader based and deeper from sides than in western *Empidonax*.

GRAY FLYCATCHER — Accidental vagrant

Empidonax wrightii (adult not shown)

Grayish; first-fall birds tinged olive above and yellow below. Bill long, pinkish with distinct black tip; outer edges of tail white. Wags tail downward.

DUSKY FLYCATCHER — Accidental vagrant

Empidonax oberholseri (adult not shown)

A grayish long-tailed *Empidonax*. First-fall birds washed olive; underparts tinged yellow; outer edges of tail whitish.

HAMMOND'S FLYCATCHER — Casual vagrant

Empidonax hammondii (adult not shown)

A small but long-winged *Empidonax* flycacther. Bill very small, almost warbler-like; thinner from sides than in Least Flycatcher. Wing projection long. Outer edges of tail whitish; wings usually duller than in Least.

FIRST-FALL *EMPIDONAX* FLYCATCHERS

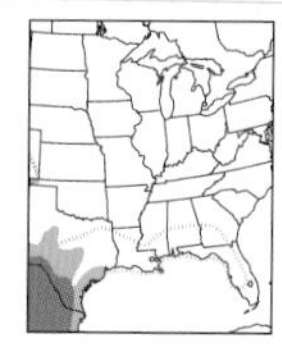

VERMILION FLYCATCHER *Pyrocephalus rubinus* Uncommon

6 in. (15 cm). *Adult male:* Crown (often raised in slight bushy crest) and underparts *flaming vermilion;* upperparts brown, tail blackish. *First-year male:* Female-like, but lower belly washed pinkish; variably gains pinkish or red mottling throughout body feathering during first year. *Female:* Breast whitish, narrowly streaked; belly washed with pinkish to salmon (adult) or pale lemon (first-year). An open-country bird, often perched on barbed wire and fenceposts, where adult males stand out. **VOICE:** *P-p-pit-zee* or *pit-a-zee.* **SIMILAR SPECIES:** Female told from Say's Phoebe by shorter tail, pale supercilium, dusky streaks on breast. Eastern Phoebe. **RANGE:** Casual vagrant to East Coast. **HABITAT:** Moist areas in arid country, such as streams, ponds, pastures, golf courses, ranches.

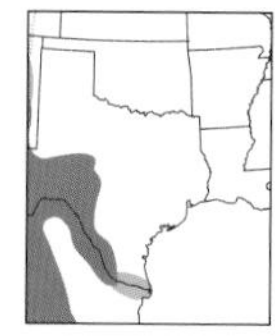

BLACK PHOEBE *Sayornis nigricans* Uncommon, local

6¾–7 in. (17–18 cm). Our only *black-breasted* flycatcher; belly white. Has typical phoebe tail-bobbing habit. *Juvenile:* Wing bars cinnamon-buff. **VOICE:** A thin, strident *fi-bee, fi-bee,* rising then dropping; also a sharp slurred *chip.* **SIMILAR SPECIES:** Eastern Phoebe, juncos (which are ground-loving birds and show very different behaviors). **RANGE:** Accidental vagrant east of range. **HABITAT:** Streams, farmyards, towns, parks; usually near water.

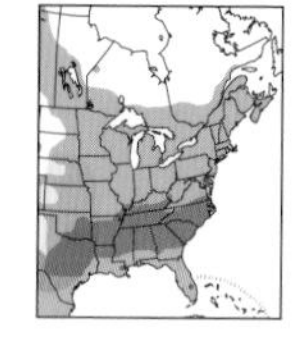

EASTERN PHOEBE *Sayornis phoebe* Fairly common

7 in. (18 cm). Note *downward tail-bobbing.* A grayish, medium-sized flycatcher *without eye-ring or strong wing bars* (thin buff wing bars on juvenile); small, *all-dark bill* and dark head; yellowish belly in fall. **VOICE:** Song a well-enunciated *phoe-be* or *fi-bree* (second note alternately higher or lower). Call a sharp *chip.* **SIMILAR SPECIES:** Eastern Wood-Pewee and smaller *Empidonax* flycatchers have conspicuous wing bars; lower mandibles partly yellowish. All *Empidonax* flick tail *upward.* **HABITAT:** Streamsides, bridges, farms, roadsides, towns. Often nests under bridges or in abandoned wooden structures.

SAY'S PHOEBE *Sayornis saya* Uncommon

7½ in. (19 cm). A midsized, pale brownish flycatcher with contrasty black tail and *orange-buff belly.* **VOICE:** A plaintive, down-slurred *pweer* or *pee-ee.* **SIMILAR SPECIES:** Eastern Phoebe, Vermilion Flycatcher, Ash-throated Flycatcher. **RANGE:** Casual vagrant to East Coast. **HABITAT:** Open country, dry scrub, canyons, ranches. Frequents barbed wire and fenceposts in open fields, where graceful and conspicuous.

PHOEBES
VERMILION FLYCATCHER
adult male
first-year male
adult female
first-year female
BLACK PHOEBE
EASTERN PHOEBE
SAY'S PHOEBE

MYIARCHUS FLYCATCHERS

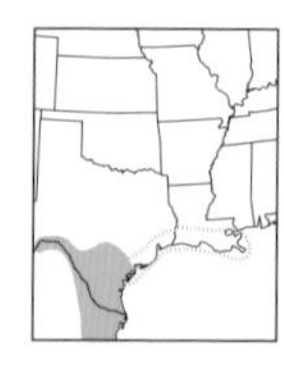

BROWN-CRESTED FLYCATCHER *Myiarchus tyrannulus* **Uncommon**

8¾ in. (22 cm). Similar to Ash-throated Flycatcher but larger, with noticeably larger bill. Underparts brighter yellow. Tail rusty, a bit less so than in Ash-throated. Voice important. **VOICE:** A sharp *whit* and rolling, throaty *purreeer.* Voice much more vigorous and raucous than Ash-throated's. **SIMILAR SPECIES:** Great Crested Flycatcher. **HABITAT:** Woodlands and well-vegetated residential areas.

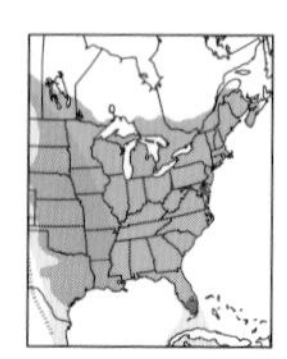

GREAT CRESTED FLYCATCHER *Myiarchus crinitus* **Fairly common**

8½–8¾ in. (21–22 cm). A kingbird-sized flycatcher with cinnamon wings and tail, dark olive back, *mouse gray breast,* and bright yellow belly. Often erects bushy crest. Note *strongly contrasting tertial pattern* and pink-based bill. **VOICE:** A loud whistled *wheeep!* Also a rolling *prrrrrreet!* **SIMILAR SPECIES:** Brown-crested and Ash-throated Flycatchers have dark lower mandibles, paler gray breasts, paler yellow bellies, and duller wings with less contrasting tertials. Vocal differences important. **HABITAT:** Woodlands, groves.

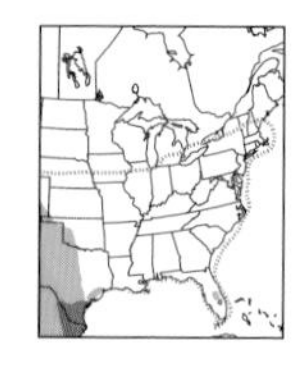

ASH-THROATED FLYCATCHER *Myiarchus cinerascens* **Fairly common**

8–8¼ in. (20–21 cm). A medium-sized flycatcher, smaller than a kingbird, grayish brown above with two pale wing bars, *whitish* throat, *pale* gray breast, *pale yellowish belly,* and *rufous tail.* Head slightly bushy. **VOICE:** *Prrt* (likened to a police whistle); also a rolling *chi-queer* or *prit-wheer.* **SIMILAR SPECIES:** Great Crested and Brown-crested Flycatchers, Say's Phoebe. **RANGE:** Rare late-fall vagrant to East Coast. **HABITAT:** Semiarid country, deserts, brush, mesquite, pinyon-juniper, chaparral, open woods.

LA SAGRA'S FLYCATCHER *Myiarchus sagrae* **Casual vagrant**

7¼–7½ in. (19 cm). Casual vagrant to FL from W. Indies. Similar to Ash-throated Flycatcher but smaller, darker above, has only a *hint of yellow on belly. Tail brownish, not rufous.* Short primaries. Often "droopy" posture. **VOICE:** A high, rapid, double *wick-wick.* **SIMILAR SPECIES:** Great Crested and Ash-throated Flycatchers. **HABITAT:** Shrubby coastal woods.

MYIARCHUS FLYCATCHERS

Most Have Extensively Rusty Tails

BROWN-CRESTED FLYCATCHER

GREAT CRESTED FLYCATCHER

ASH-THROATED FLYCATCHER

LA SAGRA'S FLYCATCHER

WESTERN KINGBIRD *Tyrannus verticalis* — Fairly common

8¾ in. (22 cm). Note *pale gray head and breast,* white throat, *yellowish belly,* smaller bill. *Black tail* has *narrow white edges.* **VOICE:** Shrill, bickering calls; a sharp *kip* or *whit-ker-whit;* dawn song *pit-PEE-tu-whee.* **SIMILAR SPECIES:** Couch's and Tropical Kingbirds. **RANGE:** Very rare vagrant to East Coast. **HABITAT:** Farms, shelterbelts, semiopen country, roadsides, fences, wires.

EASTERN KINGBIRD *Tyrannus tyrannus* — Common

8¾ in. (22 cm). Lack of yellow underparts and *white band* across tail tip mark Eastern Kingbird. Red crown mark is concealed and rarely seen. Often seems to fly quiveringly on tips of wings. Harasses crows, hawks. **VOICE:** A rapid sputter of high, bickering electric-shock notes: *dzee-dzee-dzee,* etc., and *kit-kit-kitter-kitter,* etc. Also a nasal *dzeep.* **SIMILAR SPECIES:** Gray Kingbird. **HABITAT:** Wood edges, river groves, farms, shelterbelts, roadsides, fences, wires.

GRAY KINGBIRD *Tyrannus dominicensis* — Fairly common, local

9 in. (23 cm). Resembles Eastern Kingbird but larger and much paler. Conspicuously *notched tail* has no white band. *Very large bill* gives large-headed look. Dark ear patch. **VOICE:** A rolling *pi-teer-rrry* or *pe-cheer-ry.* **SIMILAR SPECIES:** Eastern Kingbird. **RANGE:** Vagrants occur well to north up coast in spring and summer. **HABITAT:** Roadsides, wires, mangroves, edges.

TROPICAL KINGBIRD *Tyrannus melancholicus* — Uncommon, local

9¼ in. (23 cm). Nearly identical to Couch's Kingbird. Both species similar to Western Kingbird, but in Tropical, the birds that reach our area usually have *bill much larger and longer,* tail *notched* and *brownish;* bright yellow on underparts *includes breast.* **VOICE:** Repeated twittery *kip-kip-kip* calls. **SIMILAR SPECIES:** Couch's Kingbird. **RANGE:** Accidental vagrant north and east of range, to East Coast. **HABITAT:** Groves along streams and ponds, open areas with scattered trees, phone wires.

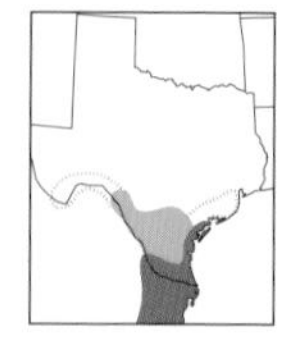

COUCH'S KINGBIRD *Tyrannus couchii* — Fairly common, local

9¼ in. (23 cm). Very similar to Tropical Kingbird and best distinguished by voice. Couch's usually has *shorter and smaller bill* than Tropicals that reach our area, and brighter green back. **VOICE:** A nasal *queer* or *beeer* (suggests Common Pauraque). Also a sharp *kip.* **SIMILAR SPECIES:** Tropical Kingbird. **RANGE:** Accidental vagrant north and east of range, to East Coast. **HABITAT:** Open wooded and brushy areas with large trees; most common in native habitat.

KINGBIRDS
EASTERN KINGBIRD
WESTERN KINGBIRD
GRAY KINGBIRD
TROPICAL KINGBIRD
COUCH'S KINGBIRD
Tropical

More TYRANT FLYCATCHERS and BECARD

FORK-TAILED FLYCATCHER *Tyrannus savana* Very rare vagrant

14½–16 in. (37–41 cm). A vagrant from Tropics. Told from Scissor-tailed Flycatcher by *black cap,* white flanks and underwing. Black tail not rigid in flight. *First-year:* Much shorter tail; might be confused with Eastern Kingbird, but note paler gray back. **VOICE:** A mechanical-sounding *tik-tik-tik.* **SIMILAR SPECIES:** Scissor-tailed Flycatcher. **RANGE:** Normal range from Mex. to S. America. Vagrant to U.S. and Canada; records widespread in East, predominantly in summer through fall. **HABITAT:** Open fields, pastures with scattered trees, wires.

SCISSOR-TAILED FLYCATCHER *Tyrannus forficatus* Common

13–15 in. (33–38 cm). A beautiful bird, pale pearly gray, adult male with *extremely long, scissorlike tail* that is usually folded. Flanks orange-buff, underwing linings salmon pink. *Female and first-year male:* Shorter tail and duller sides may suggest Western Kingbird. Hybrids are known. **VOICE:** A harsh *keck* or *kew;* a repeated *ka-leep;* also shrill, kingbirdlike bickerings and stutterings. **SIMILAR SPECIES:** Western Kingbird, Fork-tailed Flycatcher. **RANGE:** Widespread vagrant north and east of range, casually to East Coast. **HABITAT:** Semiopen country, ranches, farms, roadsides, fences, wires.

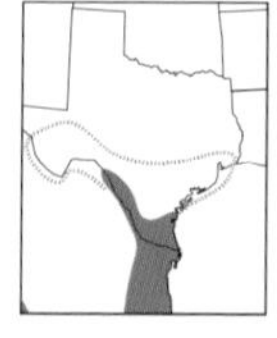

GREAT KISKADEE *Pitangus sulphuratus* Fairly common, local

9¾ in. (25 cm). A large, *big-headed* flycatcher, like Belted Kingfisher in actions, even catching small fish. Note *striking head pattern,* rufous wings and tail, *yellow underparts and crown.* **VOICE:** A loud *kiss-ka-dee;* also a loud *reea.* Often heard before it is seen. **SIMILAR SPECIES:** Tropical and Couch's Kingbirds, which share this kiskadee's limited range. **RANGE:** Accidental vagrant north of range and east to FL. **HABITAT:** Woodlands and brushy edges, usually near water.

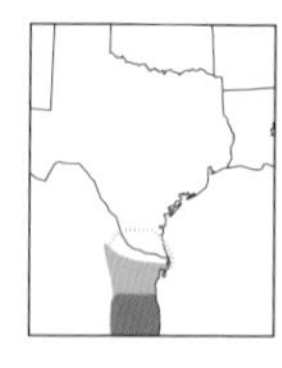

ROSE-THROATED BECARD *Pachyramphus aglaiae* Rare, local

7¼ in. (18 cm). Big-headed and thick-billed. *Adult male:* Dark gray above, pale to dusky below, with *blackish cap and cheeks* and lovely *rose-colored throat* (lacking or reduced in some males). *Female:* Brown above, with *dark cap* and *light buffy collar* around nape. Underparts strong buff. *First-year male:* Like female but with rose feathers in throat, grayish feathers in back. **VOICE:** A thin, slurred whistle, *seeoo.* **SIMILAR SPECIES:** Kingbirds, Say's Phoebe. **HABITAT:** Riparian woodlands.

FLYCATCHERS
FORK-TAILED FLYCATCHER
adult
SCISSOR-TAILED FLYCATCHER
adult male
first-year female
adult
GREAT KISKADEE
ROSE-THROATED BECARD
female
adult male

SWALLOWS Family Hirundinidae

Slim, streamlined form and graceful flight characterize swallows and martins. Pointed wings; short bill with very wide gape; tiny feet. **FOOD:** Mostly flying insects. **RANGE:** Worldwide except for polar regions, remote islands.

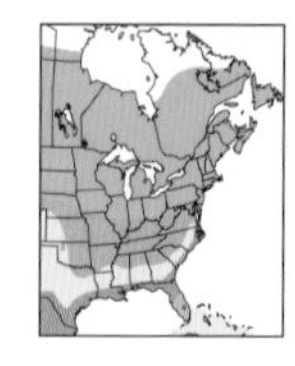

TREE SWALLOW *Tachycineta bicolor* — Common

5¾ in. (15 cm). *Adult:* Male *steely blue,* tinged green above; *white below.* Female varies from slightly duller than male to largely brown. *Juvenile:* Dusky gray-brown back and smudgy band across breast. Tree Swallows have distinctly notched tail. Flight characterized by glides followed by quick flaps and short climbs. **VOICE:** A rich *cheet* or *chi-veet;* a liquid twitter, *weet, trit, weet,* etc. **SIMILAR SPECIES:** Violet-green Swallow smaller, paler above eye, obvious white patches on sides of rump, male green and purple above. Northern Rough-winged Swallow has dingier throat, different flight style. Bank Swallow smaller, browner, has bolder dark breast-band than juvenile Tree. All N. American swallows also have different calls. **HABITAT:** Open country near water, marshes, meadows, streams, lakes. Nests in holes, in trees and birdhouses.

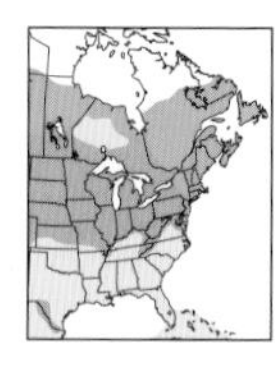

BANK SWALLOW *Riparia riparia* — Fairly common

5 in. (12 cm). *Our smallest* swallow. *Brown-backed with slightly darker wings and paler rump.* Note distinct *dark breast-band* in all plumages. White of throat *curls up behind ear. Wingbeats rapid and shallow.* Ages and sexes similar. **VOICE:** A dry, trilled chitter or rattle, *brrt* or *trr-tri-tri.* **SIMILAR SPECIES:** Northern Rough-winged Swallow and juvenile Tree Swallow. Bank's smaller size stands out in mixed flock. **HABITAT:** Near water; fields, marshes, lakes, coasts. Nests colonially in dirt or sandbanks.

NORTHERN ROUGH-WINGED SWALLOW — Fairly common
Stelgidopteryx serripennis

5¼ in. (13 cm). *Adult: Brown-backed and -rumped; throat and upper breast brownish to dusky;* no breast-band. Flight more languid than Bank Swallow; wings pulled back at end of stroke. *Juvenile:* Has cinnamon-rusty wing bars. **VOICE:** Call a low, liquid *trrit,* lower and less grating than Bank Swallow. **SIMILAR SPECIES:** Plainer than Bank Swallow and juvenile Tree Swallow. **HABITAT:** Near streams, lakes, rivers, coasts. Nests in banks, pipes, and crevices, but not colonially as Bank Swallow does.

VIOLET-GREEN SWALLOW *Tachycineta thalassina* — Casual vagrant

5¼ in. (13 cm). Note *white patches that almost meet* over base of tail. *Adult male:* Dark and shiny above; glossed with *green and purple;* clear white below. *White of face partially encircles eye. Female and first-winter male:* Duller above, white above eye tinged grayish or brownish. *Juvenile:* Brown above, with little or no green. **VOICE:** A thin *ch-lip* or *chew-chit* or rapid *chit-chit-chit wheet, wheet.* **SIMILAR SPECIES:** Tree Swallow lacks white patches on sides of rump and pale feathering above eye, has bluer back, slightly larger size and longer wings. **RANGE AND HABITAT:** Casual vagrant from w. U.S. and accidental vagrant to East Coast. Often in coniferous-forested areas; mixes with other swallows during migration and winter.

SWALLOWS
nests in tree
holes or nest
boxes
TREE
SWALLOW
adult
male
Bank Swallow
colony
juvenile
some can show
an indistinct brown
breast-band
BANK
SWALLOW
NORTHERN ROUGH-WINGED
SWALLOW
VIOLET-GREEN
SWALLOW
adult
male
Purple
Martin
(p. 232)
Barn
(p. 232)
Cliff
(p. 232)
Tree
Northern
Rough-winged
Bank
Violet-
green
Swallows on a wire

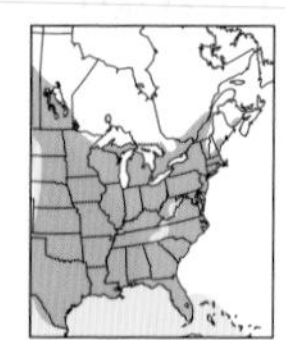

PURPLE MARTIN *Progne subis* **Fairly common**

8 in. (20 cm). The largest N. American swallow. *Adult male:* Uniformly blue-black *above and below;* no other swallow is dark-bellied. *Female and first-fall male:* Light-bellied; throat and breast grayish, often with faint gray collar; first-spring male mottled dark. Alternates quick flaps and glides; often spreads tail. **VOICE:** A throaty and rich *tchew-wew,* etc., or *pew, pew.* Song gurgling, ending in a succession of rich, low guttural notes. **SIMILAR SPECIES:** Tree and other swallows, much smaller than female Purple Martin, cleaner white below. In flight, European Starling shows similar wing shape. **HABITAT:** Towns, farms, open or semiopen country, often near water. In East, nests exclusively in human-supplied martin houses, unlike in West where natural cavities used.

CAVE SWALLOW *Petrochelidon fulva* **Uncommon**

5½ in. (14 cm). *Adult:* Similar to Cliff Swallow (rusty rump, square-cut tail), but face colors reversed: *throat and cheeks buffy* (not dark), forehead *dark chestnut* (not pale, although vagrant Cliff Swallows from sw. U.S. and elsewhere can have chestnut forehead). *Buff color sets off dark mask and cap. Juvenile:* Brown upperparts, pale brown breast, whitish belly. **VOICE:** A clear, sweet *weet* or *cheweet;* a loud, accented *chu, chu.* **SIMILAR SPECIES:** Cliff Swallow. **RANGE:** Rare but increasing vagrant north to Great Lakes and e. Canada in late fall. **HABITAT:** Open country. Cuplike nest placed in caves, culverts, and under bridges; nests colonially.

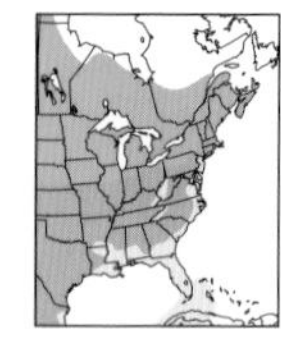

CLIFF SWALLOW *Petrochelidon pyrrhonota* **Uncommon**

5½ in. (14 cm). *Adult:* Note *rusty, orange,* or *buffy rump,* steely blue upperparts, pale hindcollar. From below, appears square-tailed, with red face and dark throat patch. *Juvenile:* Dusky above with muted head pattern; breast buff; *throat mixed with some dark.* **VOICE:** *Zayrp;* a low *chur.* Alarm call *keer!* Song consists of creaking notes and guttural gratings. **SIMILAR SPECIES:** Barn and Cave Swallows. **HABITAT:** Open to semiopen land, farms, cliffs, lakes, where breeds colonially in mud-jug or gourdlike nests under eaves and bridges.

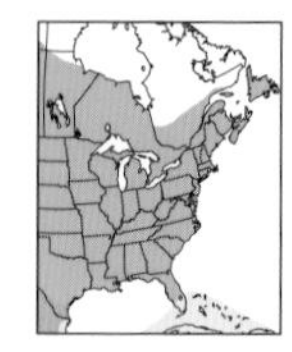

BARN SWALLOW *Hirundo rustica* **Common**

6¾ in. (17 cm). Our only swallow that is truly *swallow-tailed;* also the only one with *white tail spots. Adult:* Blue-black above; cinnamon-buff below, with darker throat; male brighter and longer tailed than female. *Juvenile and first-fall:* Duller overall and paler, more whitish below. Flight direct, close to ground; wingtips pulled back at end of stroke; not much gliding. **VOICE:** A soft *vit* or *kvik-kvik, vit-vit.* Also *szee-szah* or *szee.* Anxiety call a harsh, irritated *ee-tee* or *keet.* Song a musical twitter interspersed with guttural notes. **SIMILAR SPECIES:** Other N. American swallows have notched (not deeply forked) tail. **HABITAT:** Open or semiopen land; farms, fields, marshes, lakes; often perches on wires; usually near habitation. Builds *cuplike nest inside* barns or under eaves, not in tight colonies like Cliff Swallow.

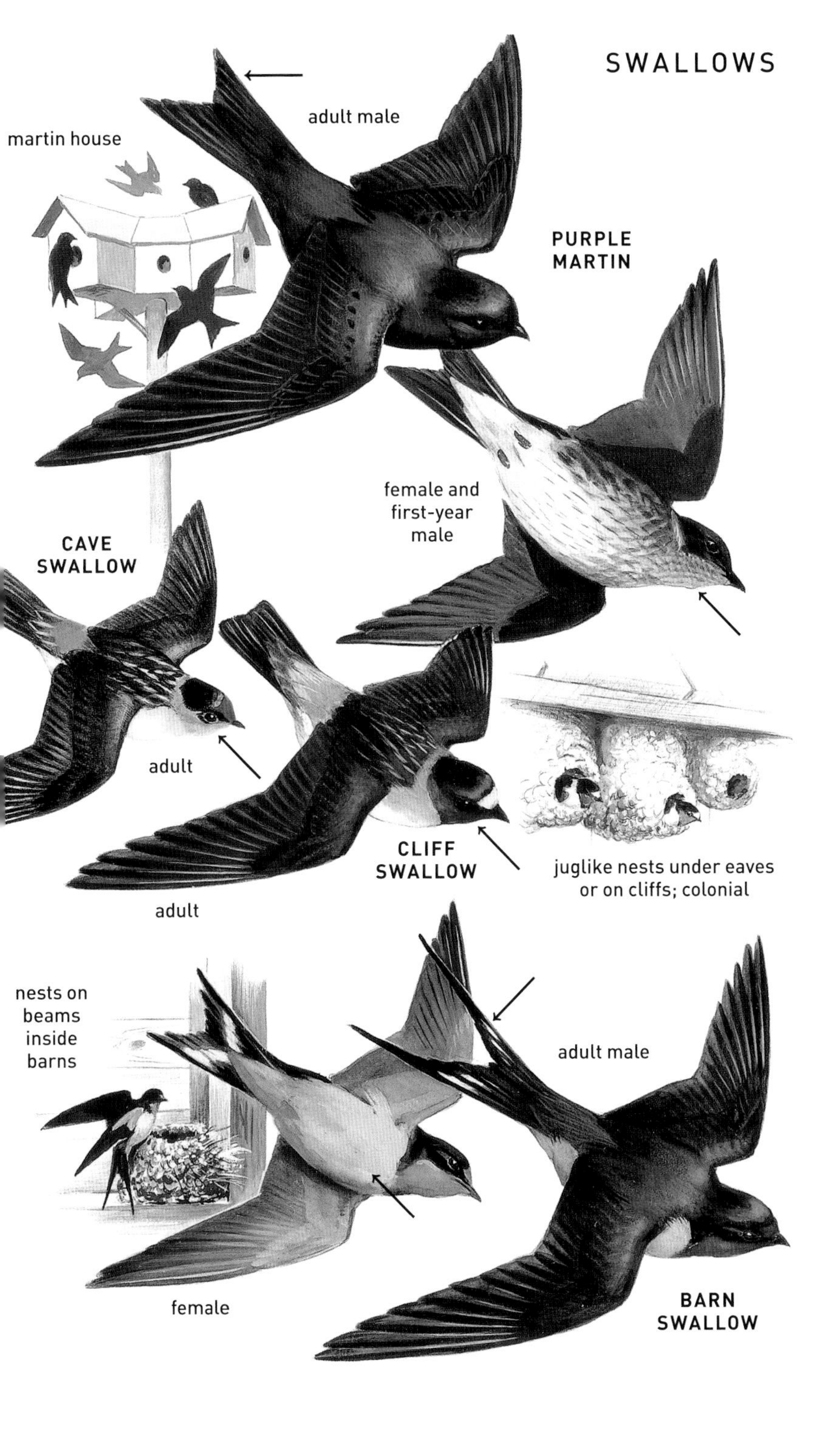
SWALLOWS
adult male
martin house
PURPLE
MARTIN
female and
first-year
male
CAVE
SWALLOW
adult
CLIFF
SWALLOW
juglike nests under eaves
or on cliffs; colonial
adult
nests on
beams
inside
barns
adult male
female
BARN
SWALLOW

OLD WORLD FLYCATCHERS Family Muscicapidae

Large and varied, primarily Old World family. Delicate sparrow-sized birds, often flicking wings and wagging tails near ground level. **FOOD:** Insects, fruit. **RANGE:** Throughout Eurasia and Africa; one species breeds in Arctic e. N. America.

NORTHERN WHEATEAR *Oenanthe oenanthe* Uncommon, local

5¾ in. (15 cm). A small, dapper bird of Arctic barrens, fanning its tail and bobbing. Note *white rump and sides of tail.* Black on tail forms *broad inverted T. Spring/summer male:* Pale gray back, black wings, and *black ear patch. Female and fall/winter male:* Variably buffier, with brown back, reduced black in face. **VOICE:** Call a hard *chak-chak* or *chack-weet, weet-chack.* **RANGE:** Very rare vagrant south of breeding range in our area. **HABITAT:** Open, stony areas; in summer, rocky tundra.

THRUSHES Family Turdidae

Large-eyed, slender-billed, usually strong-legged songbirds. Many species are brown-backed with spotted breasts; bluebirds, solitaires, and robins have speckle-breasted young. Thrushes are often fine singers, making up for their generally drab plumages. **FOOD:** Insects, worms, snails, berries, fruit. **RANGE:** Nearly worldwide.

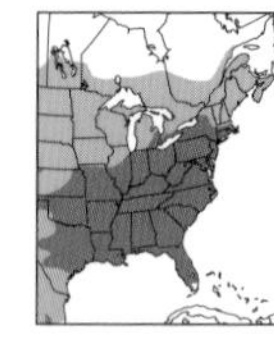

EASTERN BLUEBIRD *Sialia sialis* Fairly common

7 in. (18 cm). A blue bird with *rusty red breast;* appears round-shouldered when perched. Female duller than male, with pale rusty throat, breast, and *flanks; white belly. Juvenile:* Grayish with some telltale blue in wings and tail; *back and breast speckled.* **VOICE:** Call a musical *chur-wi.* Song three or four gurgling notes. **SIMILAR SPECIES:** Fresh female Mountain Bluebirds may be warm buff below, but flanks not as bright, and they are longer winged. **HABITAT:** Open country with scattered trees; farms, roadsides. Often nests in bluebird boxes.

MOUNTAIN BLUEBIRD *Sialia currucoides* Uncommon

7¼–7½ in. (18–19 cm). *Adult male: Turquoise blue,* paler below; belly whitish. *First-year male:* Duller with some brown. *Female:* Dull brownish gray, with touch of pale blue on rump, tail, and wings; shows pale rusty breast and sides in fresh fall/winter plumages. **VOICE:** A low *chur* or *vhew.* **SIMILAR SPECIES:** Has straighter posture than female Eastern Bluebird, with slightly longer bill and tail. Warmer colored birds in fresh plumage lack rusty-colored flanks. **RANGE:** Uncommon (mostly fall and winter) to Great Plains; accidental to coast. **HABITAT:** Open country, treeless or with with some trees.

TOWNSEND'S SOLITAIRE *Myadestes townsendi* Rare

8½ in. (22 cm). A slim gray bird with *white eye-ring, white sides on tail,* and *buffy wing patches. Juvenile:* Dark overall with light spots and scaly belly. **VOICE:** Call a high-pitched *eek,* like a squeaky bicycle wheel. **SIMILAR SPECIES:** Pattern in wing and tail might suggest Northern Mockingbird, but note eye-ring, darker breast, and especially buff wing patches. **RANGE AND HABITAT:** Rare winter visitor from w. U.S. to Great Plains, where found in open woods. Very rare to accidental vagrant as far as East Coast.

OLD WORLD FLYCATCHER
AND THRUSHES
NORTHERN
WHEATEAR
fall/winter
spring/
summer
male
juvenile
female
adult male
EASTERN
BLUEBIRD
male
MOUNTAIN
BLUEBIRD
juvenile
TOWNSEND'S
SOLITAIRE

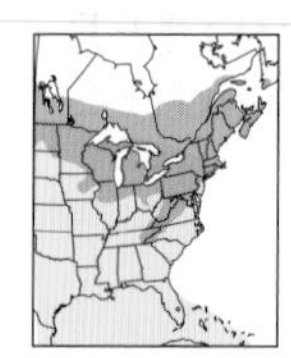

VEERY *Catharus fuscescens* Fairly common

7 in. (18 cm). *Catharus* can be difficult to separate; ages are similar (except juvenile plumage is spotted above) and sexes alike. In Veery, note *uniform rusty brown* cast above and pale grayish to whitish flanks. Grayish face with little or no eye-ring; very *indistinct* spotting on breast. **VOICE:** Song liquid, breezy, ethereal, wheeling downward: *vee-ur, vee-ur, veer, veer.* Call a down-slurred *phew* or *view.* **SIMILAR SPECIES:** Other *Catharus* thrushes are not as red above, have more distinct spotting below. **HABITAT:** Moist riparian woods, willow and alder thickets, meadows in pine forests.

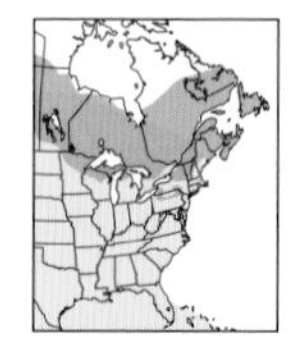

SWAINSON'S THRUSH *Catharus ustulatus* Fairly common

7 in. (18 cm). This thrush is marked by its conspicuous *buffy eye-ring* or *spectacles,* buff on cheeks and upper breast, and tail same color as back. Eastern birds are dull *olivey brown* above. Western birds are warmer, can be confused with Veery; these are causal in Great Plains and accidental farther east. **VOICE:** Song breezy, flutelike phrases, each phrase sliding *upward.* Call a liquid *whit* or *foot.* Migrants at night (in sky) give a short whistled *quee.* **SIMILAR SPECIES:** Gray-cheeked Thrush. Hermit Thrush smaller, more upright in posture, and has *contrasty rufous tail* and blackish breast spotting; posture and *vocalizations differ.* See Veery. **HABITAT:** Moist spruce and fir forests, riparian woodlands; in migration, other woods. Skulking.

BICKNELL'S THRUSH *Catharus bicknelli* Scarce, local

6½–6¾ in. (17 cm). Slightly smaller than Gray-cheeked Thrush, upperparts *warmer brown, tail dull chestnut,* breast *washed buff,* lower mandible usually more than half yellow (usually less than half in Gray-cheeked). Legs more dusky than toes (uniform pale in Gray-cheeked). **VOICE:** Melodic flutelike notes rolling from high to low to high, *whee-toolee-weee,* rising at close (falling in Gray-cheeked). Call a downward *pheu.* **SIMILAR SPECIES:** Gray-cheeked and Hermit Thrushes. **HABITAT:** Breeds in stunted mountain fir forests of Northeast to shoreline in Atlantic Provinces. In migration, forests.

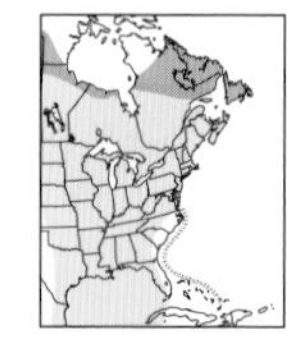

GRAY-CHEEKED THRUSH *Catharus minimus* Uncommon

7–7¼ in. (17–18 cm). A dull, "cold-colored," *gray-brown,* furtive thrush, distinguished from Swainson's Thrush by its *grayish* cheeks, *grayish,* less conspicuous, often broken eye-ring. *Little or no buff on breast.* **VOICE:** See Bicknell's Thrush. **SIMILAR SPECIES:** Other *Catharus* thrushes. **HABITAT:** Boreal forests, tundra scrub; in migration, woodlands.

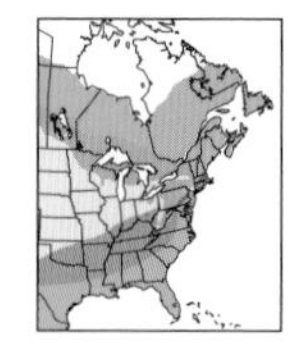

HERMIT THRUSH *Catharus guttatus* Fairly common

6¾ in. (17 cm). A smallish spot-breasted brown thrush with *rufous* tail. When perched, has habit of *flicking wings* and of *cocking tail and dropping it slowly.* Grayer interior western subspecies rare in winter to Great Plains; casual farther east. **VOICE:** Call a low *chuck;* also a scolding *tuk-tuk-tuk* and a rising, whiny *pay.* Song clear, ethereal, flutelike notes at *different pitches,* each with a *long introductory note.* **SIMILAR SPECIES:** Swainson's and Gray-cheeked Thrushes. **HABITAT:** Coniferous or mixed woods; in winter, woods, thickets, chaparral, parks, gardens.

SPOTTED THRUSHES
VEERY
SWAINSON'S THRUSH
East and interior West
Pacific Coast
BICKNELL'S THRUSH
GRAY-CHEEKED THRUSH
tail-lifting
interior West
HERMIT THRUSH
East

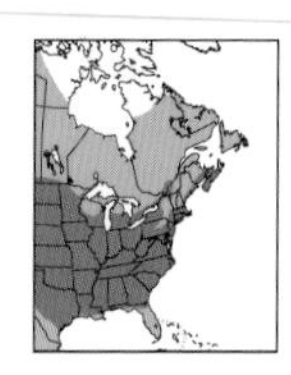

AMERICAN ROBIN *Turdus migratorius* Common

10 in. (25 cm). A very familiar bird; often seen on lawns, with an erect stance, giving short runs then pausing. Recognized by dark gray back and brick red breast. Dark stripes on white throat. Adult male has head and tail blacker and underparts solid deep reddish; these colors are duller in female, and first-year birds of each sex are slightly duller than adults. *Juvenile:* Has dark-speckled, pale rusty breast. **VOICE:** Song a clear caroling; short phrases, rising and falling, often prolonged. Calls *tyeep* and *tut-tut-tut.* **SIMILAR SPECIES:** Varied Thrush; Clay-colored Thrush and Rufous-backed Robin (both rare). **HABITAT:** Wide variety, including towns, parks, lawns, farmland, shade trees, many types of forests and woodlands; in winter, often found in berry-producing trees.

VARIED THRUSH *Ixoreus naevius* Very rare vagrant

9½ in. (24 cm). Similar to American Robin but with *orangish eye stripe, orange wing bars,* and *orange bar on underwing* visible in flight. *Male: Blue-gray above,* with wide *black breast-band. Female:* Duller gray above, with *gray breast-band.* First-year birds within each sex are duller than adults. **VOICE:** Call a quivering, low-pitched *zzzew* or *zzzeee* and a liquid *chup.* **SIMILAR SPECIES:** Orangey wing bars and eye stripe, and a breast-band, distinguish it from an American Robin, with which it only rarely mingles. **RANGE AND HABITAT:** Rare to casual winter visitor or vagrant from West, found in moist, dense woods, thickets. Vagrants (accidental as far as East Coast) often detected in gardens and near bird feeders.

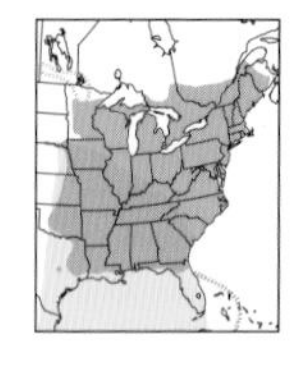

WOOD THRUSH *Hylocichla mustelina* Fairly common

7¾ in. (20 cm). *Rusty-headed.* Smaller than an American Robin; plumper than *Catharus* thrushes, distinguished by deepening rufous about head, *streaked gray cheeks,* white eye-ring, and *rounder, bolder,* more numerous *breast spots.* Ages similar (briefly held juvenile plumage with pale spots on back), sexes alike. **VOICE:** Song a pleasing series of notes with rounder phrases than other thrushes. Listen for flutelike *ee-o-lay.* Occasional guttural notes are distinctive. Call a rapid *pip-pip-pip-pip.* **SIMILAR SPECIES:** *Catharus* thrushes, juvenile American Robin. **HABITAT:** Mainly deciduous woodlands, cool moist glades. Commonly detected by distinctive song.

female
juvenile
male
AMERICAN ROBIN
juvenile
female
adult male
VARIED THRUSH
juvenile
WOOD THRUSH

FIELDFARE *Turdus pilaris* **Very rare vagrant**

10 in. (25 cm). Robinlike in size and posture, with heavily marked tawny breast. *Back rusty, contrasting with gray head and rump, dark tail.* Female and first-year birds average duller. **VOICE:** A harsh, chattering *tchak-tchak-tchak* and a quiet *see.* **SIMILAR SPECIES:** Juvenile American Robin, Redwing (vagrant, smaller in size), other spot-breasted thrushes. **RANGE:** Eurasian species; most N. American records from Northeast and n. Midwest in winter. **HABITAT:** Open country, fields, hedgerows, residential areas.

REDWING *Turdus iliacus* **Very rare vagrant**

8¼ in. (21 cm). Named for its rust-colored underwing linings (most visible in flight). Broad *pale eyebrow, heavily streaked below.* Bill two-toned, black at tip, yellow at base. Ages and sexes alike. **VOICE:** Flight call a thin, high, reedy *seeeh.* **SIMILAR SPECIES:** Fieldfare (vagrant) and juvenile American Robin both much larger. **RANGE:** Eurasian species; most N. American records from Northeast in winter. **HABITAT:** Semiopen country and young woodlands.

RUFOUS-BACKED ROBIN *Turdus rufopalliatus* **Casual vagrant**

9¼ in. (24 cm). This rare Mexican winter vagrant is like a pale American Robin but with orangier tinge below, *rufous back,* and *no white around eye.* More heavily streaked throat. *Orangier bill.* Female and first-year male duller. A timid skulker. **VOICE:** Call a soft, whistled *teeww.* **SIMILAR SPECIES:** American Robin. **RANGE AND HABITAT:** Casual vagrant to s. TX. Woods and thickets, often near water.

CLAY-COLORED THRUSH *Turdus grayi* **Scarce, local**

9 in. (23 cm). Scarce resident of southernmost TX. Warm brown above, dull tan on chest, paling to light tawny buff on belly. Throat streaked with light brown, not black. Ages and sexes similar. **VOICE:** A lower-pitched, simpler version of American Robin song. **SIMILAR SPECIES:** *Catharus* thrushes (which are smaller and less like American Robin). **HABITAT:** Tropical woodlands and well-vegetated residential areas.

MOCKINGBIRDS and THRASHERS Family Mimidae

Excellent songsters; some mimic other birds. Strong-legged; usually longer tailed than true thrushes, bill usually longer and more decurved. Ages and sexes similar. **FOOD:** Insects, fruit. **RANGE:** New World.

CURVE-BILLED THRASHER *Toxostoma curvirostre* **Fairly common**

11 in. (28 cm). Told from other thrashers by *well-curved* bill, *mottled breast.* Some individuals have narrow white wing bars. Eyes pale yellow-orange. **VOICE:** Call a sharp, liquid *whit-wheet!* Song a musical series of notes and phrases, grosbeaklike in quality but faster; little repetition. **RANGE:** Casual vagrant north and east of breeding range. **HABITAT:** Arid brush, lower canyons, ranch yards.

RARE THRUSHES

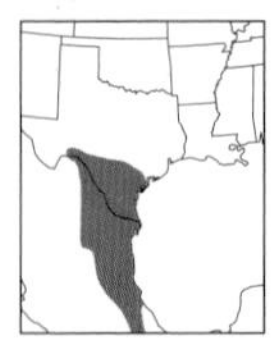

LONG-BILLED THRASHER *Toxostoma longirostre* **Uncommon, local**

11½ in. (29 cm). *Duller brown* above than Brown Thrasher, breast stripes *blacker, cheeks grayer;* bill longer, slightly more curved, and all dark. **VOICE:** Song similar to Brown Thrasher but more jumbled. Call a harsh *tchuk.* **SIMILAR SPECIES:** Brown Thrasher. **HABITAT:** Brush, mesquite.

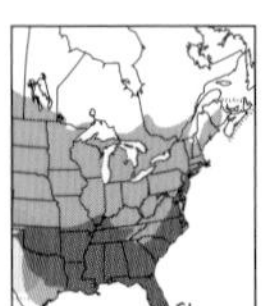

BROWN THRASHER *Toxostoma rufum* **Fairly common**

11½ in. (29 cm). Slimmer but longer tailed than an American Robin; *bright rufous* above, *heavily streaked* below. Note *wing bars,* slightly curved bill, long tail, and yellow eyes. **VOICE:** Song a succession of deliberate notes and phrases resembling Gray Catbird song, but each phrase usually *in pairs.* Call a harsh *chack!* **SIMILAR SPECIES:** The various brown thrushes have shorter tails, lack wing bars, are spotted (not striped) below, and have brown (not yellow) eyes. In s. TX see Long-billed Thrasher. **HABITAT:** Thickets, brush.

SAGE THRASHER *Oreoscoptes montanus* **Rare, local**

8½ in. (22 cm). Smaller than other thrashers. Gray-backed, with heavily streaked breast, white wing bars, and *white tail corners.* Eyes pale yellow, duller in juvenile and first-fall. **VOICE:** Song clear, ecstatic warbled phrases, sometimes repeated in thrasher fashion; more often continuous, suggestive of Black-headed Grosbeak. Call a blackbirdlike *chuck.* **RANGE AND HABITAT:** Rare winter visitor to s. TX; very rare but widespread vagrant farther east. Partial to sagebrush, mesas; in winter, also dry fields.

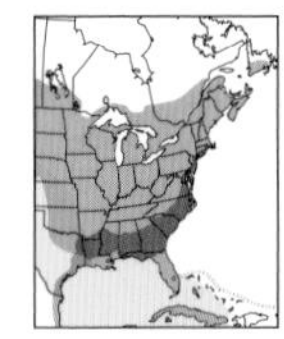

GRAY CATBIRD *Dumetella carolinensis* **Common**

8¾ in. (23 cm). Slate gray; slim. Note *black cap. Chestnut undertail coverts* (may not be noticeable). Flips tail jauntily. More often heard than seen. **VOICE:** A *catlike mewing;* distinctive. Also a grating *tcheck-tcheck.* Song disjointed notes and phrases; not repetitious compared with other mimids. **HABITAT:** Riparian undergrowth, brush.

BAHAMA MOCKINGBIRD *Mimus gundlachii* **Casual vagrant**

11 in. (28 cm). Chunkier than Northern Mockingbird and overall browner with *less white in tail* and *no white in wings.* Dark streaks on flanks, belly, and neck give this species a thrasherlike appearance. **VOICE:** Song simpler than Northern's, with two-syllable phrases. Call a sharp *tchak,* like Northern's but harsher. **RANGE:** Casual vagrant from Caribbean to s. FL, especially Dry Tortugas, FL Keys. **HABITAT:** A skulker in deep brushy cover.

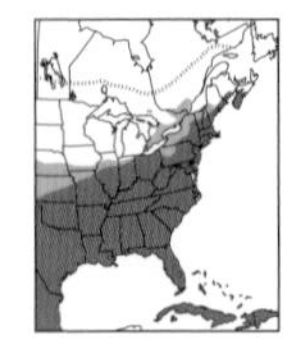

NORTHERN MOCKINGBIRD *Mimus polyglottos* **Common**

10 in. (25 cm). A familiar and conspicuous species. Gray; slimmer, longer tailed than an American Robin. Note *large white patches* on wings and tail, prominent in flight. **VOICE:** Song a varied, prolonged succession of notes and phrases, may be repeated six times or more before changing. Often heard at night. Mockingbirds are excellent mimics of other species. Call a loud *tchack;* also *chair.* **SIMILAR SPECIES:** Shrikes have dark facial mask. See Bahama Mockingbird. **HABITAT:** Towns, parks, gardens, farms, roadsides, thickets.

THRASHERS
AND MOCKINGBIRDS
LONG-BILLED
THRASHER
BROWN
THRASHER
SAGE
THRASHER
BAHAMA
MOCKINGBIRD
GRAY
CATBIRD
NORTHERN
MOCKINGBIRD
juvenile
wing-flashing

WAXWINGS Family Bombycillidae

Pointed crest may be raised or lowered. Waxy red tips on secondaries in adult and some first-year individuals; male has more waxy tips and blacker throat than female. Gregarious. **FOOD:** Berries, insects. **RANGE:** N. Hemisphere.

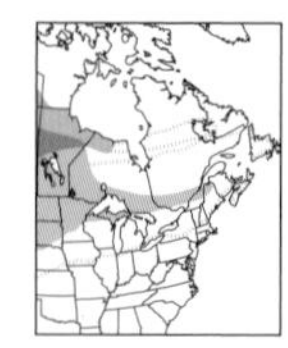

BOHEMIAN WAXWING *Bombycilla garrulus* **Uncommon, irregular**

8¼ in. (21 cm). Similar to Cedar Waxwing but larger and grayer, with *no yellow on belly;* wings with strong white or *white and yellow* markings. Note *deep rusty* undertail coverts. Juvenile larger, grayer than juvenile Cedar. Often travels in large nomadic flocks. **VOICE:** *Zrreee,* rougher than thin note of Cedar Waxwing. **RANGE:** Irruptive, occurring well south of range in some winters. **HABITAT:** In summer, boreal forests, muskeg; in winter, widespread, including towns with berries and fruiting trees.

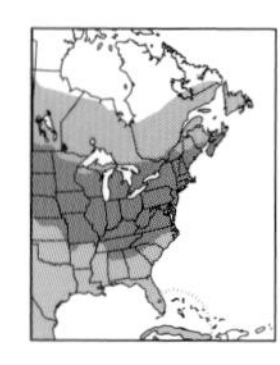

CEDAR WAXWING *Bombycilla cedrorum* **Common**

7¼ in. (18 cm). A sleek, crested, brown bird, note *yellow band* at tip of tail. *Juvenile:* Grayish olive-brown, with blurry streaks below. Waxwings are gregarious in fall/winter, flying and feeding in compact flocks in search of berries. **VOICE:** A high, thin lisp or *zeee;* sometimes slightly trilled; rather constantly given. **SIMILAR SPECIES:** Bohemian Waxwing. **HABITAT:** Open woodlands, streamside willows and alders, orchards; in winter, widespread, including towns, fruiting trees and bushes. Nomadic.

JAYS, CROWS, and ALLIES Family Corvidae

Large perching birds with strong, longish bill, nostrils covered by forward-pointing bristles. Jays are often blue. Magpies are black and white, with long tails. Crows and ravens are large and black. Sexes alike. First-year birds of most species resemble adults. **FOOD:** Almost anything edible. **RANGE:** World-wide except s. S. America, Antarctica, Oceania.

FLORIDA SCRUB-JAY **Uncommon, local, threatened**
Aphelocoma coerulescens

11–11¼ in. (29 cm). Look for this *crestless* jay in FL in stretches of oak scrub. Blue above except for pale gray back and sides; whitish throat. **VOICE:** A rough, rasping *kwesh . . . kwesh.* Also a low, rasping *zhreek* or *zhrink.* **SIMILAR SPECIES:** Blue Jay has crest and bold white spotting on wings and tail. **HABITAT:** Mainly scrub, low oaks.

WOODHOUSE'S SCRUB-JAY **Fairly common, local**
Aphelocoma woodhouseii

11–11¼ in. (29 cm). Similar to Florida Scrub-Jay but duller; throat and sides grayer, not contrasting as much with back; face plainer; note lack of blue extending down sides of breast. **VOICE:** Similar to Florida Scrub-Jay. **HABITAT:** Riparian and oak woodlands, pinyon-juniper, residential areas, parks.

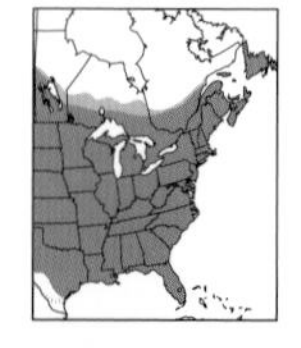

BLUE JAY *Cyanocitta cristata* **Common**

11 in. (28 cm). A showy, noisy, infamous, *crested jay;* larger than an American Robin. Bold *white spots on wings and tail;* whitish and dull gray underparts; *black necklace.* **VOICE:** A harsh, slurring *jeeah* or *jay;* a musical *queedle, queedle;* many other notes. Mimics calls of Red-shouldered and Red-tailed Hawks. **SIMILAR SPECIES:** Florida and Woodhouse's scrub-jays. **HABITAT:** Woodlands, suburban gardens, groves, towns, feeders.

WAXWINGS
AND JAYS
BOHEMIAN
WAXWING
CEDAR
WAXWING
juvenile
FLORIDA
SCRUB-JAY
WOODHOUSE'S
SCRUB-JAY
BLUE JAY

CANADA JAY *Perisoreus canadensis* **Uncommon**

11¼–11½ in. (28–29 cm). A large, fluffy, gray bird of cool northern forests; larger than an American Robin. Formerly known as Gray Jay, and nicknamed "Whiskey Jack." *Adult: Black* patch or partial cap across back of head and *white forehead* (or crown); suggests a huge overgrown chickadee. *Juvenile: Dark sooty,* almost blackish; only distinguishing mark is *whitish whisker.* **VOICE:** A soft *whee-ah;* also many other notes, some harsh. **HABITAT:** Spruce and fir forests. Becomes tame around campgrounds, picnic areas.

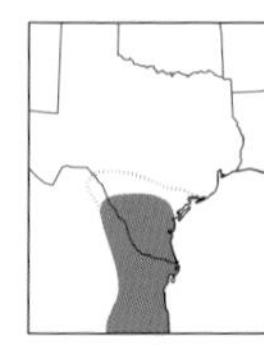

GREEN JAY *Cyanocorax yncas* **Fairly common, local**

10½ in. (27 cm). Unmistakable. The *only green-colored jay. Black throat, violet crown.* Often seen in noisy flocks. **VOICE:** Four or more harsh notes given rapidly: *cheek, cheek, cheek, cheek.* Also a variety of jaylike croaks and squeaks. **HABITAT:** Dense cover in scrubby woods. Visits feeders for fruit and seeds.

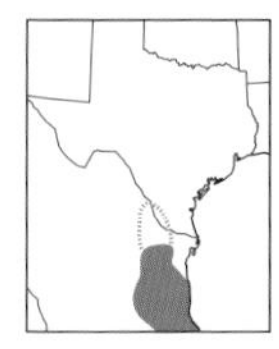

BROWN JAY *Psilorhinus morio* **Rare, local**

16½–17 in. (42–43 cm). A very large jay with *brown upperparts and pale belly.* Adult has dark bill; juvenile has yellow bill. In flight, pale belly stands out. **VOICE:** A very loud *chaa-chaa-chaa* repeated over and over. Flocks can make a loud noise. **HABITAT:** Dense scrub and brushy woods. Observations along Rio Grande, TX, have recently declined.

BLACK-BILLED MAGPIE *Pica hudsonia* **Fairly common, local**

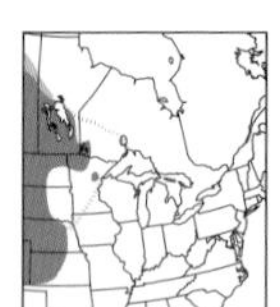

18½–19½ in. (47–49 cm); tail 9½–12 in. (24–30 cm). A large, slender, *black-and-white bird* with *long, graduated tail.* In flight, iridescent greenish black tail streams behind and large *white patches flash in wings.* **VOICE:** A harsh, rapid *queg queg queg queg* or *wah-wah-wah.* Also a querulous, nasal *maag?* or *aag-aag?* **RANGE:** Casual vagrant east of range, accidentally to East Coast. **HABITAT:** Rangeland, brushy country, conifers, streamsides, forest edges, farms. Often in flocks.

CANADA JAY
adult
juvenile
GREEN JAY
BROWN
JAY
BLACK-BILLED
MAGPIE

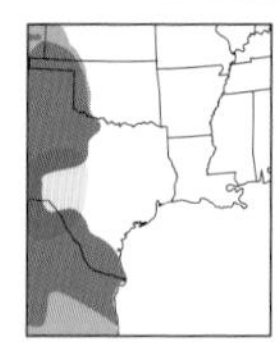

CHIHUAHUAN RAVEN *Corvus cryptoleucus* **Fairly common**

19–19½ in. (48–50 cm). A small raven of arid plains and deserts. White feather bases on neck and breast sometimes show when feathers are ruffled by the wind, hence former name White-necked Raven. **VOICE:** A hoarse *kraak,* flatter and higher than Common Raven. **SIMILAR SPECIES:** Difficult to tell from Common Raven, which is slightly larger with tail slightly more wedge-shaped, calls lower pitched, and bristles not extending as far down upper mandible. Bases of Common Raven's feathers are grayish. **HABITAT:** Arid and semiarid scrub and grasslands, deserts, yucca, mesquite, towns, dumps.

COMMON RAVEN *Corvus corax* **Common**

23½–24 in. (59–61 cm). Note long, *wedge-shaped tail.* Much larger than American Crow; has heavier voice and not inclined to be as gregarious, often solitary or in family groups. More hawklike in flight, sailing and gliding on flat, somewhat sweptback wings. Note "goiter" look created by shaggy throat feathers, and heavier "Roman-nose" bill. **VOICE:** A croaking *cr-r-ruck* or *prruk;* also a metallic *tok.* **SIMILAR SPECIES:** Chihuahuan Raven. American Crow much smaller and (especially) smaller-billed, tail shorter and rounded (not wedge-shaped). **HABITAT:** Boreal and mountain forests, desert lowlands (particularly in winter), cliffs, tundra, towns, dumps.

TAMAULIPAS CROW *Corvus imparatus* **Rare, local**

14¼–14½ in. (36–37 cm). A small crow with small bill, long tail, and slim wings. Color glossier than in other crows. **VOICE:** A "stressed," harsh, froglike *awwwk.* **SIMILAR SPECIES:** In its range this is the only small crow; next larger all-black corvid is Chihuahuan Raven. See also Great-tailed Grackle. **RANGE:** Found irregularly near Brownsville, TX; has recently declined. **HABITAT:** Arid scrub, mesquite thickets; also ranches, dumps.

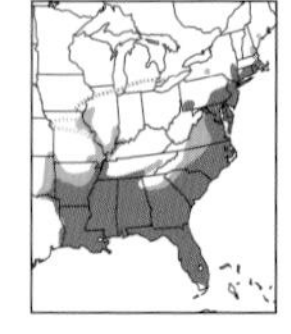

FISH CROW *Corvus ossifragus* **Fairly common**

15¼–15½ in. (38–39 cm). Slightly smaller, *glossier,* and more delicately proportioned than American Crow. Tail slightly longer and wings slightly more tapered. *Best identified by vocalizations.* **VOICE:** A short, nasal, ducklike *two*-syllable *ca-ha;* also song-note *car* or *ca.* American Crow gives a lower-pitched *caw,* lacking nasal quality, although calls of young Americans may sound like Fish Crows. **SIMILAR SPECIES:** American Crow. Tamaulipas Crow does not overlap in range. **HABITAT:** Often near tidewater, river valleys, lakes. Also farm fields, wood edges, towns and cities, dumps.

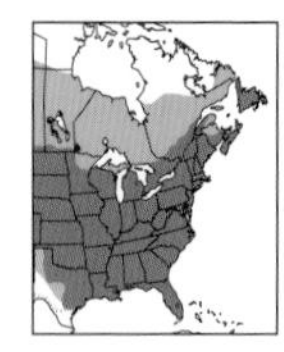

AMERICAN CROW *Corvus brachyrhynchos* **Common**

17–17½ in. (43–45 cm). A large, familiar, chunky, ebony bird. Completely black; slightly glossed with purplish in strong sunlight. Bill and feet strong and black. Often gregarious. Flies with continuous flapping and walks with distinctive gait that differs from that of Common Raven. **VOICE:** A loud *caw, caw, caw* or *cah* or *kahr.* **SIMILAR SPECIES:** Fish Crow. Common Raven larger, has longer wedge-shaped tail (shorter and more rounded in American Crow), more sweptback wings, different call. See also Chihuahuan Raven. **HABITAT:** Woodlands, farms, fields, river groves, shores, towns, dumps.

RAVENS AND CROWS
may show white on nape when feathers are ruffled
CHIHUAHUAN RAVEN
COMMON RAVEN
ravens have wedge-shaped tail
AMERICAN CROW
TAMAULIPAS CROW
FISH CROW
crows have rounded tail
American Crow
Fish Crow
Tamaulipas Crow

CHICKADEES, TITMICE, BUSHTIT, and VERDIN
Families Paridae, Aegithalidae, and Remizidae

Chickadees, titmice, and bushtits form mixed flocks; widespread in Old World. Verdins are more solitary, primarily Asian.

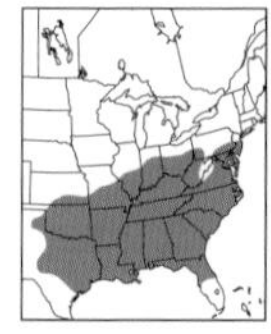

CAROLINA CHICKADEE *Poecile carolinensis* Common

4¾ in. (12 cm). Told from Black-capped Chickadee by range and voice. **VOICE:** *Chick-a-dee* call of this species higher pitched and more rapid than that of Black-capped. Whistled song a four-syllable *fee-bee, fee-bay.* **SIMILAR SPECIES:** Black-capped larger, rear cheek patch *cleaner* and whiter, *wing-covert edging more prominent, sides and flanks pinker*. Black-cappeds penetrate range of Carolinas in some winters. **HABITAT:** Woods, willow thickets, residential areas, feeders.

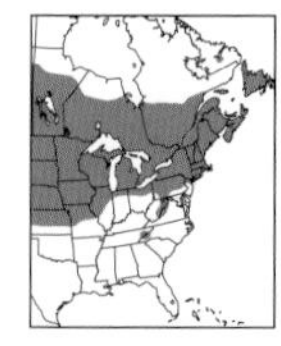

BLACK-CAPPED CHICKADEE *Poecile atricapillus* Common

5–5¼ in. (12–13 cm). A familar species in Northeast. **VOICE:** A clearly enunciated *chick-a-dee-dee-dee.* Song a clear whistle, *fee-bee-ee* or *fee-bee,* first note higher. **SIMILAR SPECIES:** Carolina Chickadee is mostly separated by range. **HABITAT:** Same as Carolina Chickadee.

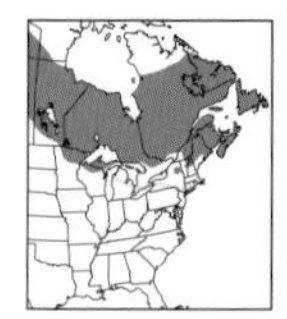

BOREAL CHICKADEE *Poecile hudsonicus* Uncommon

5½ in. (14 cm). Note *dull brown cap,* rich brown to pinkish brown flanks, extensively *grayish cheeks.* **VOICE:** A wheezy *chick-che-day-day,* slower and raspier than Black-capped Chickadee. **SIMILAR SPECIES:** Browner than Carolina and Black-capped Chickadees. **RANGE:** Occasionally moves south in winter. **HABITAT:** Coniferous forests, evergreen plantations.

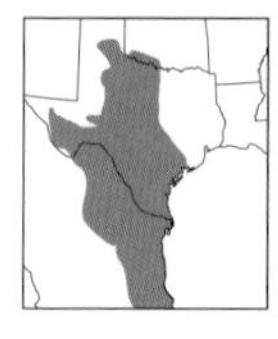

BLACK-CRESTED TITMOUSE *Baeolophus atricristatus* Fairly common

6¼ in. (16 cm). Like Tufted Titmouse, but *crown and crest black,* forehead white, sides paler. Juveniles, usually observed with parents, have mostly gray crest. **VOICE:** Chickadee-like calls. Song a whistled *peter peter peter peter* or *hear hear hear hear.* **SIMILAR SPECIES:** Tufted Titmouse. **HABITAT:** Woodlands, canyons, towns, feeders.

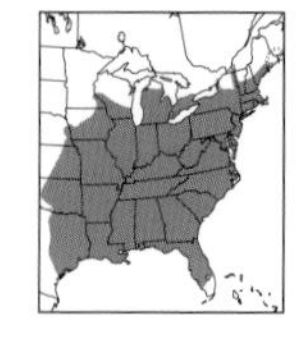

TUFTED TITMOUSE *Baeolophus bicolor* Common

6¼ in. (16 cm). A familiar, *mouse-colored bird with tufted crest. Flanks rusty buff.* Plain face, large black eyes. Inquisitive and vocal. **VOICE:** A clear, whistled *peter, peter, peter* or *here, here, here, here.* Also chickadee-like calls but more drawling, nasal, and complaining. **SIMILAR SPECIES:** Other titmice, chickadees. **HABITAT:** Woodlands, shade trees, groves, residential areas, feeders.

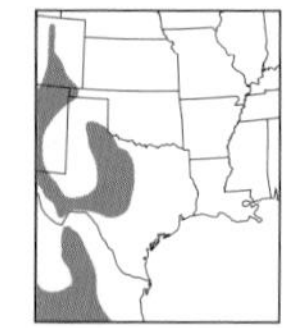

BUSHTIT *Psaltriparus minimus* Common, local

4½ in. (11 cm). A very small, plain, long-tailed bird, often found in *straggling flocks,* conversing in twittering notes. Gray back, pale underparts, brownish crown and cheeks, stubby bill. Adult male and juvenile have dark eyes, female yellow eyes. **VOICE:** Insistent light *tsits*, *pits*, and *clenks* given constantly. **HABITAT:** Mixed woods, parks, residential areas.

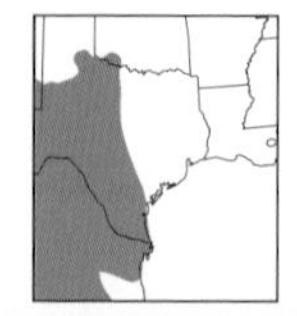

VERDIN *Auriparus flaviceps* Uncommon, local

4½ in. (11 cm). Tiny. Does not flock like Bushtit does. *Adult:* Gray, with *yellowish head, rufous bend of wing* (often hidden). Sexes similar. *Juvenile:* Plain gray. **VOICE:** An insistent *see-lip;* rapid chipping; song a *tsee see-see.* **SIMILAR SPECIES:** Bushtit longer tailed than juvenile Verdin; bill thicker, less sharp. **HABITAT:** Brushy desert and semiarid lowlands, mesquite.

CHICKADEES, TITMICE, BUSHTIT, VERDIN
CAROLINA CHICKADEE
BLACK-CAPPED CHICKADEE
BOREAL CHICKADEE
BLACK-CRESTED TITMOUSE
TUFTED TITMOUSE
BUSHTIT
females have yellow eyes
adult
VERDIN
juvenile

NUTHATCHES Family Sittidae

Small, stubby tree climbers with strong, woodpecker-like bills and strong feet. Short, square-cut tails are not braced like those of woodpeckers or creepers during climbing. Nuthatches habitually go down trees headfirst. Ages and sexes similar or differ slightly. **FOOD:** Bark insects, seeds, nuts; attracted to feeders by suet, sunflower seeds. **RANGE:** Most of N. Hemisphere.

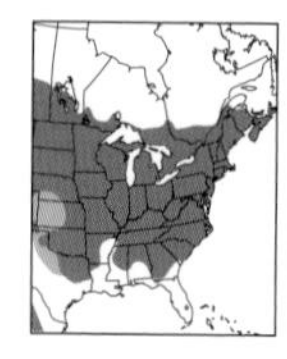

WHITE-BREASTED NUTHATCH *Sitta carolinensis* **Common**

5¾ in. (15 cm). This widespread and familiar nuthatch is known by its *black cap* (gray in female) and beady black eye on white face. Undertail coverts chestnut. **VOICE:** Song a rapid series of low, nasal, whistled notes on one pitch: *whi, whi, whi, whi, whi, whi* or *who, who, who,* etc. Call a distinctive nasal *yank, yank, yank;* also a nasal *tootoo.* **SIMILAR SPECIES:** Red-breasted Nuthatch. **HABITAT:** Forests, woodlots, groves, river woods, shade trees; visits feeders.

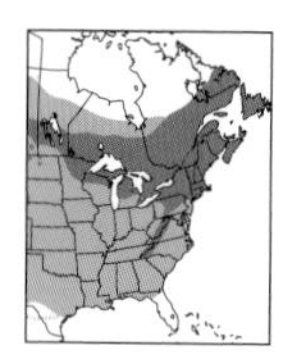

RED-BREASTED NUTHATCH *Sitta canadensis* **Common**

4½ in. (11 cm). A small nuthatch with *broad black line* through eye and white line above it. Crown black in male, gray in female; underparts washed rusty in male, paler in female. First-year birds of each sex duller than adults, crown blackish rather than black in male and duller gray in female. **VOICE:** Call higher, more nasal than White-breasted Nuthatch, a distinctive *ank* or *enk,* sounding like a baby nuthatch or tiny tin horn. **SIMILAR SPECIES:** Brown-headed Nuthatch has gray-brown or brown crown, lacks white supercilium, has very different calls. **RANGE:** Irruptive, sometimes moving well south of range in winter. **HABITAT:** Coniferous forests; in winter, also other trees, feeders.

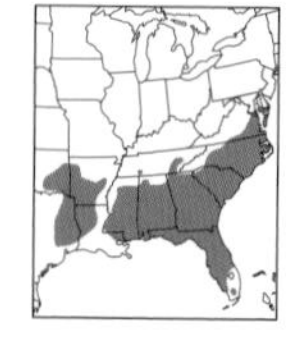

BROWN-HEADED NUTHATCH *Sitta pusilla* **Uncommon**

4½ in. (11 cm). A small nuthatch of southeastern pinelands. Smaller than White-breasted Nuthatch, with brown cap coming down to eye and a usually pale or whitish spot on nape. Travels in groups. Ages and sexes similar. **VOICE:** Sounds like a toy rubber duck: a high, rapid *kit-kit-kit;* also a squeaky piping *ki-day* or *ki-dee-dee,* constantly repeated, sometimes becoming an excited twitter or chatter. **SIMILAR SPECIES:** Other nuthatches, Brown Creeper. **HABITAT:** Open pine woods.

TREE CREEPERS Family Certhiidae

Small, slim, stiff-tailed birds, with slender, slightly curved bills used to probe bark of tree trunks, often from vertical position or even upside down. Ages and sexes alike. **FOOD:** Bark insects. **RANGE:** Cooler parts of N. Hemisphere.

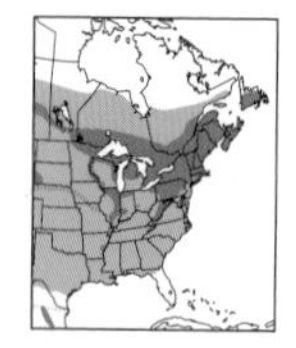

BROWN CREEPER *Certhia americana* **Uncommon**

5¼ in. (13 cm). A very small, slim, camouflaged tree climber. Brown above, whitish below, with *slender decurved bill* and *stiff* tail, which is used as a brace during climbing. Ascends trees spirally from base, hugging bark closely. **VOICE:** Call a single high, thin *seee,* similar to quick three-note call (*see-see-see*) of Golden-crowned Kinglet. Song a high, thin, sibilant *see-ti-wee-tu-wee* or *trees, trees, trees, see the trees.* **HABITAT:** Nests in variety of coniferous and mixed woodlands; in fall and winter, also in deciduous woods, groves, shade trees.

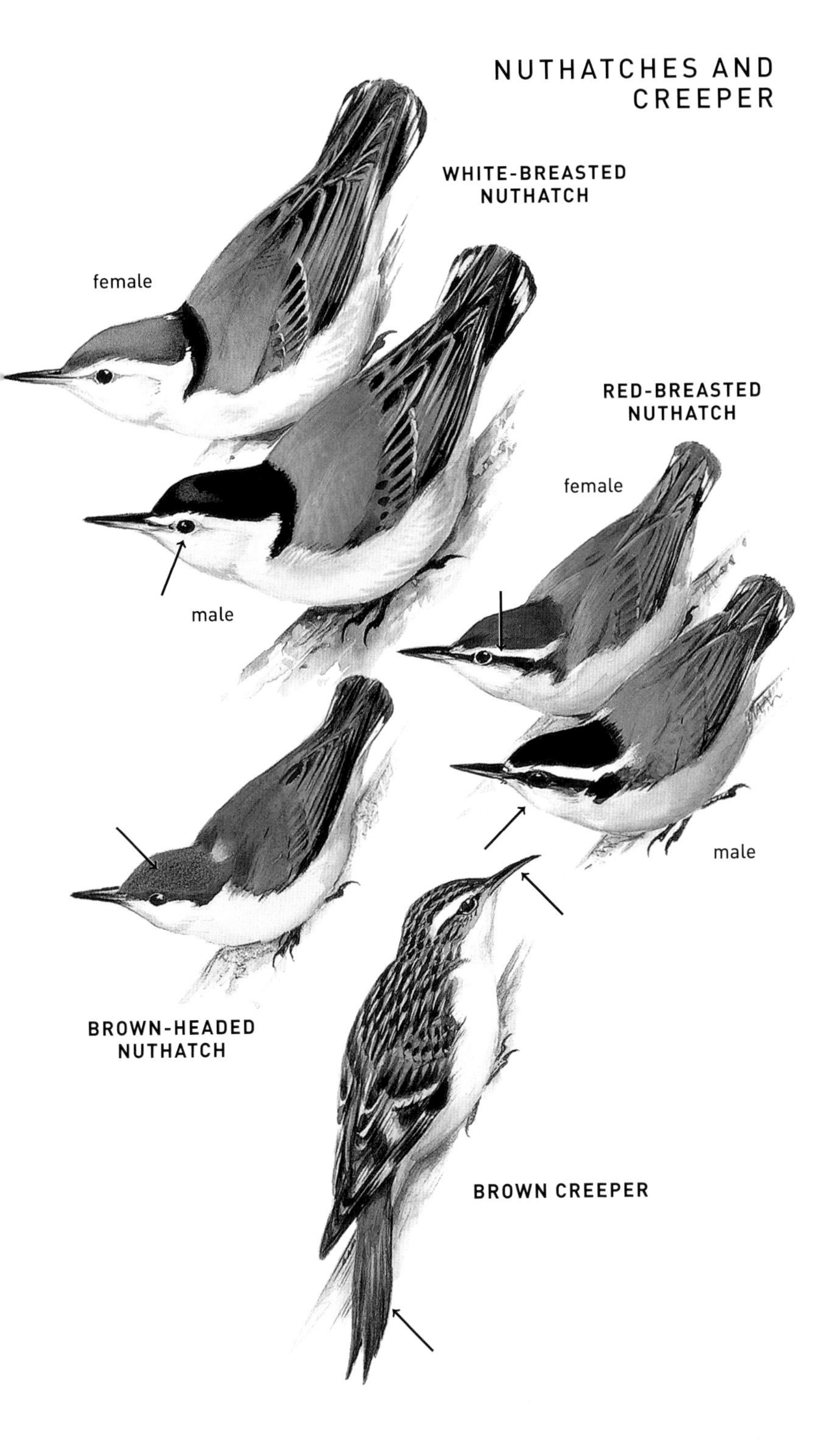
NUTHATCHES AND CREEPER
WHITE-BREASTED NUTHATCH
female
male
RED-BREASTED NUTHATCH
female
male
BROWN-HEADED NUTHATCH
BROWN CREEPER

WRENS Family Troglodytidae

Small brown birds with slim, slightly curved bills; tails often cocked. Songs often pleasing, making up for the drab plumage. Ages and sexes alike. **FOOD:** Insects, spiders. **RANGE:** Throughout Americas; one species in Eurasia.

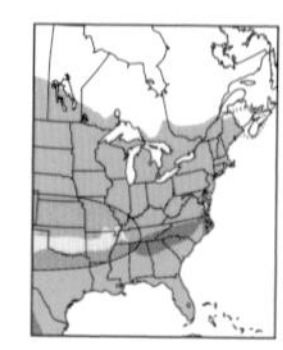

HOUSE WREN *Troglodytes aedon* **Common**

4½–4¾ in. (11–12 cm). A small, energetic, gray-brown wren with long tail, light eye-ring, and no strong eyebrow stripe. A familiar songster in residential areas and on farms. **VOICE:** A stuttering, gurgling song rises in a musical burst, then falls at end; calls a rolled *prrrrr* and harsh *cheh, cheh.* **SIMILAR SPECIES:** Winter Wren, Pacific Wren (see Winter Wren below). **HABITAT:** Open woods, thickets, towns, gardens; often nests in bird boxes.

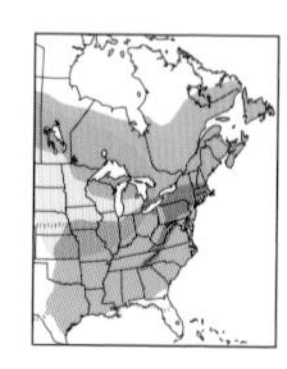

WINTER WREN *Troglodytes hiemalis* **Uncommon**

4 in. (10 cm). Similar to House Wren but smaller, with *much stubbier tail,* stronger eyebrow, and *dark, heavily barred belly.* Mouselike, staying close to ground. **VOICE:** Song a rapid succession of high tinkling warbles, trills. Call a soft, two-syllable *chemp-chemp.* **SIMILAR SPECIES:** House Wren. Recently split Pacific Wren (*T. pacificus;* not shown) of West, a vagrant to Great Plains states, averages darker, more russet; has different voice. **HABITAT:** Dense, shaded woodlands, underbrush, fallen trees.

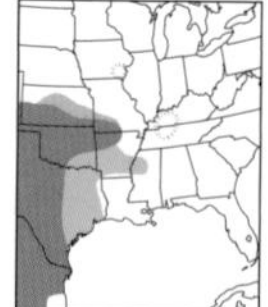

BEWICK'S WREN *Thryomanes bewickii* **Scarce**

5¼ in. (13 cm). Note longish tail with *white corners* and bold *white eyebrow stripe.* Reddish brown above, white below. **VOICE:** Song suggests Song Sparrow but thinner, ending in a trill; calls a sharp *vit, vit* and buzzy *dzzzzzt.* **SIMILAR SPECIES:** Carolina Wren. **RANGE:** More common in TX than farther East. **HABITAT:** Thickets, underbrush.

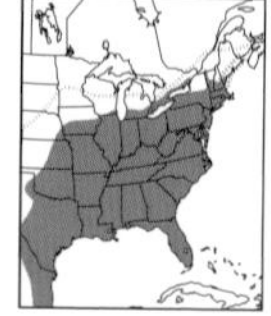

CAROLINA WREN *Thryothorus ludovicianus* **Common**

5½ in. (14 cm). A large, familiar wren, near size of a sparrow. *Warm rusty brown* above, variably buff below; conspicuous *white eyebrow stripe.* Often travels in pairs, near ground, skulking over and under logs and brush for insects. **VOICE:** A two- or three-syllable chant: *tea-kettle, tea-kettle, tea kettle,* or *chirpity, chirpity, chirpity, chirp. Chips* and *churrs.* **SIMILAR SPECIES:** Bewick's and Marsh Wrens. **RANGE:** Casual vagrant west of range. **HABITAT:** Tangles, undergrowth, woods, gardens. Sometimes nests in bird boxes.

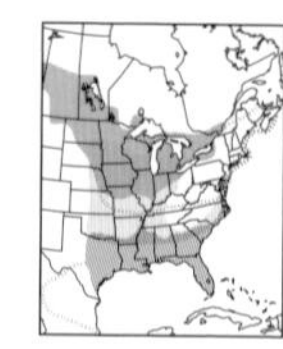

SEDGE WREN *Cistothorus platensis* **Uncommon, secretive**

4½ in. (11 cm). Stubbier than Marsh Wren; buffier, with *buffy* undertail coverts, *barred wings,* and *finely streaked* crown. **VOICE:** Song a dry staccato chattering: *chap chap chap chap chap chap chap chapper-rrrrr.* Call a single or double warblerlike *chap.* **SIMILAR SPECIES:** Smaller than Marsh Wren, crown streaked; supercilium indistinct. **HABITAT:** Grassy and sedgy marshes and meadows.

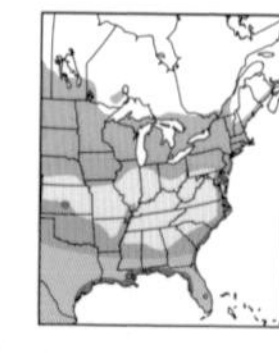

MARSH WREN *Cistothorus palustris* **Fairly common**

5 in. (13 cm). Note *white stripes on back* and white eyebrow stripe, unstreaked dark crown. **VOICE:** Song reedy, gurgling, *cut-cut-turrrrrrrr-ur,* often ending in rattle; may sing at night. Call a low *tsuck-tsuck.* **SIMILAR SPECIES:** Sedge Wren. **HABITAT:** Fresh and brackish cattail and bulrush marshes; in winter, also salt marshes. Rare outside marsh habitats during migration.

HOUSE
WREN
WINTER
WREN
BEWICK'S
WREN
CAROLINA
WREN
SEDGE
WREN
MARSH
WREN

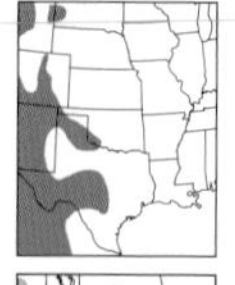

CANYON WREN *Catherpes mexicanus* — Uncommon, local

5¾–6 in. (15 cm). Primarily rusty with *white bib.* Climbs on rocks. **VOICE:** Gushing downward song; *tee tee tee tee tew tew tew tew.* Call a shrill *beet.* **RANGE:** Casual vagrant east of range. **HABITAT:** Cliffs, canyons.

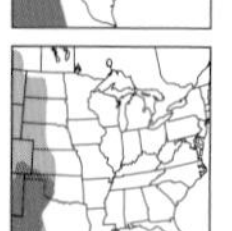

ROCK WREN *Salpinctes obsoletus* — Uncommon

6 in. (15 cm). Note *finely streaked breast, buffy terminal tail band.* Frequently bobs. **VOICE:** Song thrasherlike phrases and buzzy trills; call a loud *ti-keer.* **RANGE:** Accidental to East Coast. **HABITAT:** Rocky slopes, rubble.

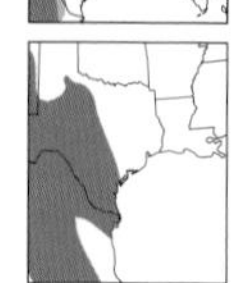

CACTUS WREN *Campylorhynchus brunneicapillus* — Uncommon, local

8½ in. (22 cm). A large wren; note *heavy spotting, clustered on upper breast.* **VOICE:** A monotonous *chug-chug-chug-chug,* on one pitch, gaining speed. **HABITAT:** Arid areas, mesquite, yucca.

KINGLETS Family Regulidae

Tiny, active birds with small slender bills, short tails, bright red crowns in males. Often found in mixed-species flocks. Ages alike, sexes differ. **FOOD:** Insects, larvae. **RANGE:** Eurasia, N. America.

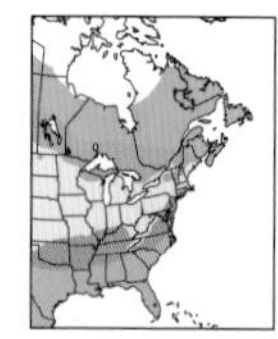

RUBY-CROWNED KINGLET *Regulus calendula* — Common

4¼ in. (11 cm). Tiny, olive-gray, smaller than warblers, *flicks wings constantly.* Note broken white eye-ring, bold wing bars bordered behind by *black bar.* Male has *scarlet crown patch;* female lacks red. **VOICE:** A husky *ji-dit.* Song high notes, lower notes, and a chant, *tee tee tee-tew tew tew—ti-didee, ti-didee, ti-didee.* **SIMILAR SPECIES:** Golden-crowned Kinglet. **HABITAT:** In summer, coniferous forests; in migration and winter, variety of other habitats, residential areas.

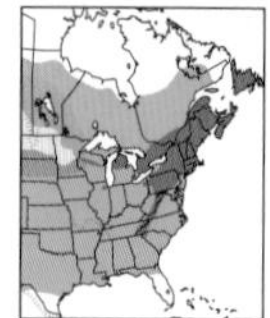

GOLDEN-CROWNED KINGLET *Regulus satrapa* — Fairly common

4 in. (10 cm). Note tiny size, *boldly striped face,* wing bars. Male but not female has red in crown. **VOICE:** A high, wiry *see-see-see.* Song a series of high thin notes, ending in a little chatter. **SIMILAR SPECIES:** Ruby-crowned Kinglet. **HABITAT:** Conifers; in winter, other trees.

GNATCATCHERS Family Polioptilidae

Active birds with slender bills. Gnatcatchers have long, mobile tails, often flipped and cocked. **FOOD:** Insects, larvae. **RANGE:** Throughout Americas.

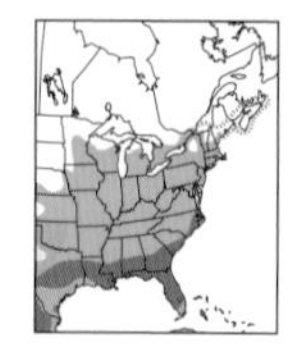

BLUE-GRAY GNATCATCHER *Polioptila caerulea* — Fairly common

4½ in. (11 cm). Tiny, slim, blue-gray above, whitish below, with narrow *white eye-ring. Long tail* is *mostly white underneath.* Adults and males average brighter than first-year birds and females; males acquire black line above eye in spring/summer. **VOICE:** Call a thin, peevish *zpee;* often doubled, *zpee-zee.* Song thin and squeaky. **SIMILAR SPECIES:** Black-tailed Gnatcatcher. **HABITAT:** Swamps, woods, brusy habitats.

BLACK-TAILED GNATCATCHER *Polioptila melanura* — Uncommon, local

4½ in. (11 cm). Similar to Blue-gray Gnatcatcher, but underside of tail *largely black.* Spring/summer male has *black cap;* in winter has black mark above eye. **VOICE:** Thin harsh *chee,* repeated two or three times; soft *chip-chip-chip* series. **HABITAT:** Dry washes, mesquite.

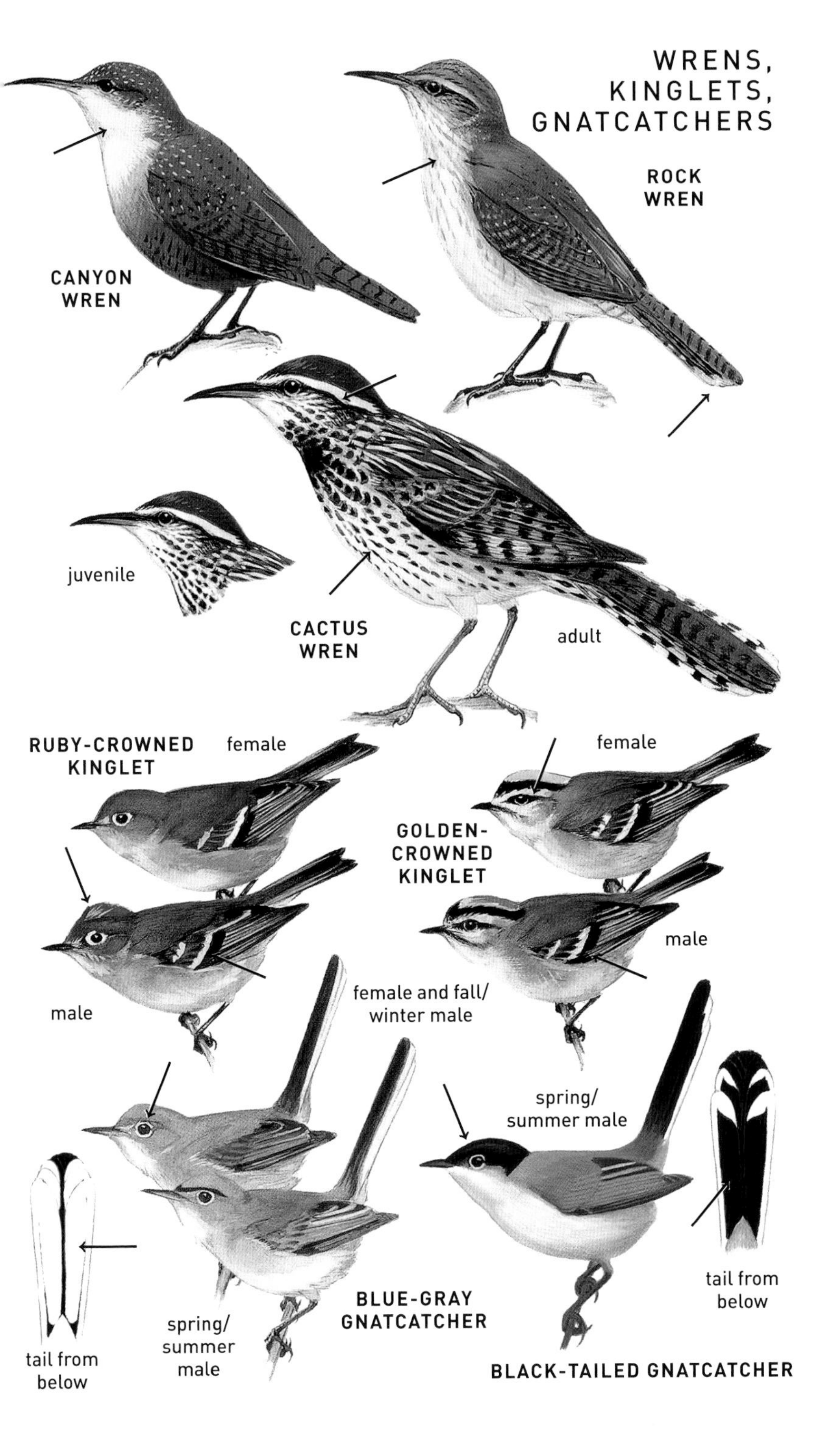
WRENS, KINGLETS, GNATCATCHERS
CANYON WREN
ROCK WREN
juvenile
CACTUS WREN
adult
RUBY-CROWNED KINGLET
female
male
GOLDEN-CROWNED KINGLET
female
male
female and fall/ winter male
spring/ summer male
tail from below
BLUE-GRAY GNATCATCHER
spring/ summer male
tail from below
BLACK-TAILED GNATCATCHER

SHRIKES Family Laniidae

Fierce songbirds, with hook-tipped bills, that perch watchfully on bush tops, treetops, wires; often impale prey on thorns, barbed wire. **FOOD:** Insects, lizards, small rodents, small birds. **RANGE:** Widespread in Old World; two species breed in N. America.

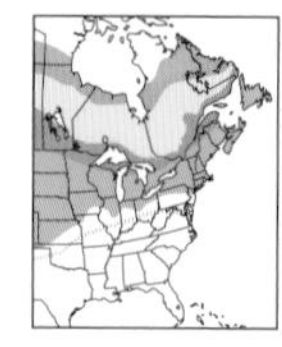

NORTHERN SHRIKE *Lanius borealis* — Scarce

10–10¼ in. (25–26 cm). Similar to Loggerhead Shrike but larger, adults paler; note *narrower dark mask with more white around eye, faintly barred* breast, and longer, more hooked bill with *pale base. Juvenile: Brown,* with weak mask and extensive *fine barring* below. Becomes mottled gray in first year. **VOICE:** Song a disjointed, thrasherlike succession of harsh and musical notes. Call *shek-shek;* a grating *jaaeg.* **SIMILAR SPECIES:** Loggerhead Shrike, Northern Mockingbird. **RANGE:** Sometimes found well south of normal range. **HABITAT:** Semiopen country, taiga, muskeg, tundra.

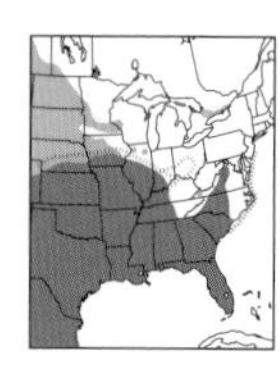

LOGGERHEAD SHRIKE *Lanius ludovicianus* — Uncommon to rare

9 in. (23 cm). Big head, slim tail; gray, black, and white, with *black mask, short hooked bill.* Flies low with flickering shallow flight, showing white patches. *Juvenile:* Has faint barring below *briefly in late summer.* **VOICE:** Song of deliberate notes and phrases, repeated 3–20 times; *queedle, queedle, tsurp-see, tsurp-see,* etc. Call *shack shack* or *jeeer jeeer.* **SIMILAR SPECIES:** Northern Shrike, Northern Mockingbird. **HABITAT:** Semiopen country, wires, fences, trees, shrubs.

VIREOS Family Vireonidae

Small olive- or gray-backed birds, less active than wood-warblers; bills slightly thicker, with more curved ridge and small hook to tip. Ages and sexes usually similar. **FOOD:** Mostly insects, also fruit in winter. **RANGE:** Canada to Argentina.

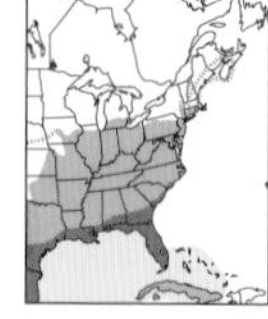

WHITE-EYED VIREO *Vireo griseus* — Fairly common

5 in. (13 cm). Distinctive combination of *yellow spectacles, whitish throat.* Also note wing bars, yellowish sides, white eye (dark through first fall). Somewhat skulking. **VOICE:** Song a sharp *CHICK-a-per-weeoo-CHICK;* variable, usually starts and ends with *chick.* **SIMILAR SPECIES:** Bell's Vireo. **HABITAT:** Wood edges, brush, brambles, dense undergrowth.

BELL'S VIREO *Vireo bellii* — Scarce

4¾ in. (12 cm). Small, nondescript. Usually stays concealed in dense cover. Thin, pale, broken eye-ring and loral stripe. One or two weak wing bars. **VOICE:** Husky phrases at short intervals: *cheedle cheedle chee? cheedle cheedle chew!* **SIMILAR SPECIES:** Warbling Vireo has plain wings, bold eyebrow. First-fall White-eyed Vireo has bolder wing bars, yellow lores. **RANGE:** Casual vagrant to East Coast. **HABITAT:** Willows, streamsides, mesquite.

BLACK-CAPPED VIREO *Vireo atricapilla* — Scarce, local, endangered

4½ in. (11 cm). Small and sprightly; cap *glossy black* in adult male, slate gray in first-fall male and female; first-spring male acquires mottled black-and-gray cap. Note wing bars, white spectacles, *red* eyes. **VOICE:** Song hurried, angry, restless phrases. Call a harsh *chit-ah.* **SIMILAR SPECIES:** Blue-headed Vireo larger and with dark eyes. **RANGE:** Accidental to East Coast. **HABITAT:** Oak scrub, brushy hills, rocky canyons.

SHRIKES AND VIREOS

VIREOS

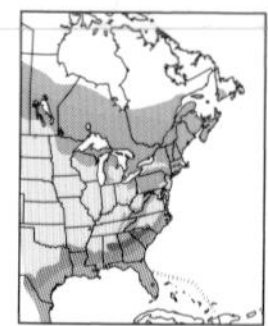

BLUE-HEADED VIREO *Vireo solitarius* **Fairly common**

5¼ in. (14 cm). Note *sharply demarcated* blue-gray cap, *bright white* spectacles and throat, *bright green* back, yellowish wash to side. **VOICE:** Song of sweet but burry phrases with deliberate pauses: *wee-ay, chweeo, chuweep* (slower than Red-eyed Vireo). Also a whiny chatter. **HABITAT:** Coniferous, mixed, and deciduous woods.

YELLOW-THROATED VIREO *Vireo flavifrons* **Fairly common**

5½ in. (14 cm). Bright yellow throat, yellow spectacles, and white wing bars. Olive back contrasts with gray rump. **VOICE:** Song similar to Blue-headed Vireo but lower pitched and burrier*: ee-yay, three-eight.* **SIMILAR SPECIES:** See Pine Warbler. **HABITAT:** Deciduous woodlands, shade trees, particularly oaks.

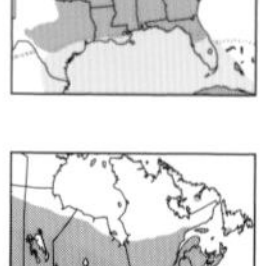

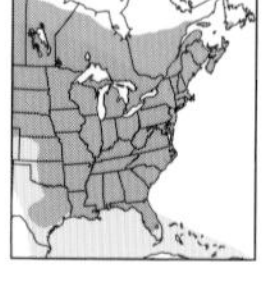

RED-EYED VIREO *Vireo olivaceus* **Common**

6 in. (15 cm). Note *gray cap* contrasting with olive back, and strong, *black-bordered white eyebrow stripe (supercilium).* Red iris of adult is brown in first-fall birds of this species and in Yellow-green and Black-whiskered Vireos. **VOICE:** Song repeated, robinlike phrases, monotonous. Call a nasal, whining *chway.* **SIMILAR SPECIES:** Warbling Vireo smaller, duller, with pale lores and dark brown eyes. See Yellow-green and Black-whiskered Vireos. **HABITAT:** Deciduous woodlands, shade trees, groves.

YELLOW-GREEN VIREO *Vireo flavoviridis* **Very rare, local**

6–6¼ in. (15–16 cm). Similar to Red-eyed Vireo but has *yellow tones* on sides, flanks, and undertail coverts; back *yellower* green; head stripes *less distinct;* bill slightly *longer* and paler. **VOICE:** Song slower than Red-eyed. **SIMILAR SPECIES:** First-fall Red-eyed may have yellow on flanks and undertail coverts. **RANGE:** Rare summer resident in lower Rio Grande Valley, TX; accidental farther East. **HABITAT:** Deciduous woods.

BLACK-WHISKERED VIREO *Vireo altiloquus* **Uncommon, local**

6¼ in. (16 cm). Narrow dark whisker on each side of throat. Otherwise similar to Red-eyed Vireo but brownish olive above, slightly longer bill. **VOICE:** Song slightly slower than Red-eyed. **RANGE:** Casual vagrant to n. Gulf Coast states. **HABITAT:** Mangroves, subtropical hardwoods.

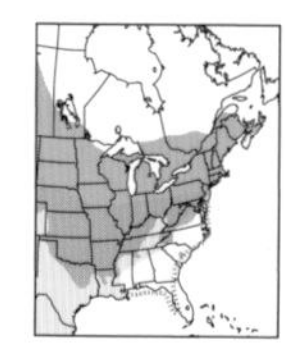

WARBLING VIREO *Vireo gilvus* **Uncommon to fairly common**

5½ in. (14 cm). Note *whitish breast, pale lores,* and *lack of black borders* on eyebrow stripe. First-fall birds have more yellow on sides. **VOICE:** Song a languid warble, suggesting Purple Finch song, but slightly burrier, less spirited. Call a querulous *twee* and short *vit.* **SIMILAR SPECIES:** Philadelphia Vireo rounder headed, smaller billed, and has "cuter" look; yellowish concentrated on throat and breast rather than sides; lores darker. Red-eyed Vireo larger, greener above, eyebrow stripe bolder. **HABITAT:** Deciduous and mixed woods, riparian woodlands, shade trees.

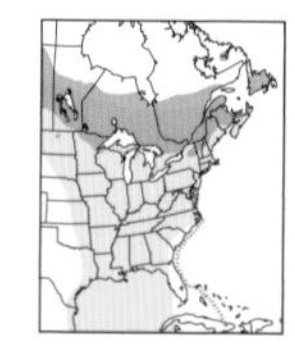

PHILADELPHIA VIREO *Vireo philadelphicus* **Uncommon**

5¼ in. (13 cm). Note more distinct dark eye line (including lores), imparting "cuter" look than Warbling Vireo, and greener back. Underparts vary from showing a small wash of pale yellow to more extensive brighter yellow in first-fall birds (p. 286). **VOICE:** Song higher and slower than Red-eyed Vireo. Call a quick, husky *niff-niff-niff-niff.* **SIMILAR SPECIES:** Bright Warbling Vireos in fall. Tennessee Warbler smaller, has finer bill, white undertail coverts, blackish legs. **HABITAT:** Second-growth woodlands, poplars, willows, alders.

BLUE-HEADED VIREO
YELLOW-THROATED VIREO
adult
RED-EYED VIREO
first-fall
adult
YELLOW-GREEN VIREO
adult
BLACK-WHISKERED VIREO
first-fall
adult
WARBLING VIREO
spring/summer
fall/winter
PHILADELPHIA VIREO

WOOD-WARBLERS Family Parulidae

Popular, active, brightly colored birds, smaller than sparrows, with thin bills. The majority have some yellow in plumage. Generally forest dwelling, being found anywhere from treetops to midstory to ground level. Ages and sexes usually differ. **FOOD:** Mainly insects, though many species also eat fruit in fall and winter. **RANGE:** N. America to n. Argentina.

BACHMAN'S WARBLER — Almost certainly extinct
Vermivora bachmanii

4¾ in. (12 cm). Bill thin and *downcurved. Male: Face and underparts yellow;* bib and crown black. *Female:* Lacks black bib; forehead yellow; crown and cheek grayish; eye-ring yellow. First-year birds average duller than adults within each sex. **VOICE:** Song a rapid series of flat mechanical buzzes rendered on one pitch: *bzz-bzz-bzz-bzz-bzz-bzz-bzz-bzz.* **SIMILAR SPECIES:** Female might recall female Yellow and Nashville Warblers. **RANGE:** Former resident of Southeast; last definite record in 1962. **HABITAT:** Swampy areas, canebrakes.

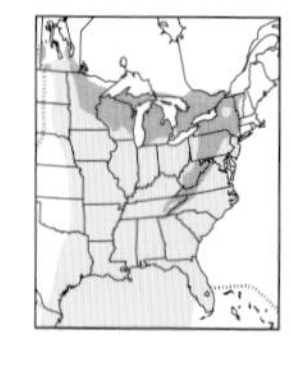

GOLDEN-WINGED WARBLER *Vermivora chrysoptera* — Uncommon

4¾ in. (12 cm). Gray above, white below. *Male:* The only warbler with combination of *yellow wing patch* and *black throat.* Note yellow forecrown, black *ear patch,* whitish underparts. *Female:* Ear and throat patches grayer. First-year birds average duller than adults within each sex. **VOICE:** Song a buzzy note followed by one to three on a lower pitch: *bee-bz-bz-bz.* Call like Blue-winged Warbler. **SIMILAR SPECIES:** See "Brewster's," "Lawrence's," and Blue-winged Warblers. **HABITAT:** Open woodlands, swampy edges, brushy clearings, undergrowth. Declining in many northeastern and southern areas.

"BREWSTER'S" AND "LAWRENCE'S" WARBLERS — Scarce

Golden-winged and Blue-winged Warblers hybridize where their ranges overlap, producing two basic types, "Brewster's" and "Lawrence's" Warblers. "Brewster's" is more variable, typically showing whitish underparts; some can have white or yellow wing bars and some are tinged yellow below. "Lawrence's" is typically bright yellow below like Blue-winged, but with black head pattern of Golden-winged. Hybrid combinations can also resemble parent species with hints of the other species. Females and first-year males duller, adding to the variation in these hybrids. **VOICE:** May sing like either parent. **HABITAT:** Same as Blue-winged and Golden-winged Warblers.

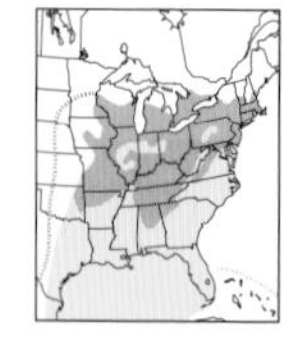

BLUE-WINGED WARBLER *Vermivora cyanoptera* — Fairly common

4¾ in. (12 cm). Note *narrow black line through eye.* Face and underparts yellow; wings *have two white bars.* Female averages duller than male, especially in crown, and first-year birds average duller than adults within each sex. **VOICE:** Song a buzzy *beeee-bzzz,* as if inhaled and exhaled. Call a sharp *tsik.* **SIMILAR SPECIES:** "Brewster's," "Lawrence's," Prothonotary, Golden-winged, and Yellow Warblers. **HABITAT:** Field edges, undergrowth, bushy edges, woodland openings.

female
BACHMAN'S
WARBLER
male
female
GOLDEN-WINGED
WARBLER
male
male
"BREWSTER'S"
WARBLER
"LAWRENCE'S"
WARBLER
male
BLUE-WINGED
WARBLER
male

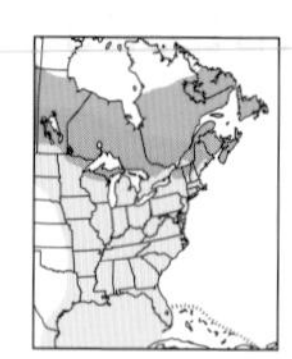

TENNESSEE WARBLER

Uncommon to fairly common

Oreothlypis peregrina

4¾ in. (12 cm). Note short tail, *bold eyebrow, white undertail coverts. Spring/summer male: Pale bluish gray head contrasting with greenish back. Female and first-fall male:* Washed with greenish on head, yellow on breast; often showing trace of a single wing bar. First-fall females duller (p. 286). **VOICE:** Song staccato, three-parted: *ticka ticka ticka ticka, swit swit, chew-chew-chew-chew-chew.* Call a sweet *chip.* **SIMILAR SPECIES:** Orange-crowned Warbler. Warbling and Philadelphia Vireos larger and thicker-billed, duller on back. **HABITAT:** Deciduous and mixed forests; in migration, a variety of woodlands.

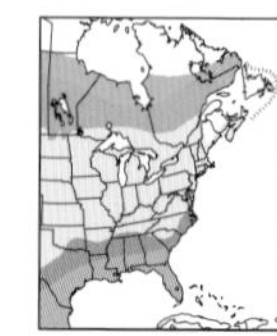

ORANGE-CROWNED WARBLER

Uncommon to fairly common

Oreothlypis celata

5 in. (13 cm). Drab *olive green* with *yellow undertail coverts* and *blurry breast streaking.* The subspecies in e. N. America (*O. c. celata*) is gray-headed; Rocky Mountain subspecies (*O. c. orestera*), rare migrant through Great Plains and accidental further east, is brighter and yellower. Adult males brighter than first-year males and females, with first-year females dullest (see p. 286). Orange crown patch largest in males but seldom visible; can be absent in first-year females. **VOICE:** Song a colorless trill, becoming weaker toward end. Call a sharp *stik.* **SIMILAR SPECIES:** Fall/winter Tennessee Warbler brighter green, with white undertail coverts, shorter tail. **HABITAT:** Open woodlands, brushy clearings, willows.

NASHVILLE WARBLER *Oreothlypis ruficapilla*

Uncommon

4¾ in. (12 cm). Note *white eye-ring* in combination with *yellow* throat. *Head gray,* contrasting with olive-green back. No wing bars. Underparts bright yellow with white vest. Adults and males brighter than first-year birds and females. Regularly bobs tail. **VOICE:** Song two-parted: *seebit, seebit, seebit, seebit, tititititi* (ends like Chipping Sparrow song). Call a sharp *pink.* **SIMILAR SPECIES:** Connecticut Warbler larger, behaves very differently, has grayish or brownish throat. Some dull first-year female Nashvilles (see p. 286) can look almost as dull as Virginia's Warbler but always have *yellow on throat.* **HABITAT:** Open mixed woods with undergrowth, forest edges, bogs; in migration, also brushy areas.

VIRGINIA'S WARBLER *Oreothlypis virginiae*

Casual vagrant

4¾ in. (12 cm). *Male:* A slim *gray* warbler with *yellowish rump* and *bright yellow undertail coverts, white eye-ring,* rufous patch in crown (usually concealed), and touch of yellow on breast. Flicks or jerks tail. *Females and first-year male:* Duller; can lack yellow on breast but always have *yellow undertail coverts.* **VOICE:** Call a sharp *pink,* like Nashville Warbler. **RANGE AND HABITAT:** Casual vagrant to East from w. U.S.; on migration, found in brushy habitats, secondary forest.

WARBLERS
first-fall
male
TENNESSEE
WARBLER
spring/
summer
male
Rocky
Mountain
female
eastern
male
ORANGE-CROWNED
WARBLER
NASHVILLE
WARBLER
female
male
female
VIRGINIA'S
WARBLER
male

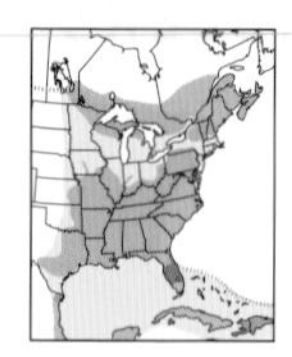

NORTHERN PARULA *Setophaga americana* Fairly common

4½ in. (11 cm). A small, short-tailed warbler, *bluish above,* with yellow throat and breast and two white wing bars, *greenish patch* on back, and *broken white eye-ring.* Bright adult male with *dark breast-band;* first-year female lacks breast-band, has greenish wash on head (see p. 284). **VOICE:** Song a buzzy trill that climbs scale and trips over the top: *zeeeeeeeee-up.* Also *zh-zh-zh-zheeeeee.* **SIMILAR SPECIES:** Tropical Parula. **HABITAT:** Breeds mainly in humid woods where lichen or Spanish moss hangs from trees (occasionally where neither is found).

TROPICAL PARULA *Setophaga pitiayumi* Rare, local

4½ in. (11 cm). Similar in size and habits to Northern Parula but limited in our area to s. TX, near Rio Grande. Dark head and *black face, lacks white eye-ring.* Two bold white wing bars; lacks distinct color bands across chest. Adults and males slightly brighter than first-year birds and female. **VOICE:** Like Northern Parula. **SIMILAR SPECIES:** Northern Parula. **HABITAT:** Oaks, dry forests.

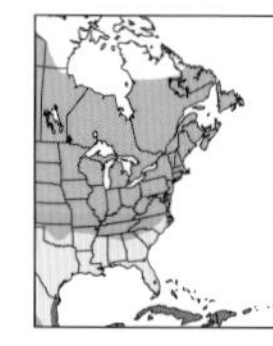

YELLOW WARBLER *Setophaga petechia* Common

5 in. (13 cm). Extensively yellow, with *yellow tail spots* and *yellow edgings to wing and tail.* Male has *rusty breast streaks* (in females and first-fall male these are faint or lacking). Note dark beady eye. First-fall female lacks breast streaks; some individuals may be quite dull, with noticeable eye-ring and brighter yellow restricted to lower vent and undertail coverts (see p. 284). **VOICE:** Song a bright cheerful *tsee-tsee-tsee-tsee-titi-wee* or *weet weet weet weet tsee-tsee wew.* Variable. Call a soft, rich *chip.* **SIMILAR SPECIES:** Wilson's Warbler longer tailed, lacks yellow tail spots. Note vocal differences. **HABITAT:** Riparian woodlands and understory, swamp edges, particularly alders and willows; also parks, gardens.

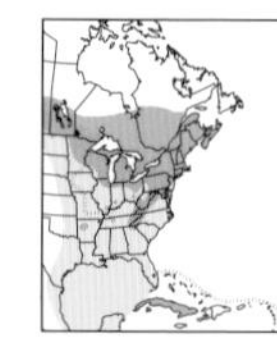

CHESTNUT-SIDED WARBLER *Setophaga pensylvanica* Fairly common

5 in. (13 cm). Usually holds tail cocked up at an angle. *Spring/summer:* Identified by combination of *yellow crown, chestnut sides.* Males brighter than females in spring. In fall, chestnut on sides lacking or reduced, upperparts plainer lime greenish with narrow white eye-ring and *two pale yellow* wing bars (p. 284). **VOICE:** Song similar to Yellow Warbler: *see see see see Miss BEECHer* or *please please pleased to MEETcha,* last note dropping. Call a rich *chip.* **HABITAT:** Undergrowth, overgrown field edges, small trees.

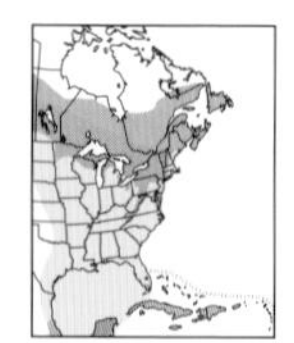

MAGNOLIA WARBLER *Setophaga magnolia* Fairly common

5 in. (13 cm). *Spring/summer male:* Upperparts blackish, with large white patches on wings and tail; underparts yellow, with heavy black stripes. *Female and fall/winter male:* Duller. First-fall female has weak stripes on sides and thin, weak grayish band across upper breast, but tail pattern distinctive (p. 284). In all ages and sexes, black tail crossed midway by *broad white band.* **VOICE:** Song suggests Yellow Warbler but shorter: *weeta weeta weetsee* (last note rising); or a Hooded Warbler–like *weeta weeta wit-chew.* Call an odd nasal note. **SIMILAR SPECIES:** Yellow-rumped and Black-throated Green Warblers. **HABITAT:** Low conifers; in migration, a variety of woodlands.

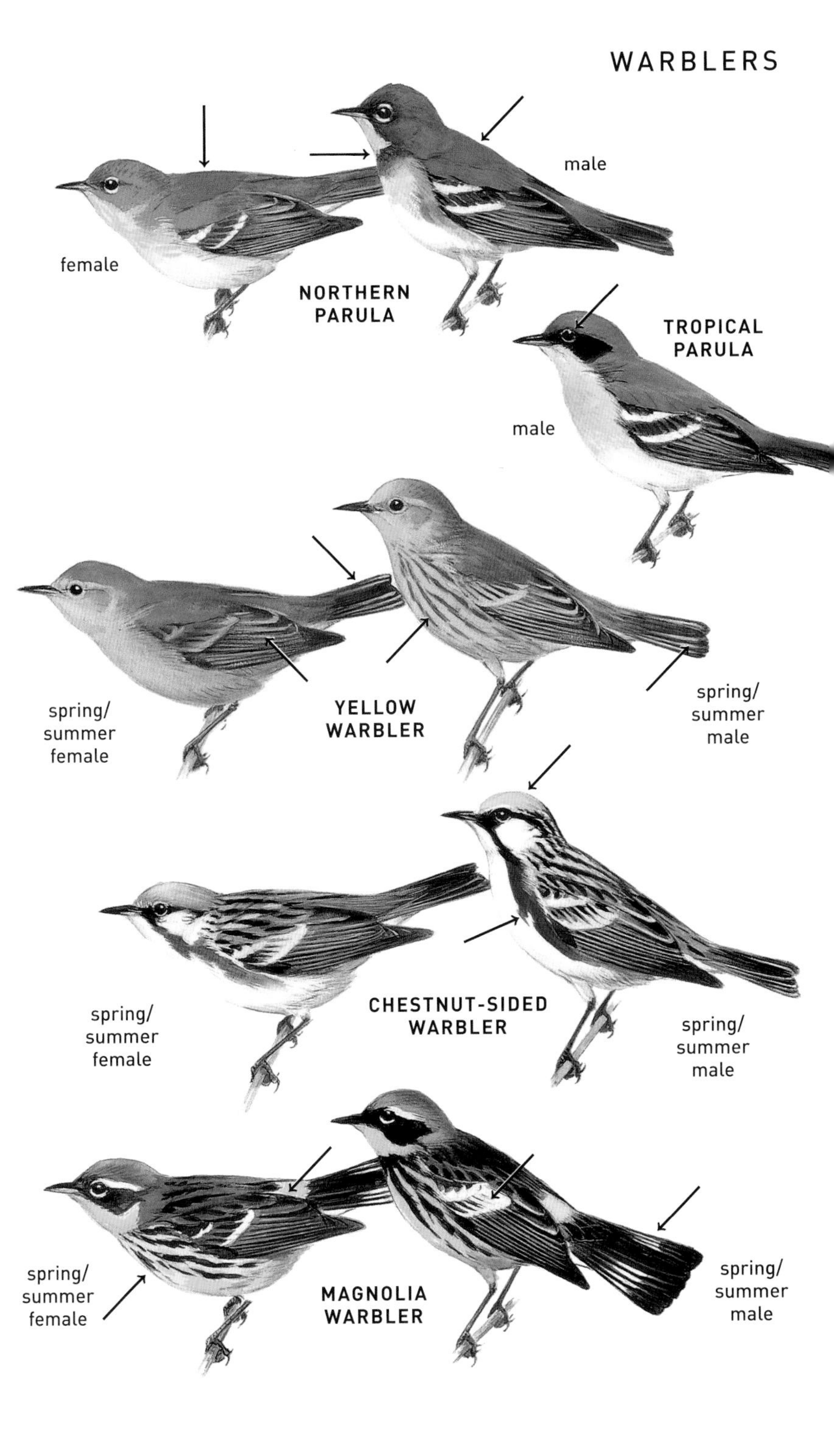

male
female
NORTHERN PARULA
TROPICAL PARULA
male
spring/ summer female
YELLOW WARBLER
spring/ summer male
spring/ summer female
CHESTNUT-SIDED WARBLER
spring/ summer male
spring/ summer female
MAGNOLIA WARBLER
spring/ summer male

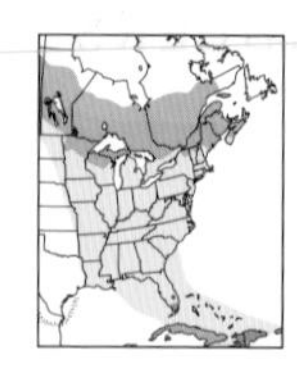

CAPE MAY WARBLER *Setophaga tigrina* Uncommon

5 in. (13 cm). *Spring/summer male:* Note *chestnut* cheeks. Yellow below, striped with black; rump yellow, crown black. *Female and fall/winter male:* Lack chestnut cheeks; duller, breast often whitish. Note dull *patch of yellow behind ear, yellowish rump,* and *one wing bar bolder than the other.* First-fall female (p. 284) distinctly *gray,* can lack yellow. **VOICE:** Song a very high, thin *seet seet seet seet.* **SIMILAR SPECIES:** Dull fall/winter females may be confused with Yellow-rumped Warbler but are plainer; have small pale patch behind ear; duller, greenish-yellow rump; and shorter tail. **HABITAT:** Spruce forests; in migration, also broadleaf trees.

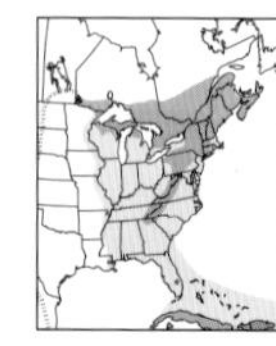

BLACK-THROATED BLUE WARBLER *Setophaga caerulescens* Fairly common

5¼ in. (13 cm). *Male:* Clean-cut; upperparts *deep blue;* throat and sides *black,* belly white; wing with large white spot at base of primaries. First-year male similar but slightly duller, wing edging greener. *Female:* Olive-backed, with light line over eye and smaller *white wing spot.* First-year female (p. 286) may lack this white spot, but note *dark cheek.* **VOICE:** Song a husky, lazy *zur, zur, zur, zreee* or *beer, beer, bree.* Call a hard *thip,* similar to call of Dark-eyed Junco. **HABITAT:** Understory of deciduous and mixed woodlands.

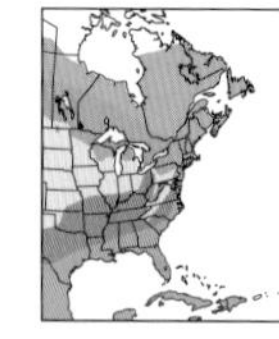

YELLOW-RUMPED WARBLER *Setophaga coronata* Common

5½ in. (14 cm). Includes "Myrtle" (*S. c. coronata*) and "Audubon's" (*S. c. auduboni*) Warblers, two subspecies groups formerly considered separate species. Note bright *yellow rump* in all subspecies, ages, and sexes. *Spring/summer male:* Blue-gray above; heavy black breast patch (like an inverted U); crown and side patches yellow. "Audubon's" (breeds in w. U.S., sw. Canada; rare vagrant to East) differs from "Myrtle" (breeds in AK, much of Canada, e. U.S.) in having *yellow throat* (which does not extend back below cheek, as white does in "Myrtle"), larger white wing patches, no white supercilium, plainer face. *Spring/summer female:* Similar but duller overall. *Fall/winter* (see also p. 284): More brownish above; whitish below; *rump yellow.* **VOICE:** Variable song, juncolike but two-parted, rising or dropping in pitch, *seet-seet-seet-seet-seet, trrrrrrrr.* Call a loud *check* or *chip* ("Myrtle") or more nasal *tchenp* ("Audubon's"). **SIMILAR SPECIES:** Cape May and Magnolia Warblers. **HABITAT:** Coniferous forests. In migration and winter, varied; open woods, coastal bushes, brush, thickets, parks, gardens, upper beaches.

BLACK-THROATED GRAY WARBLER *Setophaga nigrescens* Rare

5 in. (13 cm). *Spring/summer and adult male:* Gray above, with black throat, cheek, and crown separated by *white. Small yellow spot in lores. Female: Slaty* crown and cheek; dusky or light throat; loral spot duller yellow. *First-fall:* Male like adult female but throat mottled black; female duller, may be tinged brownish above; cheeks dull gray; loral spot pale. **VOICE:** Call a dull *tup.* **SIMILAR SPECIES:** Suggests Black-and-white Warbler but lacks white stripes on back and crown, does not crawl around on branches. **RANGE AND HABITAT:** Rare winter visitor to s. TX and casual vagrant to East Coast. In migration and winter, mixed woods.

spring/
summer
male
spring/
summer
female
CAPE MAY
WARBLER
BLACK-
THROATED BLUE
WARBLER
male
female
spring/summer
male
first-fall
"Myrtle"
Warbler
spring/summer
female
YELLOW-RUMPED
WARBLER
BLACK-
THROATED GRAY
WARBLER
spring/
summer male
spring/
summer
male
"Audubon's"
Warbler
female and
fall/winter
male
first-fall male
female

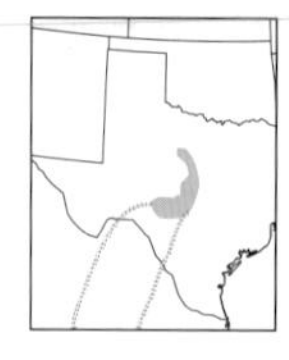

GOLDEN-CHEEKED WARBLER *Setophaga chrysoparia* Scarce, local

5¼ in. (14 cm). Like Black-throated Green Warbler, but back *black to darker olive with black streaks* and blacker line through eye; flanks lack yellow. **VOICE:** Song a hurried *tweeah, tweeah, tweesy* or *bzzzz, laysee, daysee.* Call a flat *tip.* **RANGE AND HABITAT:** Breeds in junipers, oaks, and streamside trees of Edwards Plateau, TX.

HERMIT WARBLER *Setophaga occidentalis* Casual vagrant

5 in. (13 cm). *Adult:* Note bright *yellow face* set off by *black throat and nape* and gray back. *First-fall:* Black of throat reduced or lacking, plain-looking yellow face, gray back, *unstreaked* underparts. **VOICE:** Call a soft, flat *tip.* **SIMILAR SPECIES:** Black-throated Green Warbler has green back, black mottling or streaking down sides of breast. See Townsend's Warbler. **RANGE AND HABITAT:** Casual vagrant to East Coast from w. N. America. In migration, found in coniferous and deciduous woods.

TOWNSEND'S WARBLER *Setophaga townsendi* Casual vagrant

5 in. (13 cm). Easily distinguished in all plumages by *black-and-yellow pattern of head,* with *blackish to grayish cheek patch; underparts yellow,* and heavily striped sides. **VOICE:** Call a soft, flat *tip.* **SIMILAR SPECIES:** Hermit and Black-throated Green Warblers lack dark cheeks and yellow on lower sides. See also dull Blackburnian Warblers. **RANGE AND HABITAT:** Casual vagrant to East Coast from w. N. America. In migration, found in mixed woodlands.

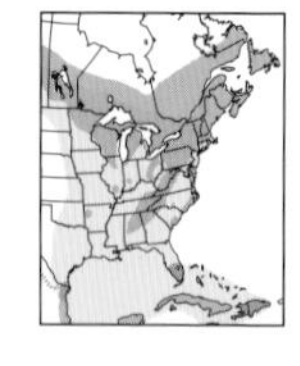

BLACK-THROATED GREEN WARBLER *Setophaga virens* Fairly common

5 in. (13 cm). *Spring/summer and adult male:* Bright *yellow face* framed by black throat and olive-green crown. *Adult female and first-fall male:* Yellow face; much less black on throat; unmarked olive-green back, *black mottling on sides of upper breast. First-fall female:* Dullest (see also p. 284). **VOICE:** A lisping, weezy, or buzzy *zoo zee zoo zoo zee* or *zee zee zee zee zoo zee.* Call a flat *tip* or *tup.* **SIMILAR SPECIES:** Hermit and Townsend's Warblers. Golden-cheeked Warbler has black line through eye. **HABITAT:** Mainly coniferous or mixed woods; in migration, variety of woodlands.

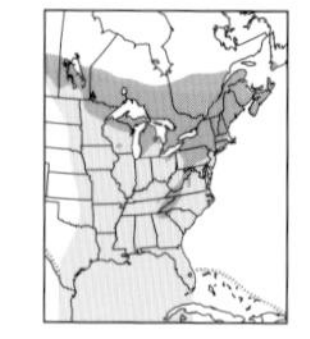

BLACKBURNIAN WARBLER *Setophaga fusca* Fairly common

5 in. (13 cm). The "fire throat." *Spring/summer and adult male:* Black and white, with *flame orange* on head and throat. *Adult female and first-fall male:* Paler orange on throat; dark cheek patch. First-fall female (p. 284) dullest, with dull yellowish to beige throat. Note head stripes, *pale back stripes.* **VOICE:** Song *zip zip zip titi tseeeeee,* ending on a very high, upslurred note (inaudible to some ears). Also a two-parted *teetsa teetsa teetsa teetsa zizizizizi,* more like Nashville Warbler. Call a rich *chip.* **SIMILAR SPECIES:** Dull Townsend's Warbler yellower in head, back greener. See also Yellow-throated and Cerulean Warblers. **HABITAT:** Woodlands; in summer, conifers.

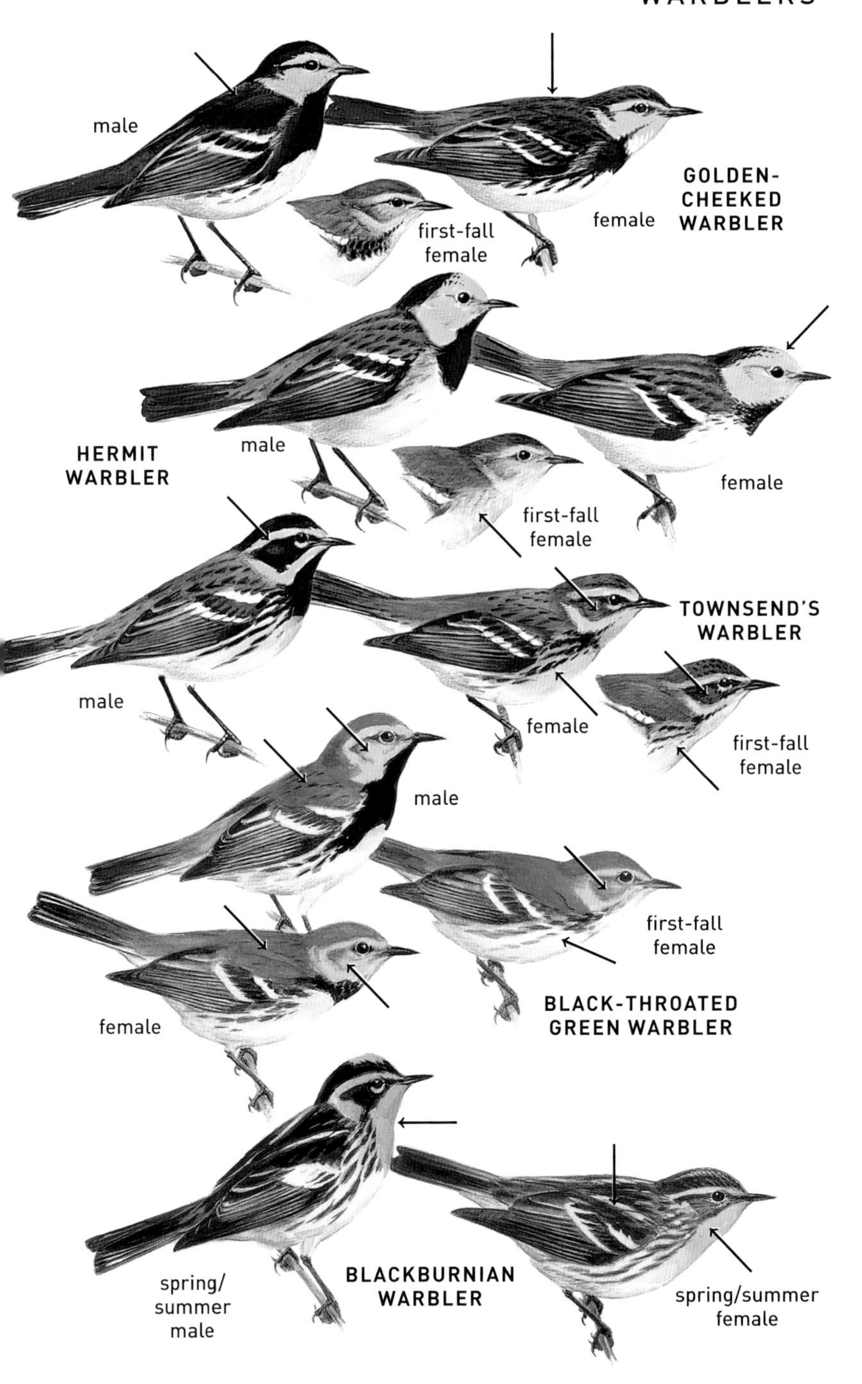
male
GOLDEN-CHEEKED WARBLER
female
first-fall female
HERMIT WARBLER
male
female
first-fall female
TOWNSEND'S WARBLER
male
female
first-fall female
male
first-fall female
female
BLACK-THROATED GREEN WARBLER
BLACKBURNIAN WARBLER
spring/summer male
spring/summer female

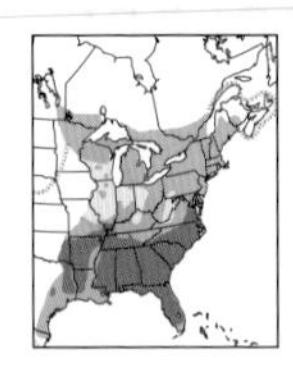

PINE WARBLER *Setophaga pinus* **Common**

5½ in. (14 cm). All plumages have dark cheeks, blurry streaking on sides of breast, unstreaked back, white tail spots, and dark feet. *Adult male:* Yellow-breasted, with olive-green back, *white wing bars. Adult female and first-year male:* Duller; brownish olive above. *First-fall female:* Can lack yellow (see also p. 284). **VOICE:** Song a trill on one pitch, slower and more musical than Chipping Sparrow song. Call a sweet *chip.* **SIMILAR SPECIES:** Fall/winter Blackpoll and Bay-breasted Warblers have more black streaking on back, less prominent dark cheeks; Blackpoll has *yellow feet.* **HABITAT:** Pine woods. In winter, sometimes in fields with bluebirds.

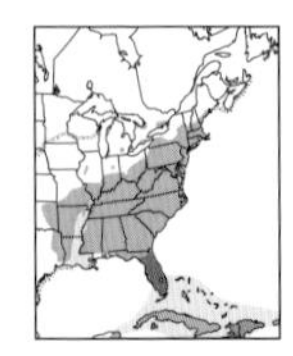

PRAIRIE WARBLER *Setophaga discolor* **Fairly common**

4¾ in. (12 cm). This warbler *bobs its tail* (as does Palm Warbler); underparts yellow, paling on undertail coverts; black stripes *confined to sides; two black face marks,* one through eye, one below. At close range, chestnut marks may be seen on back of male (reduced in female). First-fall birds (p. 284) duller, especially female. **VOICE:** Song a thin *zee zee zee zee zee zee zee zee,* ascending the chromatic scale. Call a sharp *tschip.* **SIMILAR SPECIES:** Pine, Palm, and Yellow Warblers. **HABITAT:** Brushy pastures, low pines, mangroves.

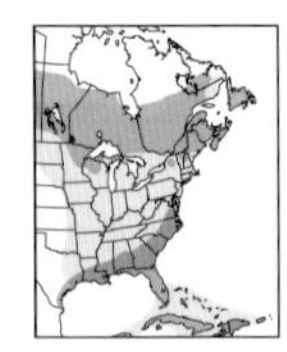

PALM WARBLER *Setophaga palmarum* **Common**

5¼ in. (14 cm). Note constant *bobbing* of tail. Both sexes brownish or olive above; yellowish or dirty white below, narrowly streaked; *bright yellow* undertail coverts, white spots in tail corners. In spring/summer has *chestnut cap;* ages and sexes rather similar. Two subspecies: eastern breeders have more yellow below and on eyebrow; western breeders duller (p. 284), may have yellow restricted to undertail coverts in fall. **VOICE:** Song weak, repetitious notes: *zhe-zhe-zhe-zhe-zhe-zhe.* Call a distinctive sharp *tsup.* **SIMILAR SPECIES:** Yellow-rumped and Prairie Warblers. **HABITAT:** In summer, wooded borders of muskeg, bogs. In migration and winter, bushes, weedy fields. A ground-loving warbler.

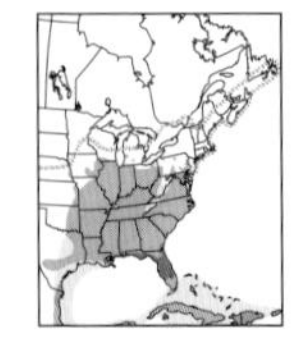

YELLOW-THROATED WARBLER *Setophaga dominica* **Fairly common**

5½ in. (14 cm). A gray-backed warbler with *yellow throat. Black eye mask,* white wing bars, black stripes on sides. Ages and sexes similar; first-fall female slightly duller. Creeps about branches of trees. "Sutton's" Warbler is a very rare hybrid of Yellow-throated Warbler and Northern Parula. **VOICE:** Song a series of clear slurred notes dropping slightly in pitch: *tee-ew, tew, tew, tew, tew, tew wi* (last note rising). Call a rich *chip.* **SIMILAR SPECIES:** Female Blackburnian Warbler. **HABITAT:** Open woodlands, especially sycamores, live oaks, pines. In winter, palms.

KIRTLAND'S WARBLER *Setophaga kirtlandii* **Rare, local, endangered**

5¾ in. (15 cm). Bluish gray above, *streaked with black;* yellow below, with black spots or streaks confined to sides. *Male:* Has *blackish mask. Female:* Duller, lacks mask; first-fall female browner. Persistently wags tail. **VOICE:** Song, loud and low-pitched, starts with three or four low staccato notes, continues with rapid ringing notes on higher pitch, and ends abruptly. **SIMILAR SPECIES:** Yellow-rumped, Yellow-throated, and Magnolia Warblers. **HABITAT:** Groves of young jack pines 5–18 ft. (1½–5½ m.) high with ground cover of berries and fern.

first-fall
female
PINE
WARBLER
male
female
PRAIRIE
WARBLER
male
fall/
winter
spring/
summer
male
Western
PALM
WARBLER
spring/
summer
male
Eastern
"Sutton's"
Warbler
YELLOW-THROATED
WARBLER
male
KIRTLAND'S
WARBLER

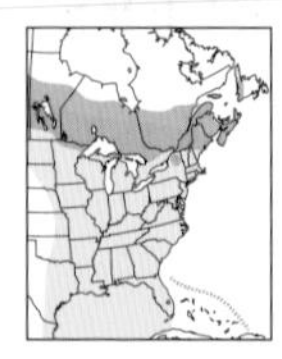

BAY-BREASTED WARBLER *Setophaga castanea* Uncommon

5½ in. (14 cm). *Spring/summer male: Chestnut throat, upper breast,* and sides. Note *buff patch* on neck. *Spring/summer female:* Paler, with whitish throat. *Fall/winter* (see also p. 284): Olive green above; two white wing bars; pale *buff to chestnut flanks, dark feet.* **VOICE:** A high, sibilant *tees teesi teesi;* thinner and shorter than Black-and-white Warbler. Call a sharp *chip,* like Blackpoll Warbler. **SIMILAR SPECIES:** See fall/winter Blackpoll Warbler and Pine Warbler. **HABITAT:** Woodlands; in summer, conifers.

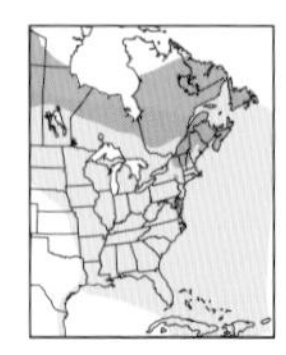

BLACKPOLL WARBLER *Setophaga striata* Common

5½ in. (14 cm). *Spring/summer male:* Note *black cap, white cheeks, distinct pale legs. Spring/summer female: Greenish gray above,* whitish below, *streaked. Fall/winter* (see also p. 284): Olive and greenish, *faintly streaked* on back and on breast; two wing bars; *whitish undertail coverts;* bright *yellow legs or feet.* **VOICE:** Song a thin, very high-pitched *zi-zi-zi-zi-zi-zi-zi-zi-zi* on one pitch, becoming stronger, then diminishing. Call a sharp *chip.* **SIMILAR SPECIES:** Fall/winter Bay-breasted Warbler has buff wash on flanks and undertail coverts, dark feet. See also Pine Warbler. **HABITAT:** Conifers; in migration, broadleaf trees.

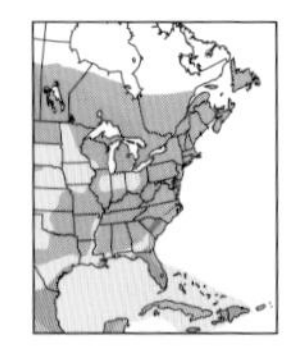

BLACK-AND-WHITE WARBLER *Mniotilta varia* Common

5¼ in. (13 cm). *Creeps along trunks* and branches of trees. Note *lengthwise black-and-white stripes,* white stripes on crown and back. *Spring/summer and adult male:* Black throat partly or mostly lost in winter. *Female and first-fall male:* Paler cheeks, fainter streaks below, buffy wash on flanks. **VOICE:** Song a thin *weesee weesee weesee weesee,* sometimes dropping in pitch midway. Call a sharp *chip.* **SIMILAR SPECIES:** Blackpoll and Black-throated Gray Warblers. **HABITAT:** Woods.

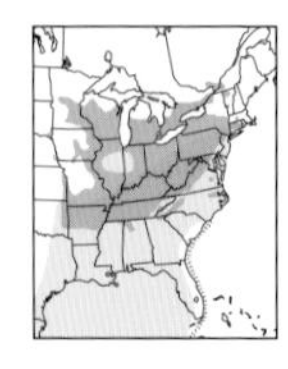

CERULEAN WARBLER *Setophaga cerulea* Uncommon

4¾ in. (12 cm). A small, short-tailed warbler, often high up in large trees. *Spring/summer and adult male: Blue* above, white below, *narrow black band* across chest. *Adult female and first-fall male: Olive green above* with dull blue crown and rump and *broad whitish eyebrow;* whitish below (breast sometimes tinged yellow); two white wing bars. *First-fall female:* Green with white wing bars, can lack blue. **VOICE:** Buzzy notes on same pitch, followed by longer note on a higher pitch: *zray zray z-z-z zeeeee.* Call a rich, slurred *chip.* **SIMILAR SPECIES:** Dull Tennessee Warbler lacks bold white wing bars. Dull female Blackburnian Warbler has darker streaked back pattern. **HABITAT:** High in deciduous forests, especially in river valleys and ridges.

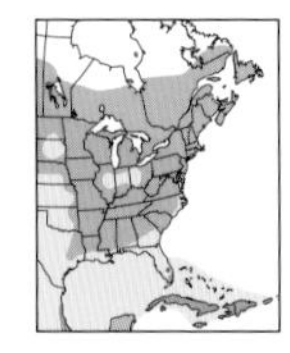

AMERICAN REDSTART *Setophaga ruticilla* Common

5¼ in. (13 cm). Butterfly-like; actively flitting, with drooping wings and spread tail. *Adult male:* Black; *bright orange patches* on wings and tail. *Female:* Gray-olive above; *yellow patches* on wings and tail. *First-year male:* Like female but tinged with orange on chest patches; acquires black splotches on face in spring. **VOICE:** Songs *zee zee zee zee zwee* or *tsee tsee tsee tsee tsee-o;* also *teetsa teetsa teetsa teetsa teet* (notes paired). Call a slurred, rich *chip.* **HABITAT:** Second-growth woods, riparian woodlands.

WARBLERS
fall/winter
BAY-BREASTED
WARBLER
spring/
summer male
spring/
summer
female
spring/
summer male
fall/
winter
BLACKPOLL
WARBLER
spring/
summer
female
female and
first-fall male
adult and
spring male
BLACK-AND-
WHITE
WARBLER
adult
male
adult and
spring male
female
and
first-fall
male
first-year
male
female
CERULEAN
WARBLER
AMERICAN
REDSTART

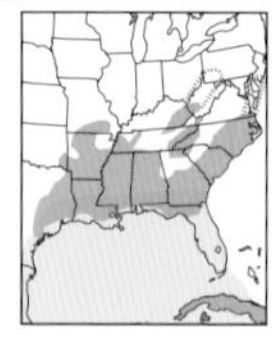

SWAINSON'S WARBLER *Limnothlypis swainsonii* Uncommon

5½ in. (14 cm). A plain brown skulker, difficult to see. Long bill. Olive-brown above and buffy white below, with *brown crown* and *light eyebrow stripe*. Ages and sexes similar. **VOICE:** Song suggests Louisiana Waterthrush but shorter (five notes: two slurred notes, two lower notes, and a higher note): *wee-wee-chip-poor-will.* Call a sharp, loud *chip.* **SIMILAR SPECIES:** Ovenbird, Worm-eating Warbler, waterthrushes. **HABITAT:** Cane thickets, swamps, stream bottoms, thick woodland brush; locally in rhododendron-hemlock tangles in Appalachians.

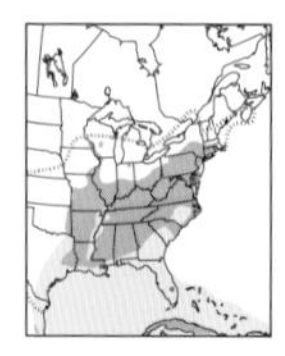

WORM-EATING WARBLER *Helmitheros vermivorum* Uncommon

5¼ in. (13 cm). An unobtrusive forager of wooded slopes and thick understory. Often probes dead-leaf clusters. *Dull olive,* with *black stripes on buffy head.* Breast *rich buff.* Ages and sexes similar. **VOICE:** Song a series of thin dry notes; resembles trill or rattle of Chipping Sparrow but thinner, more rapid, and insectlike. Call a flat *chip,* also a distinctive *zeet-zeet* in flight. **SIMILAR SPECIES:** Ovenbird, Swainson's Warbler, waterthrushes. **HABITAT:** Wooded hillsides, undergrowth, ravines.

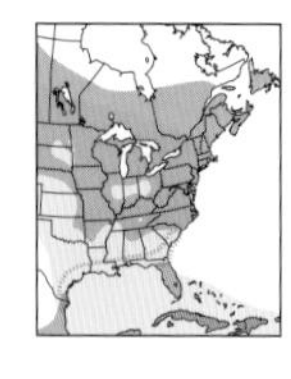

OVENBIRD *Seiurus aurocapilla* Common

6 in. (15 cm). When breeding, less often seen than heard. When seen, usually walking on leafy floor of woods. Suggests a small thrush but *striped* rather than spotted beneath. *Orangish patch on crown bordered by blackish stripes. White eye-ring.* Ages and sexes similar. **VOICE:** Song an emphatic *TEACHer, TEACHer, TEACHer,* etc., in crescendo. In some areas, a monosyllabic *TEACH, TEACH, TEACH,* etc. Call a loud, sharp *tshuk.* **SIMILAR SPECIES:** Waterthrushes. See also spotted thrushes (p. 236). **HABITAT:** Near or on ground in leafy and pine-oak woods; in migration, also thickets.

NORTHERN WATERTHRUSH *Parkesia noveboracensis* Common

5¾ in. (15 cm). Suggests a small thrush. *Walks* along water's edge and *teeters* like a Spotted Sandpiper. Brown-backed, often tinged olive, with *striped* underparts, strong eyebrow stripe; both eyebrow and underparts vary from whitish to pale yellow. *Throat striped.* Ages and sexes similar. **VOICE:** Call a sharp *chink.* Song a vigorous, rapid *twit twit twit sweet sweet sweet chew chew chew* (*chews* drop in pitch). **SIMILAR SPECIES:** Louisiana Waterthrush, Ovenbird. **HABITAT:** Swamps, bogs, wet woods with standing water, streamsides, pond shores; in migration, also marsh edges, puddles, mangroves.

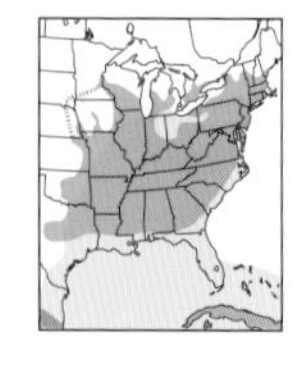

LOUISIANA WATERTHRUSH *Parkesia motacilla* Uncommon

6 in. (15 cm). Similar to Northern Waterthrush, but underparts *white on breast, pinkish buff on flanks and undertail coverts.* Bill slightly larger. *Eyebrow stripe pure white and flares noticeably behind eye.* Throat usually *lacks stripes.* Legs pinkish. Ages and sexes similar. **VOICE:** Song musical and ringing; three clear, slurred whistles, followed by a jumble of twittering notes dropping in pitch. **SIMILAR SPECIES:** Some fall/winter Northern Waterthrushes (particularly of western populations) have whitish eyebrow stripe. Northern has small spots or stripes on throat and *even-toned* (yellow to off-white) underparts, without buff flanks as in Louisiana. Song of Swainson's Warbler somewhat similar. **HABITAT:** Streams, brooks, ravines, wooded swamps. Bobs when walking, more exaggerated than in Northern Waterthrush.

SWAINSON'S
WARBLER
WORM-EATING
WARBLER
OVENBIRD
fresh fall
NORTHERN
WATERTHRUSH
worn
spring
LOUISIANA
WATERTHRUSH

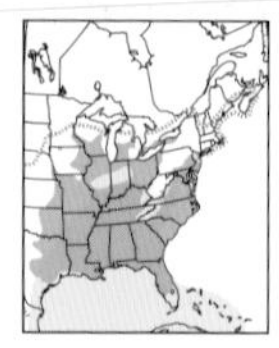

PROTHONOTARY WARBLER *Protonotaria citrea* **Fairly common**

5½ in. (14 cm). A golden bird of wooded swamps. *Male:* Entire head and breast deep *yellow to orangey.* Wings blue-gray *with no bars. Female:* Duller, fewer white spots in tail. First-year birds duller within each sex (p. 286). **VOICE:** Song *zweet zweet zweet zweet zweet zweet,* on one pitch. Call a loud *seep.* **SIMILAR SPECIES:** Yellow and Blue-winged Warblers. **HABITAT:** Wooded swamps, backwaters, river edges.

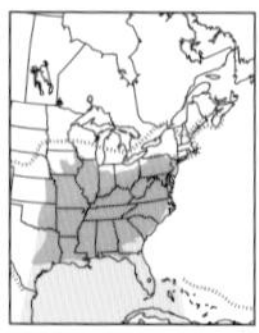

KENTUCKY WARBLER *Geothlypis formosa* **Uncommon**

5¼ in. (13 cm). Note *broad black mustaches* extending down from eye and *yellow spectacles.* Female and first-year male duller but retain distinctive mask pattern. **VOICE:** Song a rapid rolling chant, *tory-tory-tory-tory* or *churry-churry-churry-churry,* less musical than Carolina Wren. Call a rich, low *tup.* **SIMILAR SPECIES:** Common Yellowthroat lacks spectacles. See also Hooded Warbler. **HABITAT:** Woodland undergrowth, swamps.

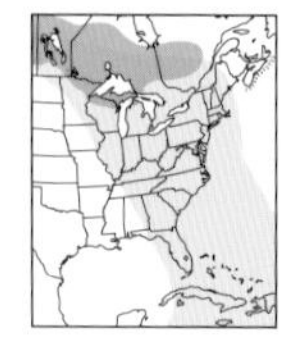

CONNECTICUT WARBLER *Oporornis agilis* **Uncommon**

5¾–6 in. (15 cm). Shy and skulking. Similar to MacGillivray's and Mourning Warblers but slightly larger and plumper; note *walking behavior* and *complete white eye-ring, long undertail coverts* reaching almost to tail tip. *Spring/summer and adult:* Hood gray in male, gray-brown in female. *First-fall:* Duller, with brownish hood, paler throat (see also p. 286). **VOICE:** A repetitious *chip-chup-ee, chip-chup-ee, chip-chup-ee, chip* or *sugar-tweet, sugar-tweet, sugar-tweet.* **SIMILAR SPECIES:** First-fall Mourning Warbler has broken eye-ring (rarely looking complete); yellow throat. Also, Mourning is faster and twitchier, hops rather than walks. Nashville Warbler also has eye-ring but is smaller, has yellow throat, and actively feeds in trees. **HABITAT:** Poplar bluffs, muskeg, mixed woods; in migration, undergrowth. Feeds mostly on ground.

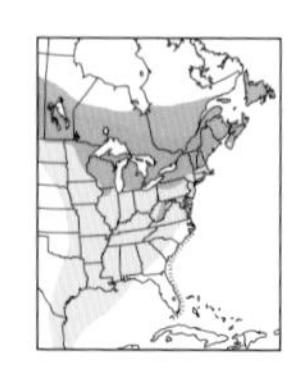

MOURNING WARBLER *Geothlypis philadelphia* **Uncommon**

5¼ in. (13 cm). Shy and skulking. Olive above, yellow below, with slate-gray hood encircling head and neck. *Spring/summer and adult male:* Has irregular black bib. *Female and first-fall male:* May have thin eye-ring that is barely broken. Some spring/summer females and all first-fall birds have yellow throat, in some extending through middle breast, in others separated by thin band. Yellow undertail coverts shorter than in Connecticut Warbler. **VOICE:** Song *chirry, chirry, chorry, chorry* (*chorry* lower). Considerable variation. Call a hard, buzzy, wrenlike *chack.* **SIMILAR SPECIES:** MacGillivray's and Connecticut Warblers. **HABITAT:** Thickets, undergrowth.

MACGILLIVRAY'S WARBLER *Geothlypis tolmiei* **Casual vagrant**

5¼ in. (13 cm). Similar to Mourning Warbler, but in all plumages partial eye-ring is *thicker, clearly broken fore and aft, forming crescents.* Spring/summer and adult males show less black on bib. Hood of females paler. First-year birds with *buff to grayish-white throat* (rather than yellow); overall, slightly duller. **VOICE:** Call a low, hard *chik,* given often. **SIMILAR SPECIES:** Mourning Warbler. **RANGE AND HABITAT:** Casual vagrant from w. N. America to East Coast. In migration, in low dense undergrowth; shady thickets.

PROTHONOTARY WARBLER
female
male
KENTUCKY WARBLER
male
spring/summer male
CONNECTICUT WARBLER
first-fall female
Nashville Warbler (p. 264) for comparison
spring/summer female
spring/summer female
MOURNING WARBLER
spring/summer male
first-fall
MACGILLIVRAY'S WARBLER
spring/summer male
spring/summer female
first-fall

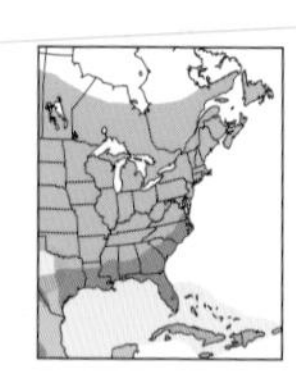

COMMON YELLOWTHROAT *Geothlypis trichas* Common

5 in. (13 cm). Wrenlike. *Spring/summer and adult male:* Distinctive *black mask,* yellow throat and upper breast. *First-fall male:* Has reduced and duller dusky mask. *Female:* Olive-brown, with rich yellow throat (can be buff in first-fall; p. 286), duller below, but brighter yellow undertail coverts; lacks black mask. **VOICE:** A bright, rapid chant, *witchity-witchity-witchity-witch;* sometimes *witchy-witchy-witchy-witch.* Call a husky *tchep.* **SIMILAR SPECIES:** Female distinguished from first-fall and female Mourning Warblers by whitish belly, smaller size. **HABITAT:** Swamps, marshes, wet thickets, woodland edges.

GRAY-CROWNED YELLOWTHROAT *Geothlypis poliocephala* Very rare

5½ in. (14 cm). Male has partial mask *not extending to forehead or cheeks; gray crown.* Both sexes have *thick bill* with *pale lower mandible;* broken white eye-ring. First-year birds duller than adults of each sex. **VOICE:** A burbling warble. Call *chlee-dee.* **SIMILAR SPECIES:** Common Yellowthroat slightly smaller and smaller-billed, has different vocalizations. **RANGE:** Very rare visitor from Mex. to s. TX, where it formerly bred. **HABITAT:** Reeds and weedy vegetation near water.

GOLDEN-CROWNED WARBLER Casual vagrant

Basileuterus culicivorus

5 in. (13 cm). Yellow crown and gray eyebrow stripe bordered by black. Broken eye-ring. Dusky yellow below. Drab olive above. Ages and sexes similar. **VOICE:** Song a series of slurred whistles. Call a short, sharp *tuk.* **SIMILAR SPECIES:** Common Yellowthroat, Orange-crowned Warbler. **RANGE:** Casual vagrant from Mex. to s. TX. **HABITAT:** Dense woodland understory.

RUFOUS-CAPPED WARBLER *Basileuterus rufifrons* Very rare

5 in. (13 cm). *Rufous cap and cheek* separated by white eyebrow stripe. Throat and upper breast bright yellow, upperparts olive. Long, spindly tail often held cocked up at angle. Ages and sexes similar. **VOICE:** An accelerating series of whistled, musical chips and warbles. Call *tick.* **SIMILAR SPECIES:** Common Yellowthroat. **RANGE:** Very rare visitor from Mex. to TX. **HABITAT:** Thick brush, oak woodlands near water.

WARBLERS
male
COMMON
YELLOWTHROAT
female
GRAY-CROWNED
YELLOWTHROAT
male
female
GOLDEN-CROWNED
WARBLER
RUFOUS-CAPPED
WARBLER

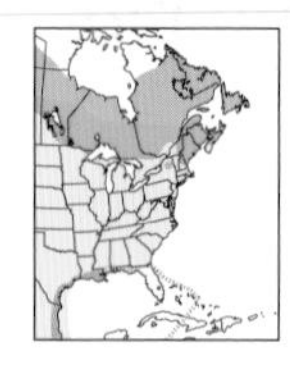

WILSON'S WARBLER *Cardellina pusilla* Uncommon

4¾ in. (12 cm). *Male:* Golden yellow with *round black cap. Female:* Has smaller cap, located closer to forecrown, or no cap. First-year birds have smaller caps than adults, sex for sex, usually lacking in first-year females (p. 286). Otherwise back olive, underparts yellow, supercilium indistinct yellow, *lores yellow,* tinged olive. Constantly moving and flitting about. **VOICE:** Song a thin, rapid little chatter, dropping in pitch at end: *chi chi chi chi chi chet chet.* Call a flat *timp.* **SIMILAR SPECIES:** Female Hooded Warbler has white spots in tail, dark lores. Yellow Warbler has yellow spots in shorter tail, yellow edging in wings. Note vocal differences. **HABITAT:** Thickets and trees along streams, moist tangles, low shrubs, willows, alders.

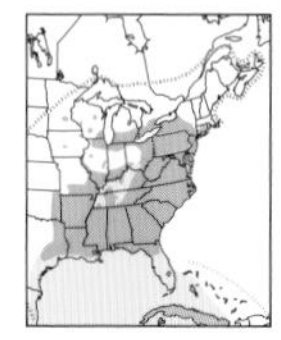

HOODED WARBLER *Setophaga citrina* Fairly common

5¼ in. (13 cm). *Male: Black hood* or cowl encircles yellow face and forehead; ages similar. *Female:* Has variable amount of black in head, from a partial hood to none in most first-year females (p. 286); yellow face usually distinctively outlined, and note *white tail spots.* **VOICE:** Song a loud whistled *weeta wee-tee-o.* Also other arrangements; slurred *tee-o* is a clue. Call a sharp *chink,* like waterthrushes. **SIMILAR SPECIES:** Female Wilson's Warblers without black cap lack tail spots and any suggestion of Hooded's face pattern. **HABITAT:** Wooded undergrowth, laurels, wooded swamps.

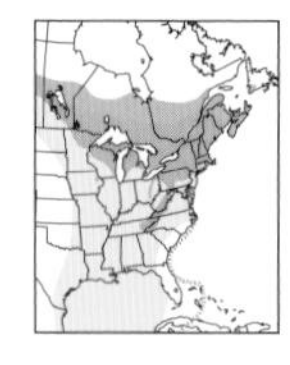

CANADA WARBLER *Cardellina canadensis* Uncommon

5¼ in. (13 cm). The "necklaced warbler." *Spring/summer and adult male: Solid gray above;* bright yellow below, with *necklace of short black stripes;* white vent. *First-year male:* Duller, more female-like. *Female:* Similar, but necklace fainter, upperparts may be washed with brownish; first-fall female (p. 286) may have only hint of necklace. All have *spectacles of white eye-ring and yellow loral stripe.* No white in wings or tail. **VOICE:** Song a staccato burst, irregularly arranged. *Chip, chupety swee-ditchety.* Call *tchip.* **SIMILAR SPECIES:** Magnolia and Yellow-throated Warblers. **HABITAT:** Forest undergrowth, shady thickets.

YELLOW-BREASTED CHAT Family Icteriidae

Traditionally placed with wood-warblers, Yellow-breasted Chat was recently afforded its own family, Icteriidae (as opposed to Icteridae for blackbirds and orioles), because of its larger body and bill size and other factors.

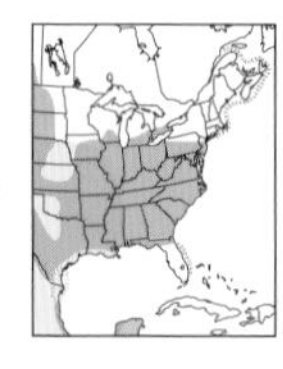

YELLOW-BREASTED CHAT *Icteria virens* Uncommon

7½ in. (19 cm). Larger than our warblers, with *heavy bill* and *long tail.* Note *white* spectacles, *bright yellow* throat and breast. No wing bars. Habitat and voice suggest a thrasher or mockingbird. **VOICE:** Repeated whistles, alternating with harsh notes and soft *caws.* Suggests Northern Mockingbird, but repertoire more limited; much longer pauses between phrases. Single notes: *whoit, kook, zhairr,* etc. Often sings in short, awkward courtship display flight. **SIMILAR SPECIES:** Common Yellowthroat (much smaller). **HABITAT:** Brushy tangles, briars, stream thickets, where it skulks.

WARBLERS
AND CHAT
female
first-fall
female
WILSON'S
WARBLER
male
adult
female
HOODED
WARBLER
male
female
CANADA
WARBLER
male
YELLOW-
BREASTED
CHAT

FALL WARBLERS

Most of these have streaks or wing bars.

RUBY-CROWNED KINGLET *Regulus calendula* p. 256
(Not a warbler.) Broken eye-ring, pale wing bars, wing-flicking behavior.

CHESTNUT-SIDED WARBLER *Setophaga pensylvanica* p. 266
First-fall: Green above, grayish white below; eye-ring; tail cocked at angle. Sexes overlap in plumage.

PINE WARBLER *Setophaga pinus* p. 272
Differs from Blackpoll and Bay-breasted in heavier bill, unstreaked back, dark legs. First-fall female can lack yellow.

BAY-BREASTED WARBLER *Setophaga castanea* p. 274
Note dark legs and feet, buff flanks and undertail coverts, unstreaked breast. First-fall males can have some richer bay on flanks.

BLACKPOLL WARBLER *Setophaga striata* p. 274
Very similar to Bay-breasted Warbler but slimmer. Note streaked back and breast, yellowish legs and especially feet.

NORTHERN PARULA *Setophaga americana* p. 266
First-fall: Small and short-tailed. Bluish head, broken eye-ring, and yellow throat; wing bars. Female lacks marks on breast.

MAGNOLIA WARBLER *Setophaga magnolia* p. 266
First-fall: Broad white band at midtail. Note yellow rump. Faint dusky band across yellow breast. Side streaking. Sexes similar in first fall.

PRAIRIE WARBLER *Setophaga discolor* p. 272
First-fall: Jaw stripe, side streaks. Bobs tail. Female duller than male.

YELLOW WARBLER *Setophaga petechia* p. 266
Yellow edging on wings and tail. Beady dark eye. Some drab females may resemble Orange-crowned Warbler, but note lack of face pattern.

BLACKBURNIAN WARBLER *Setophaga fusca* p. 270
First-fall: Yellow or yellow-orange throat, dark cheek; broad supercilium, dark blackish back with pale stripes, upper wing bar triangular.

BLACK-THROATED GREEN WARBLER *Setophaga virens* p. 270
First-fall: Yellow cheek framed by olive. Note black on sides of upper breast, plain greenish back.

PALM WARBLER *Setophaga palmarum* p. 272
Brownish back, yellowish undertail coverts. Bobs tail.

YELLOW-RUMPED WARBLER *Setophaga coronata* p. 268
First-fall: Bright yellow rump, streaked back; brownish above. Flightier behavior than most other warblers; hawks insects; eats berries in winter.

CAPE MAY WARBLER *Setophaga tigrina* p. 268
First-fall: Streaked breast, greenish yellow rump. Female grayer than dull Yellow-rumped Warbler.

SELECTED FALL WARBLERS
With Streaks or Wing Bars
RUBY-CROWNED KINGLET
emale
not a warbler
CHESTNUT-SIDED WARBLER
first-fall
fall male
PINE WARBLER
first-fall
BAY-
BREASTED
WARBLER
BLACKPOLL WARBLER
first-fall
PINE WARBLER
first-fall
female
first-fall
NORTHERN PARULA
MAGNOLIA WARBLER
first-fall
male
PRAIRIE WARBLER
first-fall
male
YELLOW WARBLER
rst-fall
emale
first-fall
female
BLACKBURNIAN WARBLER
first-fall
female
BLACK-THROATED GREEN WARBLER
PALM WARBLER
western
first-fall
first-fall
female
YELLOW-RUMPED WARBLER
"Myrtle" Warbler
first-fall
female
CAPE MAY WARBLER

FALL WARBLERS

Most of these lack streaks or wing bars.

ORANGE-CROWNED WARBLER *Oreothlypis celata* p. 264

First-year: Dingy breast with faint dusky streaks, yellow undertail coverts, faint eye line. First-fall birds of drab greenish with gray head.

TENNESSEE WARBLER *Oreothlypis peregrina* p. 264

First-fall: Similar to Orange-crowned Warbler but has white undertail coverts; more conspicuous eyebrow stripe; greener above; trace of a pale yellowish wing bar; shorter tail. Note also needle-thin bill.

PHILADELPHIA VIREO *Vireo philadelphicus* p. 260

(Not a warbler.) "Vireo" song and actions. Note also thicker vireo bill. Compare with female Tennessee Warbler.

HOODED WARBLER *Setophaga citrina* p. 282

First-year female: Yellow eyebrow stripe, mostly yellow face, dark lores, bold white tail spots. (First-fall male resembles adult male.)

WILSON'S WARBLER *Cardellina pusilla* p. 282

First-year: Smaller and slimmer than Hooded Warbler with yellow lores, mostly olive cheeks, slimmer tail with no white. First-fall male has partial black cap; many first-fall females lack black.

BLACK-THROATED BLUE WARBLER *Setophaga caerulescens* p. 268

Female: Dark cheek, white wing spot. In some first-fall females this white spot is obscured, but note dark cheek and dull olive back. (First-fall male resembles adult male.)

CONNECTICUT WARBLER *Oporornis agilis* p. 278

First-fall: Large size. Plump. Brownish hood; complete, bold eye-ring. Walks.

MOURNING WARBLER *Geothlypis philadelphia* p. 278

First-fall female: Suggestion of hood; broken eye-ring. Brighter yellow below than Connecticut Warbler, including throat, contrary to grayish white throat of MacGillivray's Warbler.

NASHVILLE WARBLER *Oreothlypis ruficapilla* p. 264

First-fall: Yellowish (male) to buff (female) throat; sides of breast and undertail coverts tinged yellow, dull in females; eye-ring white to dingy pale; crown and nape grayish. Short tail, which it can bob.

COMMON YELLOWTHROAT *Geothlypis trichas* p. 280

First-fall female: Yellowish to buff throat, dull yellow breast and undertail coverts; brownish sides; white belly. Large bill and behavior help separate this from similar dull warblers.

PROTHONOTARY WARBLER *Protonotaria citrea* p. 278

First-year: Golden head tinged greenish on crown; dark eye stands out on plain face. Gray wings, white undertail, long bill.

CANADA WARBLER *Cardellina canadensis* p. 282

First-fall: Lores yellow, eye-ring white. Grayish to brownish gray above, yellow below, trace of necklace (nearly absent on duller females).

SELECTED FALL WARBLERS

Without Streaks or Wing Bars

ORANGE-CROWNED WARBLER
first-fall eastern

first-fall female
TENNESSEE WARBLER

PHILADELPHIA VIREO
not a warbler

HOODED WARBLER
rst-fall emale

WILSON'S WARBLER
first-fall female

BLACK-THROATED BLUE WARBLER
first-fall female
adult female

CONNECTICUT WARBLER
first-fall

MOURNING WARBLER
first-fall female

NASHVILLE WARBLER
first-fall female

COMMON YELLOWTHROAT
first-fall female

PROTHONOTARY WARBLER
female

CANADA WARBLER
first-fall female

PIPITS Family Motacillidae

Streaked ground birds with white outer tail feathers, long hind claws, thin bills; they walk and most wag tail. **FOOD:** Insects, seeds. **RANGE:** Nearly worldwide.

AMERICAN PIPIT *Anthus rubescens* **Uncommon to fairly common**

6½ in. (17 cm). A *slim-billed, sparrowlike* bird of open country. *Bobs tail* as it *walks; outer tail feathers white;* legs dark. Spring/summer birds grayer above, pinker and less streaked below; fall/winter birds washed olive and buff. Ages and sexes alike. **VOICE:** Call a distinctive, thin *jeet* or *jee-eet.* Aerial flight song: *chwee chwee chwee chwee chwee chwee chwee.* **SIMILAR SPECIES:** Sprague's Pipit. **HABITAT:** In summer, Arctic and alpine tundra; in migration and winter, fields, short-grass habitats, shores.

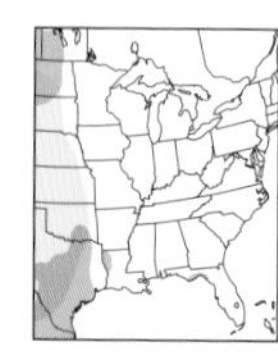

SPRAGUE'S PIPIT *Anthus spragueii* **Uncommon, secretive**

6½ in. (17 cm). A furtive species. Note *pinkish legs, striped back,* white outer tail feathers. *Plain face with beady dark eye.* Often towers high, then drops like a rock. Does *not* wag tail. **VOICE:** Aerial song a thin *shiing-a-ring-a-ring-a-ring-a.* Call a distinctive *squeet* or *squeet-squeet.* **SIMILAR SPECIES:** American Pipit differs in facial pattern and voice, has darker legs, wags tail. See juvenile Horned Lark. **RANGE:** Casual vagrant to East Coast. **HABITAT:** Short- to medium-grass prairies and fields.

LARKS Family Alaudidae

Brown terrestrial birds with long hind claws. Often sing in high display flights. **FOOD:** Seeds, insects. **RANGE:** Mainly Old World.

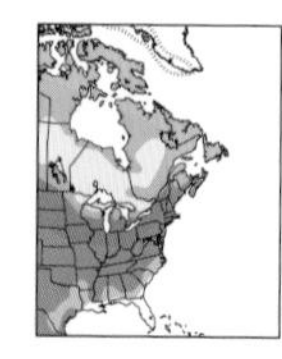

HORNED LARK *Eremophila alpestris* **Uncommon to common**

7–7¼ in. (18–19 cm). *Male:* Note head pattern. Larger than a sparrow, with *black mustache,* two small *black "horns,"* and black breast splotch. *Walks* on short legs. From below, white with *black* tail. *Female:* Similar but duller. *Juvenile:* Very different, *streaked below.* **VOICE:** Song tinkling, often prolonged; from ground or in air. Call a clear *tsee-titi.* **SIMILAR SPECIES:** Juvenile can be misidentified as Sprague's Pipit, longspurs, sparrows. **HABITAT:** Prairies, short-grass and dirt fields, airports, shores, tundra.

LONGSPURS and SNOW BUNTINGS
Family Calcariidae

A recently split family now consisting of five species in our area. Short legs; appear to feed on their bellies. Birds of open country; often found in winter flocks with pipits and larks. **FOOD:** Seeds, insects on breeding grounds.

SNOW BUNTING *Plectrophenax nivalis* **Uncommon**

6¾ in. (17 cm). Note extensive white. Browner in fall and winter, but flashing *white wing patches* in flight diagnostic. *Spring/summer male:* Black back contrasting with white head and underparts; bill black. *Females and first-fall birds:* Duller, less white, bill dull yellowish brown. Becomes whiter in spring due to plumage wear (not molt). **VOICE:** Call a sharp, whistled *teer* or *tew;* also a rough, purring *brrt,* similar to Lapland Longspur calls. Song a musical *ti-ti-chu-ree.* **SIMILAR SPECIES:** Beware leucistic sparrows. **RANGE:** Vagrant well south of normal winter range. **HABITAT:** Prairies, fields, dunes, shores. In summer, tundra.

PIPITS, LARK, AND BUNTING
AMERICAN PIPIT
fall/
winter
spring/
summer
SPRAGUE'S PIPIT
Sprague's
overhead
towering
flight
prairie
northern
juvenile
overhead
HORNED LARK
winter
female
winter
summer
SNOW BUNTING
winter
male
summer
male

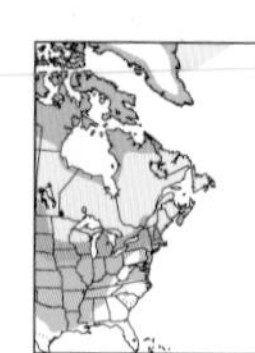

LAPLAND LONGSPUR Uncommon to fairly common
Calcarius lapponicus

6¼ in. (16 cm). The most widespread N. American longspur. *Spring/summer male: Black face outlined with white* is distinctive. Rusty collar. *Fall/winter male:* Sparse black streaks on sides, dull rusty nape, and smudge across breast help identify it. *Female:* Resembles fall/winter male, first-year duller. In all fall/winter plumages note *dark frame to rear cheek, rufous-brown edging on wing coverts,* tail pattern. **VOICE:** In flight, a dry rattle, also a musical *teew;* when perched, a soft *pee-dle.* Song in display flight is vigorous, musical. **SIMILAR SPECIES:** Smith's Longspur has similar tail pattern but buffier below; note face pattern, white checks in wings. Other longspurs have more white in tail. **HABITAT:** In summer, tundra; in winter, open fields, short-grass prairies, shores.

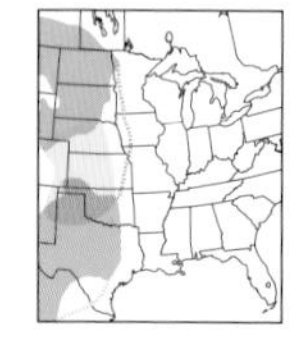

CHESTNUT-COLLARED LONGSPUR *Calcarius ornatus* Uncommon

6 in. (15 cm). *Spring/summer male:* Solid *black* below, except on throat and lower belly; nape *chestnut. Fall/winter male:* Colors muted by brown feather edging. *Female:* Very plain; best field marks are tail pattern (dark triangle on white tail), dark bill, and flight call. **VOICE:** Song short, feeble, but musical; suggests Western Meadowlark. Call a finchlike or turnstonelike *ji-jiv* or *kittle-kittle,* unique among longspurs. **SIMILAR SPECIES:** McCown's Longspur. **RANGE:** Casual vagrant to East Coast. **HABITAT:** Plains, native-grass prairies; generally prefers some cover. Winter flocks may disappear in grass until flushed.

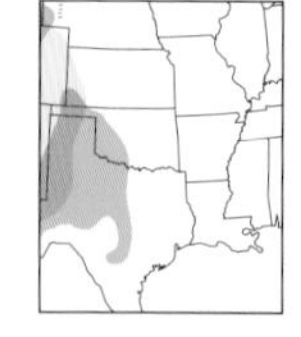

MCCOWN'S LONGSPUR *Rhynchophanes mccownii* Uncommon, local

6 in. (15 cm). *Spring/summer male:* Crown and patch on breast black, tail largely white. Hindneck *gray* (brown or chestnut in other longspurs). *Female and fall/winter male:* Rather plain; note tail pattern (inverted T of black on white) and *swollen-looking, pinkish, fleshy bill.* Some birds are especially *plain looking,* reminiscent of female House Sparrow. **VOICE:** Song in display flight is clear sweet warbles, suggestive of Lark Bunting (see p. 308). Call a dry rattle, softer than Lapland Longspur. Also a soft *pink.* **SIMILAR SPECIES:** Female Chestnut-collared Longspur darker, usually more heavily marked below, has slightly smaller and *darker (not pinkish)* bill, different call. **RANGE:** Casual vagrant to Midwest. **HABITAT:** Plains, prairies, short-grass and dirt fields.

SMITH'S LONGSPUR *Calcarius pictus* Scarce, local

6¼ in. (16 cm). This secretive longspur prefers enough grassy cover to disappear. It is *warm buff on entire underparts.* Tail edged with white, as in Vesper Sparrow and Lapland Longspur. *Spring/summer male: Deep buff;* ear patch with *white spot,* strikingly outlined by *black triangle. Female and fall/winter male:* Less distinctive; *buffy breast* lightly streaked; small pale spot on side of neck; most have white patch in wing coverts (absent or obscure in some females). **VOICE:** Rattling or clicking notes in flight (likened to winding a cheap watch). Song sweet, warblerlike, terminating in *WEchew.* Does not sing in flight. **SIMILAR SPECIES:** Lapland and Chestnut-collared Longspurs, Vesper Sparrow, Sprague's Pipit. **RANGE:** Casual vagrant to East Coast. **HABITAT:** Prairies, fields, airports; in summer, tundra with scattered bushes.

LONGSPURS
winter
female
LAPLAND
LONGSPUR
winter male
summer
male
female
CHESTNUT-
COLLARED
LONGSPUR
winter male
summer
male
female
MCCOWN'S
LONGSPUR
winter male
summer
male
female
winter
male
SMITH'S
LONGSPUR
summer
male
see Horned Lark
(p. 288) and pipits
(p. 288)

SPINDALISES Family Spindalidae

A Caribbean family, considered tanagers (Thraupidae), but genetic evidence indicates they are distinct. Found in forests. **FOOD:** Primarily fruit.

WESTERN SPINDALIS *Spindalis zena* **Casual vagrant**

6¾ in. (17 cm). *Male:* Breast burnt orange and orange-yellow, *bold black-and-white head stripes and shoulder. Female:* Plain gray-brown overall, breast sometimes washed yellow; thick bill; *pale patch at base of primaries.* **VOICE:** A series of thin, high notes, *tzee-tzee-tzee,* often given with buzzy phrase toward end. **SIMILAR SPECIES:** Tanagers. **RANGE:** Vagrant to s. FL from W. Indies. **HABITAT:** Brushy woodlands, fruit trees.

TROPICAL TANAGERS Family Thraupidae

A large, diverse, and colorful family. N. American tanagers have been recently moved to Cardinalidae (p. 316). **FOOD:** Nectar, insects. **RANGE:** Neotropics.

BANANAQUIT *Coereba flaveola* **Very rare vagrant**

4½ in. (11 cm). Small, short-tailed, with decurved bill and bold white supercilium. *Adult:* Dark above, white below with white throat and yellow wash across belly. Sexes similar. *Juvenile:* Paler overall, supercilium indistinct, bill shorter and yellow at base. **VOICE:** An explosive series of buzzy notes and sneezy squeaks. **RANGE:** Very rare visitor to s. FL from W. Indies. **HABITAT:** Open brushy areas, nectar- and fruit-bearing trees.

MORELET'S SEEDEATER *Sporophila morelleti* **Rare, local**

4½ in. (11 cm). Tiny, with stubby bill. *Adult male:* Dark cap, incomplete light collar, white wing spot; duller in fall/winter. *Female:* Buffy with eye-ring, wing bars. *First-year male:* Variably intermediate. **VOICE:** A high, then low *sweet, sweet, sweet, cheer, cheer, cheer.* Call a high *wink.* **HABITAT:** Tall, thick stands of grass and other similar deep cover.

OLD WORLD SPARROWS Family Passeridae

Differ from our native sparrows (family Passerellidae) by having a curved bill culmen. **FOOD:** Mainly insects, seeds. **RANGE:** Widespread in Old World.

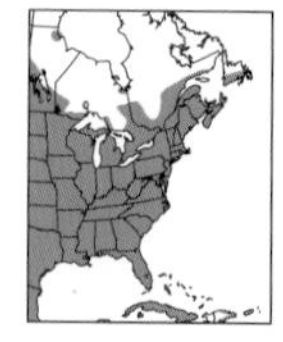

HOUSE SPARROW *Passer domesticus* **Common, introduced**

6¼ in. (16 cm). Introduced from Europe. *Male: Black throat, white cheeks, chestnut nape. Female and juvenile:* Lack black throat, have dingy breast and dull eye stripe *behind eye only;* note *single wing bar.* **VOICE:** Hoarse *chirp* and *shillip* notes, also a rising *sweep.* Song a series of such notes. **SIMILAR SPECIES:** Female Dickcissel, buntings, sparrows, Eurasian Tree Sparrow. **HABITAT:** Cities, towns, farms, feeders.

EURASIAN TREE SPARROW **Uncommon, local, introduced**
Passer montanus

6 in. (15 cm). Both sexes resemble male House Sparrow, but black throat patch smaller; note *black ear spot.* Crown brown. **VOICE:** Higher pitched than House Sparrow. A metallic *chik* or repeated *chit-tchup.* In flight, a hard *tek, tek.* **SIMILAR SPECIES:** House Sparrow. **RANGE:** Introduced from Europe around St. Louis in 1870. Some northward expansion since then. **HABITAT:** Farmland, weedy patches, locally in residential areas, feeders.

MISCELLANEOUS PASSERINES

NORTH AMERICAN TOWHEES, SPARROWS, and JUNCOS Family Passerellidae

Formerly lumped with Old World seed-eating birds, this family now consists of our familiar sparrows and allies. Juveniles are more heavily streaked; otherwise, ages and sexes similar in most species. **FOOD:** Seeds, insects, fruit, varying seasonally. **RANGE:** Throughout Americas.

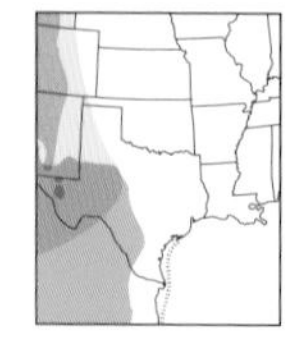

GREEN-TAILED TOWHEE *Pipilo chlorurus* **Scarce, local**

7¼ in. (18 cm). A slender towhee; note *rufous cap,* conspicuous *white throat,* black mustache, gray chest, and plain *olive-green upperparts* (sexes similar). Juvenile plainer, streaked on breast. **VOICE:** Call a catlike mewing note. **RANGE:** Casual winter vagrant to East. **HABITAT:** In winter, brushy and riparian woods, on ground under feeders.

OLIVE SPARROW *Arremonops rufivirgatus* **Uncommon, local**

6¼ in. (16 cm). Olive above, gray below, with two dull brown stripes on crown (sexes similar). Juvenile plainer and streaked. **VOICE:** Song composed of dry notes on one pitch going into Chipping Sparrow–like rattle; reminiscent of Field Sparrow. Call a sharp *chip* like Orange-crowned Warbler; also a hissing trill. **HABITAT:** Bushy thickets.

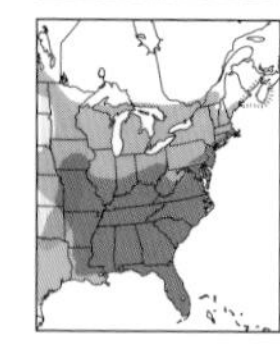

EASTERN TOWHEE *Pipilo erythrophthalmus* **Fairly common**

8 in. (20 cm). Smaller and more slender than an American Robin; rummages among leaf litter. Readily recognized by rufous sides. *Male:* Head and upperparts black; sides rufous rust, belly white. Flashes white patches at base of primaries and on tail corners. Eye usually red (but white in birds of s. Atlantic Coast and FL). *Female:* Similar, but brown where male is black. *Juvenile:* Streaked below like a large sparrow, but with diagnostic towhee wing and tail patterns. **VOICE:** Song *drink-your-tea,* last syllable higher, wavering. Call a loud *chewink!* Southern white-eyed subspecies (*P. e. alleni*) gives a more slurred *shrink* or *zree;* song *cheet cheet cheeeeee.* **SIMILAR SPECIES:** Spotted Towhee. **HABITAT:** Open woods, undergrowth, brushy edges, hedgerows, feeders.

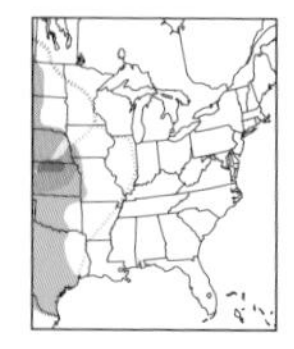

SPOTTED TOWHEE *Pipilo maculatus* **Uncommon to rare**

8 in. (20 cm). Similar to Eastern Towhee, but *back heavily spotted with white,* lacks visible white bases to primaries. Female has dusky grayish-black to brownish-black head. **VOICE:** Song a drawn-out, buzzy *chweeeeee.* Sometimes *chup chup chup zeeeeeeee;* variable. Call a catlike *gu-eeee?* **SIMILAR SPECIES:** Eastern Towhee overlaps slightly on Great Plains and in TX in winter. **RANGE:** Casual fall/winter vagrant to Midwest and Gulf Coast. **HABITAT:** Open woods, undergrowth, chaparral, brushy edges, gardens.

CANYON TOWHEE *Melozone fusca* **Uncommon, local**

8¾ in. (22 cm). Dull brown, with moderately long, dark tail; suggests a very plain, slim, overgrown sparrow. Note rufous-washed crown, faint dusky necklace, pale *rusty undertail coverts.* Juvenile streaked on breast. **VOICE:** Call an odd *shed-lp* or *kedlp.* Song an accelerating string of call notes. **HABITAT:** Brushy areas in canyons and deserts, residential areas, feeders.

GREEN-TAILED
TOWHEE
OLIVE
SPARROW
male
white-eyed
variant
female
juvenile
EASTERN
TOWHEE
male
male
juvenile
SPOTTED
TOWHEE
female
CANYON
TOWHEE

BLACK-THROATED SPARROW *Amphispiza bilineata* Uncommon, local

5½ in. (14 cm). *Adult:* Note *white face stripes* and *jet-black throat and chest.* White corners on *distinct black tail.* Sexes alike. *Juvenile:* Seen into fall; *lacks* black throat but has similar head pattern; breast weakly streaked. **VOICE:** Song a sweet *cheet cheet cheeeeeeee;* calls are light tinkling notes. **RANGE:** Casual vagrant north and east of breeding range, largely first-fall birds. **HABITAT:** Arid brush, juniper hillsides; in winter, open areas.

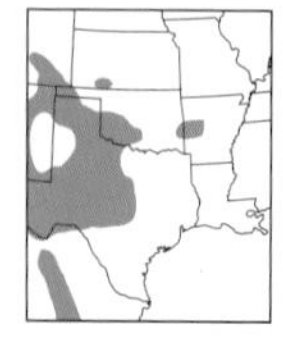

RUFOUS-CROWNED SPARROW *Aimophila ruficeps* Uncommon, local

6 in. (15 cm). Dark, with plain dusky breast, rufous cap and line behind eye, rounded tail. Note *black whiskers* bordering throat and *distinct circular whitish eye-ring.* Juvenile has streaked breast. **VOICE:** Song suggests thin House Wren song. Call *dear, dear, dear.* **SIMILAR SPECIES:** Chipping Sparrow paler, more slender, smaller billed, lacks distinct whiskers. **HABITAT:** Grassy or rocky slopes with sparse low bushes; open pine-oak woods.

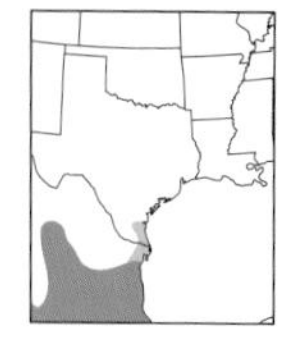

BOTTERI'S SPARROW *Peucaea botterii* Uncommon, local

6 in. (15 cm). Nondescript. Buffy breast, plain brown tail lacking white corners. Juvenile has sparse streaks on breast. *Best identified by voice.* Bill culmen slightly curved. **VOICE:** Song a constant tinkling and "pitting," sometimes ending with dry trill. Very unlike Cassin's Sparrow song. **SIMILAR SPECIES:** Cassin's Sparrow very similar but usually grayer, has small white corners on tail, straighter bill culmen; upperparts often more patterned, less streaked. **HABITAT:** Desert grasslands and bunch grass (particularly sacaton grass).

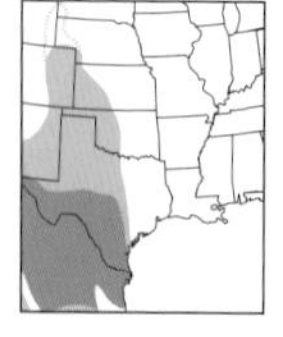

CASSIN'S SPARROW *Peucaea cassinii* Fairly common

6 in. (15 cm). A large, drab sparrow of open arid country; underparts dingy, unmarked or with faint streaking on flanks. Upperparts appear patterned, with anchor-shaped markings on individual feathers. Individuals can be either more rufous or grayer than shown. *Pale or whitish corners* on *rounded,* gray-brown tail. Juvenile has dark breast streaking. **VOICE:** Song *ti ti tseeeeeee tay tay.* Often "skylarks" in air, giving trill at climax; also flicks wings and tail in flight. **SIMILAR SPECIES:** Botteri's Sparrow. Savannah Sparrow smaller, streakier overall, and shorter tailed than Cassin's. **RANGE:** Casual to accidental fall vagrant to n. Midwest. **HABITAT:** Desert grasslands and semiarid prairies, bushes.

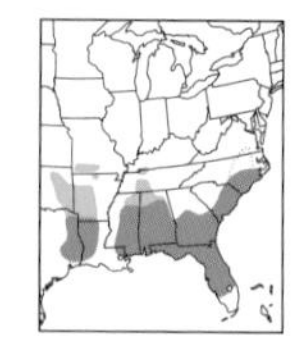

BACHMAN'S SPARROW *Peucaea aestivalis* Scarce

6 in. (15 cm). A large sparrow, with long, rounded tail. Striped with reddish brown above, washed with dingy buff across plain breast, with gray bill. Juvenile has streaked breast. **VOICE:** Song variable; usually a clear liquid whistle followed by loose trill or warble on a different pitch, e.g., *seeeee, slip slip slip slip slip.* **SIMILAR SPECIES:** Field Sparrow smaller, with pink bill. Grasshopper Sparrow has light crown stripe and short tail. Juvenile Bachman's can suggest Lincoln's Sparrow (not found in summer in Bachman's range), which has smaller bill and bolder streaks on breast. **RANGE:** Casual vagrant north of range. **HABITAT:** Restricted primarily to open pine woods with grass and palmetto scrub, where it flushes reluctantly, then drops back into cover.

SPARROWS
BLACK-THROATED SPARROW
juvenile
adult
RUFOUS-CROWNED SPARROW
BOTTERI'S SPARROW
CASSIN'S SPARROW
BACHMAN'S SPARROW

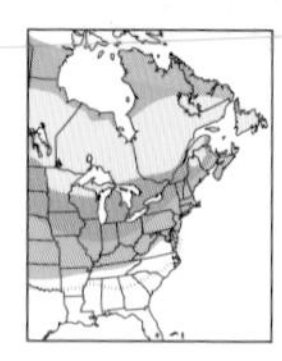

AMERICAN TREE SPARROW *Spizelloides arborea* Fairly common

6¼ in. (16 cm). Note *dark "stickpin"* on breast, and *red-brown cap. Bill dark above, yellow below;* white wing bars; rufous wash on flanks. Ages and sexes rather similar (juvenile streaked). **VOICE:** Song sweet, variable, opening on one or two high, clear notes. Call *tseet;* feeding call a musical *teelwit.* **SIMILAR SPECIES:** Field and Chipping Sparrows. **RANGE:** Casual vagrant well south of winter range. **HABITAT:** Arctic and taiga scrub, willow thickets; in winter, brushy roadsides, weedy edges, cattail marshes, feeders.

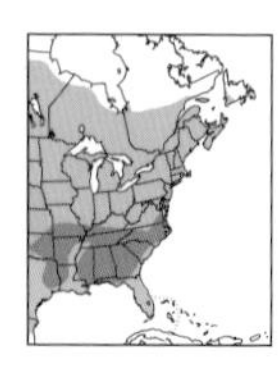

CHIPPING SPARROW *Spizella passerina* Common

5½ in. (14 cm). *Spring/summer:* A small, slim, long-tailed, plain-breasted sparrow with bright *rufous cap, black eye line, white eyebrow. Fall/winter:* Duller; note *dark eye line,* dirty grayish breast, *gray rump. Juvenile:* Has fine streaks on breast, rump not as gray; this plumage may be held through fall migration. **VOICE:** Song a dry chipping rattle on one pitch. Call a thin *tseet.* **SIMILAR SPECIES:** Clay-colored and Swamp Sparrows. **HABITAT:** Open woods, especially pine, oak; orchards, farms, towns, lawns, feeders. Often forms flocks in fall and winter.

CLAY-COLORED SPARROW *Spizella pallida* Fairly common

5½ in. (14 cm). Paler and buffier than fall/winter Chipping Sparrow, with paler lores, *more sharply outlined face pattern,* more contrasting grayer nape, browner rump, whiter underparts; rather indistinct, broken eye-ring. Juvenile streaked below (held only briefly on breeding grounds). **VOICE:** Insectlike; three or four low, flat buzzes: *bzzz, bzzz, bzzz.* Call a thin *tseet,* like Chipping Sparrow but higher. **SIMILAR SPECIES:** Chipping Sparrow. **RANGE:** Scarce migrant or vagrant to East Coast. **HABITAT:** Scrub, brushy prairies, jack pines, weedy areas.

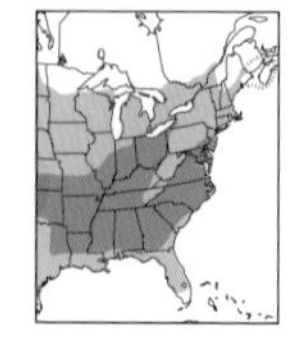

FIELD SPARROW *Spizella pusilla* Fairly common

5¾ in. (15 cm). A small, slim, rusty-capped sparrow. Note small *pink bill,* white eye-ring, plain buffy breast; rusty upperparts, and weak face striping. Juvenile has finely streaked breast; plumage not held long. **VOICE:** Song a distinctive accelerating trill, *psew-psew-psew-see-see-see-see* (ascending, descending, or on one pitch). Call *tseew.* **SIMILAR SPECIES:** American Tree and Chipping Sparrows. **HABITAT:** Overgrown fields, pastures, brush, feeders.

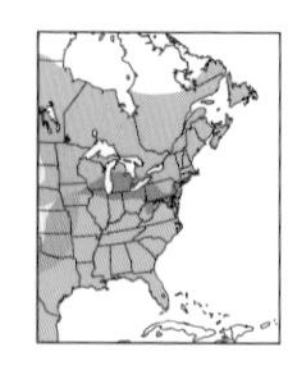

SWAMP SPARROW *Melospiza georgiana* Fairly common

5¾ in. (15 cm). A rather plump, dark, *rusty-winged* sparrow with tawny flanks and *broad black back striping. Adult male: White throat, rusty cap, blue-gray neck and breast. Female and first-year:* Variably average duller; *blackish* or dark rust crown, *olive-gray neck and breast;* dim flank streaking. *Juvenile:* Heavily streaked. **VOICE:** Song a loose trill, similar to Chipping Sparrow but slower, sweeter, and stronger. Call a hard *cheep,* similar to Eastern Phoebe. **SIMILAR SPECIES:** Song Sparrow slightly larger, larger billed, longer tailed, has *heavier breast streaks,* lacks tawny flanks. Lincoln's Sparrow has buff breast with fine sharp streaks, finer bill. *Spizella* sparrows slimmer, longer tailed, have wing bars. **HABITAT:** Nests in freshwater marshes with bushes, cattails, sedges, willows; winters in fresh and salt marshes, pond edges, weedy ditches.

AMERICAN
TREE
SPARROW
CHIPPING
SPARROW
fall/
winter
spring/
summer
juvenile
CLAY-COLORED
SPARROW
spring/
summer
fall/winter
juvenile
FIELD
SPARROW
adult
SWAMP
SPARROW
adult
first-year
female

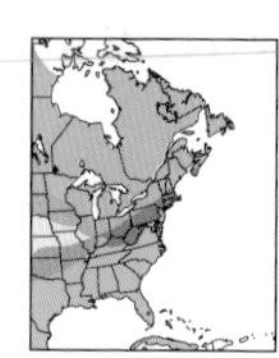

SAVANNAH SPARROW *Passerculus sandwichensis* Common

5½–5¾ in. (14–15 cm). This streaked, open-country sparrow suggests a small, short-tailed Song Sparrow, but it usually has *yellowish on front of eyebrow, whitish stripe through crown,* and pinker legs. Ages and sexes similar. Note especially the short, notched tail, with palish but not bright white outer feathers. "Ipswich" subspecies (*P. s. princeps*), which breeds on Sable I., NS, and winters along Atlantic Coast beaches, is paler overall and slightly larger than other eastern subspecies. **VOICE:** Song a lisping, buzzy *tsit-tsit-tsit, tseeee-tsaaay* (last note lower); similar to Grasshopper Sparrow except for lower last note. Call a short *tseep* or light *tsu.* **SIMILAR SPECIES:** Song Sparrow. See also Vesper Sparrow. **HABITAT:** Open fields, farms, meadows, salt marshes, prairies, dunes.

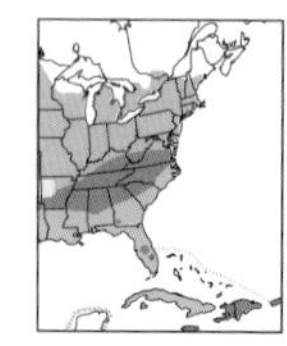

GRASSHOPPER SPARROW *Ammodramus savannarum* Uncommon

5 in. (13 cm). A small, compact-bodied sparrow, with large, flat head and short, sharp tail, found in taller grasslands. Flight feeble. *Adult:* Crown with pale median stripe; *yellow lores; whitish eye-ring,* purplish-edged upperpart feathers; note relatively *unstriped buffy breast.* Endangered resident FL subspecies (*A. s. floridanus*) darker and browner above, paler below, larger billed. *Juvenile* (found on migration): Less colorful, has dusky streaks on breast. **VOICE:** A distinctive, very thin, dry, insectlike *pi-tup zeeeeeeeeeeee.* **SIMILAR SPECIES:** LeConte's Sparrow slimmer, longer tailed, smaller billed; adult with orangier eyebrow; juvenile buffier, plainer faced. See Savannah Sparrow, female Bobolink. **HABITAT:** Grasslands, hayfields, pastures, prairies.

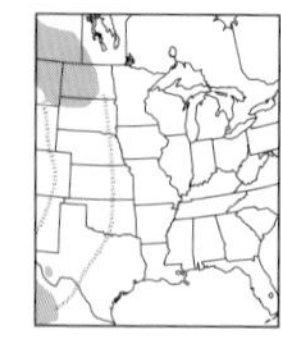

BAIRD'S SPARROW *Centronyx bairdii* Scarce, local

5½ in. (14 cm). An elusive, skulking prairie sparrow. *Adult:* Light breast crossed by *narrow band* of fine black streaks. Head ocher buff, streaked, with broad *ocher* median crown stripe, *double mustache stripes.* Flat head. **VOICE:** Song begins with two or three high musical *zips*, ends with trill on lower pitch; more musical than Savannah Sparrow. **SIMILAR SPECIES:** Savannah Sparrow has smaller bill, more extensively streaked below, lacks dark marks at rear of auriculars and double mustache stripes. See Henslow's Sparrow (ranges do not normally overlap). **RANGE:** Accidental vagrant east of range to OH. **HABITAT:** Native prairies, scattered bushes used as song perches.

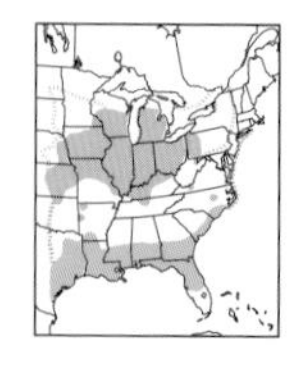

HENSLOW'S SPARROW *Centronyx henslowii* Scarce, local

5 in. (13 cm). A secretive sparrow best located by its odd song. Short-tailed and flat-headed, with large, pale bill. *Adult:* Has fine stripes across breast. Olive-colored head, double mustache stripes, spots behind "ear," and reddish wings. *Juvenile:* Back and underparts dull olive and without breast streaking (breeding grounds only). **VOICE:** Song a hiccuping *tsi-lick.* May sing on quiet, windless nights. **SIMILAR SPECIES:** Juvenile Henslow's Sparrow (without breast streaks) might resemble adult Grasshopper Sparrow, whereas juvenile Grasshopper has breast streaks but lacks Henslow's olive and russet tones. **RANGE:** Casual vagrant to Great Plains states. **HABITAT:** Breeding habitat very specific: partially overgrown fields with dead or dried vegetation and dense leaf litter; has adapted to hayfields. Winters in dense cover in southern pine forests.

OPEN-COUNTRY SPARROWS
typical
"Ipswich"
SAVANNAH
SPARROW
juvenile
GRASSHOPPER
SPARROW
adult
BAIRD'S
SPARROW
juvenile
HENSLOW'S
SPARROW
adult

SALTMARSH SPARROW *Ammospiza caudacuta* **Uncommon**

5¼ in. (13 cm). A short-tailed, often shy sparrow of coastal marshes. Note deep ocher yellow or orange on face, which completely surrounds gray ear patch. Distinct streaks on mostly whitish or light buff breast, flat-headed appearance. Juvenile similar but washed *pale buff*. **VOICE:** Song a weak, varied jumble of buzzy hisses and clicks; not distinctly two-parted like Nelson's Sparrow. **SIMILAR SPECIES:** Nelson's and LeConte's Sparrows. Juvenile Seaside Sparrow on same breeding grounds grayer, loral area yellowish. Savannah Sparrow smaller, smaller headed, shorter notched tail. **HABITAT:** Coastal salt marshes.

NELSON'S SPARROW *Ammospiza nelsoni* **Uncommon**

5 in. (13 cm). A shy marshland skulker with three widely separated breeding populations. Note bright *orange on face,* completely surrounding gray ear patch. *Breast warm buff with faint blurry streaks,* stronger streaks on flanks. Gray central crown and *unmarked gray nape.* Back sharply striped with white. Ages and sexes similar. Birds breeding in New England and Maritime Provinces grayer with less distinct stripes. Juvenile similar but washed *orange-buff*. **VOICE:** Song a buzzy, two-parted *shleeee-tup.* **SIMILAR SPECIES:** Saltmarsh Sparrow has heavier breast streaking, and orange on breast, if present, is *paler* than orange on face (breast and face equally bright in Nelson's); juveniles paler, less orangish. LeConte's Sparrow has white median crown stripe, purplish chestnut streaks on nape. **RANGE:** Casual inland migrant or vagrant. **HABITAT:** In summer, prairie and coastal marshes, muskeg; in winter, coastal marshes.

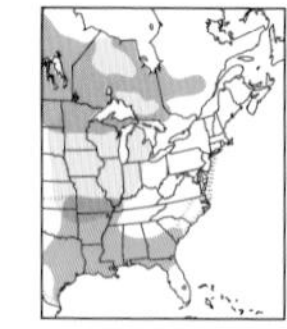

LECONTE'S SPARROW *Ammospiza leconteii* **Uncommon**

5 in. (13 cm). A skulking sharp-tailed sparrow of prairie marshes, boggy fields. Note *bright orange* eyebrow and buffy breast (with streaks *confined to sides*), *purplish-chestnut streaks on nape,* white median crown stripe, strong stripes on back. Juvenile, which can be found on migration, buffier overall and has streaked breast. **VOICE:** Song consists of two extremely thin, grasshopper-like hisses. **SIMILAR SPECIES:** Nelson's Sparrow. Adult and juvenile Grasshopper Sparrow. **RANGE:** Casual vagrant to East Coast. **HABITAT:** Grassy marshes, tallgrass fields, weedy hayfields.

SEASIDE SPARROW *Ammospiza maritima* **Fairly common**

6 in. (15 cm). A dark, *gray* sparrow of salt marshes. *Adult:* Short *yellow area above lores. Whitish throat* and white above dark malar. *Juvenile:* Similar but duller, browner. Shares marshes with Nelson's and Saltmarsh Sparrows along Atlantic Coast. "Cape Sable" Seaside Sparrow (*A. m. mirabilis*) is an endangered subspecies confined to s. FL (the only Seaside Sparrow breeding there); more greenish than typical birds and with *much heavier breast streaking.* **VOICE:** Song *cutcut ZHE-eeeeeeee;* much stronger than Saltmarsh Sparrow. Call *chack.* **HABITAT:** Salt marshes.

MARSH SPARROWS
SALTMARSH SPARROW
coastal
NELSON'S SPARROW
interior
LECONTE'S SPARROW
juvenile
adult
"Cape Sable"
SEASIDE SPARROW

LARK SPARROW *Chondestes grammacus* Uncommon, local

6½ in. (17 cm). *Adult:* Note *black tail with white corners;* also dark *central breast spot* on clean whitish underparts, *quail-like head pattern. Juvenile:* Head pattern duller, a few dusky streaks on breast. **VOICE:** Clear buzzing and churring trills, with pauses between. Call a sharp *tsip.* **SIMILAR SPECIES:** Vesper Sparrow. **RANGE:** Rare vagrant to East Coast. **HABITAT:** Open country with bushes, trees; pastures, farms, roadsides.

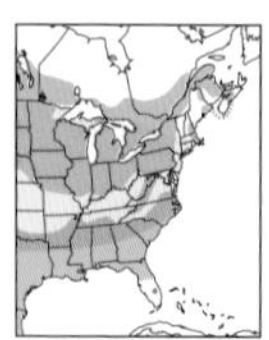

VESPER SPARROW *Pooecetes gramineus* Uncommon

6¼ in. (16 cm). *White outer tail feathers* are conspicuous when bird flies; note also prominent *whitish eye-ring, chestnut* bend of wing *(sometimes difficult to see),* white malar stripe, and lack of central crown stripe. **VOICE:** Song similar to Song Sparrow but throatier; usually begins with two clear minor notes, followed by two higher ones. Call a brief *tseet.* **SIMILAR SPECIES:** Savannah Sparrow lacks bright white outer tail feathers and distinct eye-ring. **HABITAT:** Meadows and prairies with scattered trees or bushes (such as sage), roadsides, farm fields.

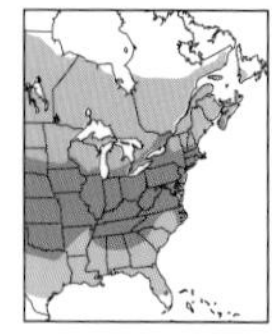

SONG SPARROW *Melospiza melodia* Common

5¾–6½ in. (15–17 cm). This common, midsized sparrow has a *long rounded tail* and *heavy breast streaks* that merge into a *large central spot.* Broad grayish eyebrow. Juvenile more finely streaked, often lacks central spot. **VOICE:** Song a variable series of notes, some musical, some buzzy; usually starts with three or four bright repetitious notes, *sweet sweet sweet,* etc., and ends in lower buzzy trill. Call a low, nasal *tchep.* **SIMILAR SPECIES:** Savannah Sparrow has yellowish over eye, shorter notched tail, pinker legs. See Lincoln's and Swamp Sparrows. **HABITAT:** Thickets, brush, marshes, roadsides, gardens, feeders.

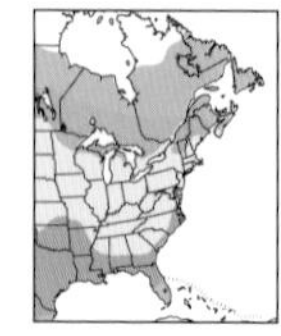

LINCOLN'S SPARROW *Melospiza lincolnii* Uncommon

5¾ in. (15 cm). A somewhat skulking species, preferring to be near cover. Similar to Song Sparrow but smaller and trimmer, side of face grayer, sharp breast streaks *much finer* and overlaid on *creamy buff* breast. **VOICE:** Song sweet and gurgling; starts with low passages, rises abruptly, drops. Calls a hard *tik* and buzzy *zzzeeet.* **SIMILAR SPECIES:** First-year Swamp Sparrow has duller breast with blurry streaks, rustier wing. Juvenile Swamp, Song, and Lincoln's Sparrows very similar: check differences in breast streaking, malar stripe, and bill size and shape. **HABITAT:** Willow and alder thickets, muskeg, brushy bogs; in winter, wet fields, brush, thickets, sometimes feeders.

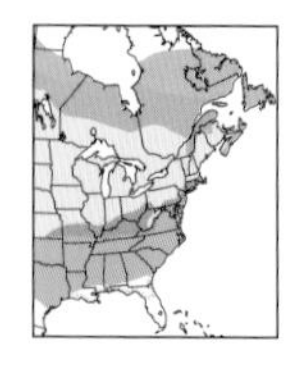

FOX SPARROW *Passerella iliaca* Uncommon

7 in. (18 cm). A large, plump sparrow; most subspecies have *rusty rump and tail.* Action towhee-like, kicking among dead leaves and other ground litter. *Breast heavily streaked* with triangular spots, often clustered in a large blotch on upper breast. Ages and sexes similar. Fox Sparrows vary widely; many subspecies, but only one is found regularly in East: "Red" (*P. i. iliaca*), which is bright rusty with rusty back stripes; grayer western subspecies scarce winter vagrant to East. **VOICE:** Song brilliant and musical; a varied arrangement of short clear notes and sliding whistles. Call a flat *chup* or *chick.* **SIMILAR SPECIES:** Song Sparrow much smaller. **HABITAT:** Wooded undergrowth, brush, feeders.

STREAKED SPARROWS
LARK SPARROW
juvenile
adult
VESPER SPARROW
SONG SPARROW
LINCOLN'S SPARROW
FOX SPARROW
"Red" (East)
"Gray" (West)

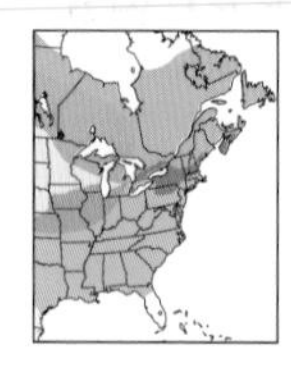

WHITE-THROATED SPARROW *Zonotrichia albicollis* Common

6¾ in. (17 cm). *Spring/summer:* A gray-breasted sparrow with distinct white throat and yellow lores. Bill grayish. Polymorphic, with head stripes varying in shades of black, brown, and tan; some adults have bold black-and-white head stripes, others duller brown and tan. Tan morph may be moderately streaked on breast; throat duller. Ages and sexes similar, although first-fall birds and females average duller than adults and males within each morph. **VOICE:** Song several clear pensive whistles, often rendered *old sam peabody peabody peabody.* Call a hard *chink;* also a thin, slurred *tseet.* **SIMILAR SPECIES:** White-crowned Sparrow. **HABITAT:** Thickets, brush, undergrowth of coniferous and mixed woodlands. Regularly visits feeders, preferring to stay on ground.

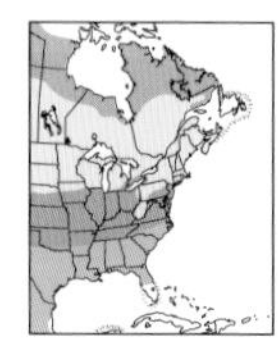

WHITE-CROWNED SPARROW *Zonotrichia leucophrys* Uncommon

7 in. (18 cm). This species comprises two subspecies in East, one exhibiting whitish lores and tawny bill (*Z. l. gambelii,* a rare migrant from w. N. America eastward), the other black lores and yellowish bill (*Z. l. leucophrys,* eastern breeding and wintering). *Adult:* Clear breast, crown *striped with black and white. First-fall/-winter:* Head stripes dark red-brown and light buff. **VOICE:** Song one or more clear, plaintive whistles, often *chew-chee-tzip-tzip-tzip tseew* but variable by subspecies. Call a sharp *pink.* **SIMILAR SPECIES:** White-throated Sparrow has well-defined white throat, yellow spot before eye, grayish bill. First-fall/-winter Golden-crowned Sparrow slightly larger, has *duskier bill and underparts,* more muted head pattern, usually with *dull yellowish forehead.* **HABITAT:** Brush, forest edges, thickets; in winter, also farms, desert washes, gardens, parks, feeders.

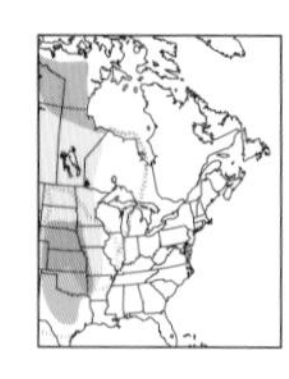

HARRIS'S SPARROW *Zonotrichia querula* Uncommon

7½ in. (19 cm). Large; almost size of Fox Sparrow. *Adult: Black crown, face, and bib encircling pink bill* (sexes similar). *First-fall/-winter:* Has *white on throat,* less black on crown, buffy brown on rest of head; blotched and streaked on breast. Plumage varies. **VOICE:** Song has quavering quality of other *Zonotrichia* sparrows: one to three clear whistles, *peee* to *peee-pee-pee,* on same pitch. Alarm call *wink.* **RANGE:** Casual vagrant to East Coast. **HABITAT:** Stunted boreal forests; in winter, brush, hedgerows, open woods. May mix with White-crowned Sparrows in winter.

GOLDEN-CROWNED SPARROW Very rare vagrant
Zonotrichia atricapilla

7¼ in. (18 cm). *Spring/summer:* Note *yellow central crown stripe,* bordered broadly with black. Dusky bill. *First-fall/-winter:* May look like large female House Sparrow but usually with dull yellow suffusion on forehead; fall/winter adult similar but with blacker head stripes. **VOICE:** Call a sharp *tsew.* **SIMILAR SPECIES:** White-crowned Sparrow. **RANGE AND HABITAT:** Very rare vagrant or winter visitor to Midwest from West, casual vagrant to East Coast. Winter habitat similar to that of White-crowned Sparrow but favors denser shrubs; vagrants often found at feeders.

SPARROWS
tan-striped morph
dull
WHITE-THROATED SPARROW
white-striped morph
first-fall/ winter
interior
adult
WHITE-CROWNED SPARROW
Eastern
adult
first-fall/ winter
GOLDEN-CROWNED SPARROW
spring/ summer
adult
first-fall/ winter
HARRIS'S SPARROW

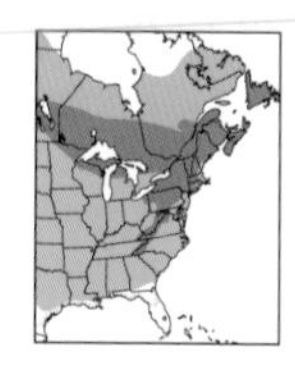

DARK-EYED JUNCO *Junco hyemalis* **Common**

6–6½ in. (15–16 cm). This familiar winter songbird is characterized by a variably dark-hooded appearance and *white outer tail feathers* that flash conspicuously in flight. Bill and belly white to whitish. Adult male has dark hood; first-year male and adult female slightly less so; first-year female drabbest. *Juvenile:* Finely streaked on head and breast; note extensively white outer tail feathers typical of juncos. A complex species of distinct subspecies groups (often intergrading) as follows.

"Oregon" Junco (*J. h. oreganus*) is the widespread subspecies in West and a casual fall and winter vagrant in East. Male has *rusty brown back* with *blackish hood* and *buffy, brownish, or rusty sides* (side color variable). Female duller, but note contrast between paler gray hood and brown back, convex shape on lower border of hood.

"Pink-sided" Junco (*J. h. mearnsi*) breeds in n. Great Basin and is a rare winter vagrant in East. Male has gray hood, pink flanks, and black lores; female duller. (Not shown.)

"Slate-colored" Junco (*J. h. hyemalis*) is the common northern and eastern breeding subspecies that winters throughout East. Gray with *gray back and sides,* white belly. Female and first-years of both sexes duller gray tinged brownish on back. The more uniform coloration, lacking rusty areas, is distinctive. Some particularly brownish young birds may be confused with "Oregon" Junco but usually have gray rather than brown or buff sides.

"White-winged" Junco (*J. h. aikeni*) breeds in Black Hills region and winters in prairie states; casual vagrant farther east. Large and dark (resembling "Slate-colored") with gray back; usually has *two whitish wing bars* and exhibits considerably more white in tail (four outer feathers on each side). Some female "White-winged" Juncos and some "Slate-colored" Juncos can have thin, weak, or broken wing bars, so caution is warranted.

VOICE: Song a loose trill, suggestive of Chipping Sparrow but more musical. Call a light *smack;* also clicking or twittering notes. **HABITAT:** Coniferous and mixed woods. In fall and winter, open woods, undergrowth, roadsides, brush, parks, gardens, feeders; often in flocks.

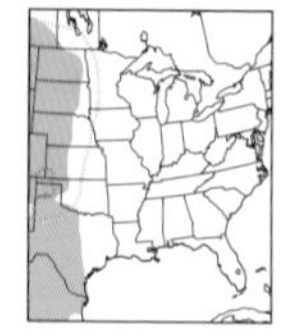

LARK BUNTING *Calamospiza melanocorys* **Uncommon, local**

7 in. (18 cm). A plump, short-tailed prairie sparrow. Gregarious in fall and winter. Note rather *heavy, blue-gray bill. Spring/summer male: Black,* with *large white wing patches. Female and first-year and fall/winter adult male:* Brown, streaked; pattern suggests female Purple Finch but with *whitish* or *buffy white wing patches* and *tail corners.* Fall/winter adult males can retain some black on face, wings, and belly. **VOICE:** Song, given in flight display, composed of cardinal-like slurs, unmusical chatlike *chugs*, piping whistles and trills; each note repeated 3–11 times. Call a flat, mellow *heew.* **SIMILAR SPECIES:** Male Bobolink has yellow nape patch and white rump. Beware leucistic blackbirds with odd patches of white in wings. **RANGE:** Widespread vagrant to East, casually to coast. **HABITAT:** Plains, prairies; in winter, also weedy desert lowlands and farm fields.

DARK-EYED JUNCO
adult female
adult male
"Oregon" West
juvenile
"Slate-colored"
adult male East
adult female
"White-winged" Black Hills region
LARK BUNTING
female and fall/winter male
spring/summer male

FINCHES Family Fringillidae

Plump, small to medium-small birds with seed-cracking bills; relatively short, notched tails; often undulating flight. Sexes usually differ. More arboreal than sparrows. **FOOD:** Seeds, insects, small fruit. **RANGE:** Worldwide.

BRAMBLING *Fringilla montifringilla* Casual vagrant

6¼ in. (16 cm). *Tawny* or *orangey buff* breast and shoulders, *whitish rump distinctive in flight. Spring/summer male:* Black head and back. *Female and fall/winter male:* Gray cheek bordered by dark, flanks streaked or spotted. **VOICE:** Call a whiny *zweee;* in flight, a distinctive nasal, hollow *eck.* **RANGE:** Eurasian species; casual but widespread records from Canada and n. U.S. to East Coast, primarily in winter and often at feeders.

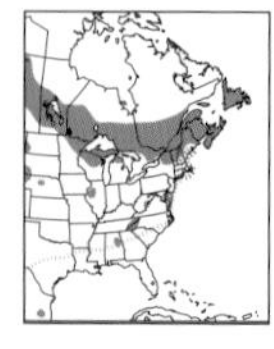

RED CROSSBILL *Loxia curvirostra* Uncommon, irregular

5¾–7 in. (14–17 cm). Note *crossed mandibles* and *plain wings.* Usually found in *flocks. Adult male: Dull red,* brighter on rump. *Second-year male:* Often washed orange. *Female and first-year male:* Dull olive-gray to mustard yellow; yellowish on rump. *Juvenile:* Streaked above and below, suggesting a large Pine Siskin, but note bill. **VOICE:** Call a hard *jip-jip* or *kip-kip-kip* (in some populations, *kwit-kwit* or *kewp-kewp*). Song consists of finchlike warbled passages, *jip-jip-jip-jeeaa-jeeaa;* trills, *chips.* **SIMILAR SPECIES:** White-winged Crossbill. **RANGE:** Erratic and irruptive wanderings throughout range, especially in winter. **HABITAT:** Variety of conifers; rarely at feeders.

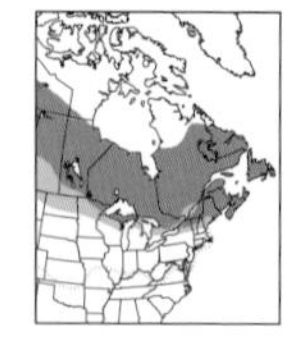

WHITE-WINGED CROSSBILL *Loxia leucoptera* Uncommon, irregular

6½ in. (17 cm). Told from Red Crossbill by *bold white wing bars* and white tertial tips in all plumages. *Adult male:* Dull *rose pink. Female and first-year male:* Olive-gray, with yellowish rump (see Red Crossbill). *Juvenile:* Heavily streaked. **VOICE:** Calls a liquid *peet* and a dry *chif-chif.* Song a succession of loud trills on different pitches. **SIMILAR SPECIES:** Red Crossbill more scarlet (less rosy) red; may have a single weak wing bar, but not two broad ones, and it lacks white tertial tips. **RANGE:** Irruptive winter visitor south of normal range. **HABITAT:** Spruce and fir forests, hemlocks; very rarely at feeders.

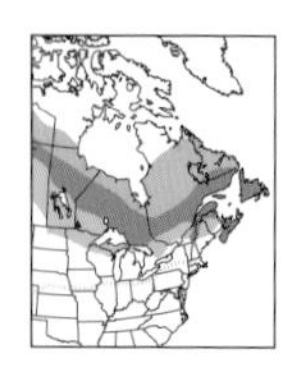

PINE GROSBEAK *Pinicola enucleator* Scarce, irregular

8¾–9 in. (23 cm). A large, plump, tame finch with dark, stubby bill, longish tail, dark wings with *two white wing bars.* Flight undulating. *Adult male:* Dull *rose red. Female and first-fall/-winter male:* Gray; head and rump tinged with dull mustard yellow. First-spring/-summer male can molt in scattered red feathers. **VOICE:** Song a rich, rapid warbling: *richy-rich-chew-twee-chur-chur.* Call a *pe-pew-pew.* **SIMILAR SPECIES:** Crossbills, Purple Finch. **RANGE:** Irruptive winter visitor south of normal range. **HABITAT:** Conifers, particularly larches; in winter, also crabapples and other fruiting trees, ashes.

BRAMBLING, CROSSBILLS, PINE GROSBEAK
female
BRAMBLING
fall/winter male
spring/summer male
adult male
adult female
juvenile
RED CROSSBILL
juvenile
adult male
WHITE-WINGED CROSSBILL
female and first-year male
PINE GROSBEAK
adult male

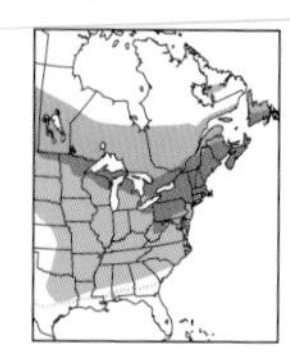

PURPLE FINCH *Haemorhous purpureus* **Uncommon**

6 in. (15 cm). *Adult male:* Like a sparrow dipped in raspberry juice. Dull rose red, brightest on head, chest, and rump. Sides and flanks unstreaked. *Female and first-year male:* Heavily streaked, brown to olive-brown; undertail coverts usually lack streaks. **VOICE:** Song a fast, lively warble recalling Warbling Vireo but bubblier; call a dull, flat, metallic *pik* or *tick*. **SIMILAR SPECIES:** House Finch. See also female Rose-breasted Grosbeak. **HABITAT:** Woods, groves, riparian thickets, suburbs, feeders.

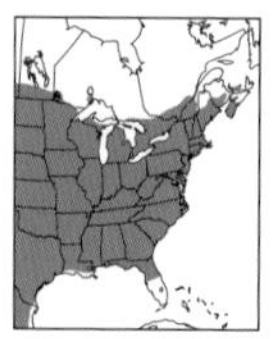

HOUSE FINCH *Haemorhous mexicanus* **Common**

5¾–6 in. (14–15 cm). Slimmer than Purple and Cassin's Finches; tail longer, square-tipped. *Male:* Breast, forehead, stripe over eye, and rump vary from *red to orange to dull mustard yellow* (diet-related). Note *dark streaks* on sides and belly. *Female:* Streaked brown; told from female Purple Finch by paler brown overall, slimmer body, longer tail, smaller head and bill; *plainer face;* undertail coverts usually streaked. **VOICE:** Song bright, loose, and disjointed finchlike notes; often ends in a nasal *wheer.* Call a nasal, finchlike *chirp*. **SIMILAR SPECIES:** Purple Finch. **RANGE:** Native to w. N. America. Introduced to New York City in 1941 and spread throughout East but subsequently has declined in some areas. **HABITAT:** Cities, suburbs, farms, feeders; prefers drier habitats.

COMMON REDPOLL *Acanthis flammea* **Uncommon, irregular**

5¼ in. (13 cm). A small finch, often found in flocks. Note *bright red forehead* and *black chin. Adult male:* Has noticeable *pink breast. Adult female and first-year male:* Usually show a pink tinge. *First-year female:* Lacks pink. **VOICE:** Song a trill, followed by a rattling *chet-chet-chet,* the latter also given in flight. **SIMILAR SPECIES:** Hoary Redpoll, Pine Siskin. **RANGE:** Irruptive winter visitor south of normal range. **HABITAT:** Birches, tundra scrub. In winter, weeds, brush, thistle feeders.

HOARY REDPOLL *Acanthis hornemanni* **Rare, irregular**

5¼–5½ in. (13–14 cm). Can be found in winter flocks of Common Redpolls. Look for a "frostier" bird, with whiter rump containing *little or no streaking; bill stubbier;* streaks on flanks and undertail coverts reduced. Adult males whitest; females and first-year birds duller, can overlap with adult male Commons, and can be *very difficult to identify.* **VOICE:** Like Common Redpoll. **SIMILAR SPECIES:** Common Redpoll, Pine Siskin. **RANGE:** Irruptive in winter, though not as much as Common. **HABITAT:** Birches, tundra scrub. In winter, weeds, brush, feeders.

adult
male
PURPLE
FINCH
female and
first-year male
HOUSE FINCH
orange
variant
male
female
female
male
male
COMMON REDPOLL
HOARY
REDPOLL

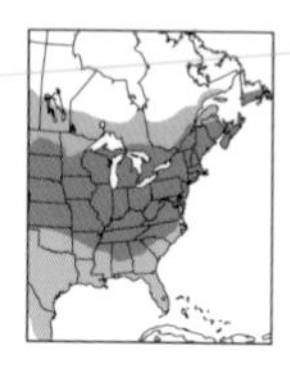

AMERICAN GOLDFINCH *Spinus tristis* **Common**

5 in. (13 cm). Goldfinches are distinguished from other small, olive-yellow birds (such as warblers) by their short, conical bill and behavior. *Spring/summer male: Bright yellow with black forehead and wings;* tail also black; bill pale. *Spring/summer female:* Dull yellow-olive; darker above, with brownish-black wings and conspicuous wing bars. *Fall/winter:* Both sexes much like spring/summer female, but bill dark; wings blacker in males. **VOICE:** Song clear, light, canary-like. Call given in undulating flight, each dip punctuated by *ti-DEE-di-di* or *per-chik-o-ree* or *po-ta-to-chip.* **SIMILAR SPECIES:** Lesser Goldfinch, Pine Siskin. **HABITAT:** Patches of thistles and weeds, dandelions on lawns, sweetgum balls, roadsides, open woods, edges; in winter, also feeders, where often in flocks.

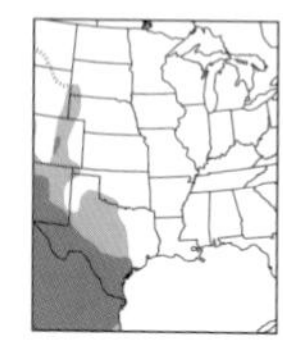

LESSER GOLDFINCH *Spinus psaltria* **Uncommon, local**

4½ in. (11 cm). *Male:* A very small finch with *black cap* and yellow underparts; white on wings. Males of eastern subspecies (*S. p. psaltria*) have *black* back; others can have *greenish* back. *Female:* Similar to fall/winter American Goldfinch but usually yellower below, has *less contrasting wing bars, yellowish* (not white) *undertail coverts,* and *dark rump.* Calls differ. First-year female plain dull greenish overall. **VOICE:** Sweet, plaintive, whiny notes, *tee-yee* (rising) and *tee-yer* (dropping). Song more phrased than American Goldfinch. **SIMILAR SPECIES:** American Goldfinch. **RANGE:** Casual vagrant east of range, accidentally to East Coast. **HABITAT:** Dry brushy and weedy country, open woods, wooded streams, towns, parks, gardens, feeders.

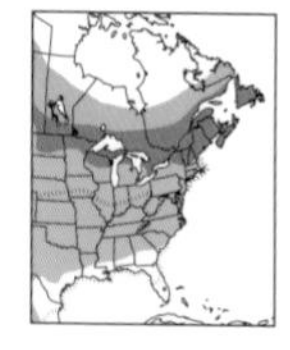

PINE SISKIN *Spinus pinus* **Fairly common, irregular**

5 in. (13 cm). Size of a goldfinch. A small, dark, *heavily streaked* finch with deeply notched tail, sharply pointed bill. *Yellow bases to wings and tail* (more prominent in male; less evident in female). Often first detected by voice, flying over. **VOICE:** Call a loud, finchy *jjeee-ip;* also a light *tit-i-tit;* a buzzy *shreeeee.* Song suggests goldfinch but coarser, wheezy. **SIMILAR SPECIES:** Fall/winter American Goldfinch lacks streaks. Female House Finch much larger, has stubbier bill. Common Redpoll has red forehead. All lack yellow in wings and tail. **HABITAT:** Conifers, mixed woods, alders, sweetgum balls, weedy areas, feeders.

EUROPEAN GOLDFINCH *Carduelis carduelis* **Scarce, exotic**

5½ in. (14 cm). Occasional reports, mostly at feeders. Assumed to be escaped captive birds. Note red face, yellow wing patches.

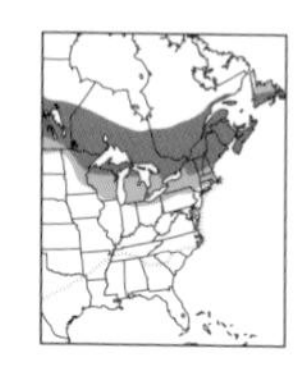

EVENING GROSBEAK *Coccothraustes vespertinus* **Uncommon, irregular**

8 in. (20 cm). Size of a starling. A *chunky, short-tailed* finch with *very large, pale, conical bill* (sometimes tinged greenish). *Male:* Deep yellow, with darker head, *yellow eyebrow,* and black-and-white wings. *Female:* Silver gray, with yellow sides of neck, patterned black-and-white wings and tail; suggests an overgrown female American Goldfinch. Gregarious. In flight, overall shape and *large white wing patches* identify this species. **VOICE:** Song repeated short trills. Calls a distinctive, ringing, finchlike *clee-ip* and a high, clear *thew.* **SIMILAR SPECIES:** American Goldfinch (much smaller), female crossbills. **RANGE:** Shows decadal irruptive tendencies to winter south of normal range. **HABITAT:** Coniferous and mixed forests; in winter, box elders, fruiting shrubs, feeders.

YELLOW FINCHES
fall/winter male
spring/summer male
spring/summer female
AMERICAN GOLDFINCH
female
adult male
adult male
green-backed
black-backed
LESSER GOLDFINCH
male
female
PINE SISKIN
EUROPEAN GOLDFINCH
female
EVENING GROSBEAK
male

NORTH AMERICAN TANAGERS, CARDINALS, BUNTINGS, and ALLIES Family Cardinalidae

Medium-sized songbirds with heavy, fruit-eating or seed-crushing bills. Now includes N. American tanagers as well as the crested cardinals, heavy-billed grosbeaks, smaller *Passerina* buntings, and Dickcissel. **FOOD:** Seeds, fruit, insects. **RANGE:** New World.

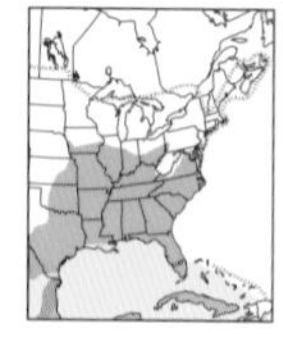

SUMMER TANAGER *Piranga rubra* — Fairly common

7¾ in. (20 cm). *Adult male: Rose red all over,* with *pale* bill. *Female and first-winter male:* Olive above, *mustard yellow* below; pale bill. Adult female has orangey throat and undertail coverts. *First-spring/-summer males:* Patched with red, yellow, and green. **VOICE:** Song is short phrases; similar to American Robin in form, but richer. Call a staccato *pi-tuk* or *pik-i-tuk-i-tuk.* **SIMILAR SPECIES:** Female Scarlet Tanager more yellow-green color and has darker wings, *whiter underwing lining,* and smaller, duskier bill. **RANGE:** Rare to casual vagrant well north of range. **HABITAT:** Riparian woodlands, oaks.

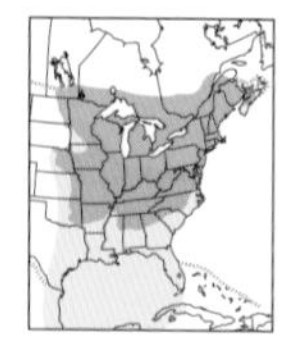

SCARLET TANAGER *Piranga olivacea* — Fairly common

7 in. (18 cm). *Spring/summer male: Flaming scarlet,* with *jet-black* wings and tail. *Female and fall/winter male: Greenish olive* above, variably *yellowish* below; dark *brownish to black wings;* normally no wing bars, but young birds may have single faint bar. **VOICE:** Song four or five short phrases, robinlike but hoarse (suggesting an American Robin with a sore throat): *hurry-worry-flurry-blurry.* Call a distinctive *chip-burr.* **SIMILAR SPECIES:** Summer and Western Tanagers. **HABITAT:** Deciduous and mixed forests, shade trees, especially oaks. Often stays high in trees.

WESTERN TANAGER *Piranga ludoviciana* — Casual vagrant

7¼ in. (18 cm). Our only tanager with *strong wing bars. Adult male:* Yellow with black back, wings, and tail, two wing bars, and *reddish head. Female and first-fall/-winter male:* Variably yellow below, with white belly but yellow undertail coverts; dull olive above, dull grayish "saddle" may be apparent on back, white wing bars. **VOICE:** Calls a dry *pr-tee* or *pri-ti-tic* and breathy *whee?* **SIMILAR SPECIES:** Female orioles have longer tails, more pointed bills. **RANGE AND HABITAT:** Casual vagrant to East Coast, primarily in fall and winter. Vagrants occur in widespread habitats during migration; often in fruit trees and at feeders in winter.

CRIMSON-COLLARED GROSBEAK *Rhodothraupis celaeno* — Casual vagrant

8½ in. (22 cm). *Adult male:* A blackish grosbeak with *dark red collar and underparts* encircling throat and chest. Red underparts often spotted or blotched with black. *Female and first-fall male:* Similar but *dark olive* replaces red. First-spring male acquires red on nape and breast. **VOICE:** Song similar to Black-headed Grosbeak; a hoarse, bouncy warble, ending in up-slurred note: *zwee!* **SIMILAR SPECIES:** Female tanagers and orioles. **RANGE:** Mexican species, casual vagrant (mostly in winter) to s. TX. **HABITAT:** Brushy woods, second growth.

TANAGERS, GROSBEAK
SUMMER TANAGER
adult male
female
first-spring male
molting
female
spring/ summer male
fall/winter male
orange variant
SCARLET TANAGER
female
spring/ summer male
first-fall/ winter male
WESTERN TANAGER
female
CRIMSON-COLLARED GROSBEAK
adult male

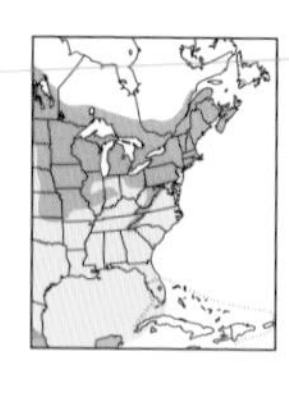

ROSE-BREASTED GROSBEAK *Pheucticus ludovicianus* Fairly common

8 in. (20 cm). *Adult male:* Black and white, with large triangle of rose red on breast and thick pale bill. In flight, pattern of black and white flashes across upperparts. Underwing linings rose pink. Plumage fringed brown in fall/winter. *Female:* Streaked, like a large sparrow or female Purple Finch; recognized by large, pink, grosbeak bill, broad white wing bars, striped crown, and broad white eyebrow stripe. Underwing linings yellow. *First-year male:* In first fall like female but has pink underwing lining; attains partial adult plumage by first spring/summer. **VOICE:** Song consists of rising and falling passages; resembles American Robin song, but more melodic. Call a squeaky, metallic *kick* or *eek*. **SIMILAR SPECIES:** Purple Finch (see p. 312), Black-headed Grosbeak. **HABITAT:** Deciduous woods, orchards, groves, thickets, sometimes at feeders in spring.

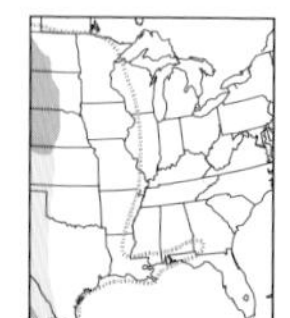

BLACK-HEADED GROSBEAK Uncommon, local
Pheucticus melanocephalus

8¼ in. (21 cm). A stocky bird, larger than a sparrow, with outsized bill. *Adult male:* Breast, collar, and rump *dull orange-brown.* Otherwise, black head and bold black-and-white wing and tail pattern are similar to those of Rose-breasted Grosbeak. *Female and first-fall male:* Similar to Rose-breasted Grosbeak female but breast *washed with yellow-buff, ocher-buff, or butterscotch;* dark streaks on sides *fine,* nearly absent across middle of chest. Underwing linings yellow in all ages and sexes. *Maxilla dark. First-spring/-summer male:* Adultlike, but head variably mottled buff and wings brown rather than black. **VOICE:** Similar to Rose-breasted Grosbeak. **SIMILAR SPECIES:** Female and first-fall male Rose-breasted Grosbeak have *heavier streaks on paler* breast, paler bill; underwing pink in male. Beware some intermediates, and hybrids also occur that can be difficult to identify. **RANGE:** Rare vagrant to East Coast, primarily in fall and winter, when often observed at feeders. **HABITAT:** Deciduous and riparian woods.

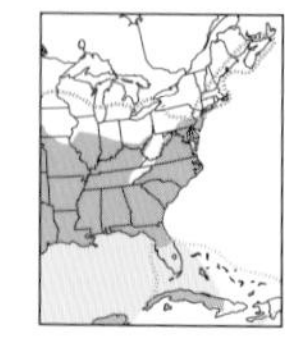

BLUE GROSBEAK *Passerina caerulea* Uncommon

6¾ in. (17 cm). *Adult male:* Deep *dull blue,* with thick bill, *two broad rusty or chestnut wing bars.* Often *flips or twitches tail.* Head mottled brown in fresh fall/winter plumage. *Female and first-fall/-winter male:* About size of Brown-headed Cowbird; warm or tawny brown, slightly lighter below, with two *rusty buff wing bars;* rump or tail may be tinged with blue. First-year male begins acquiring mottled blue plumage on winter grounds and by first spring/summer is a variable mixture of brown and blue, as in Indigo Bunting. **VOICE:** A warbling song, phrases rising and falling; suggests Purple or House Finch, but slower, more guttural. Call a sharp *chink,* in flight a flat *bzzzt.* **SIMILAR SPECIES:** Female and first-year male Indigo Bunting usually paler (less tawny) brown, show *weaker wing bars* and are *much smaller* and smaller-billed. **RANGE:** Casual vagrant well north of range. **HABITAT:** Thickets, hedgerows, riparian undergrowth, brushy hillsides, weedy ditches.

ROSE-BREASTED
GROSBEAK
adult
male
female and
first-fall male
BLACK-HEADED
GROSBEAK
female and
first-fall male
adult male
BLUE
GROSBEAK
female and
first-fall
male
adult male

BLUE BUNTING *Cyanocompsa parellina* Casual vagrant

5½ in. (14 cm). *Adult male:* Deep blue-black; brighter blue on crown, shoulders, and rump. *Female and first-fall/-winter male:* Richer brown than female Indigo Bunting; no bars or streaks; *bill blacker.* **VOICE:** Song a high and sweet jumble of warbled phrases. Call a metallic *chink!* **SIMILAR SPECIES:** Indigo Bunting, Blue Grosbeak. **RANGE:** Mexican vagrant, casual to s. TX, mostly in winter. **HABITAT:** Brushy woods with dense cover.

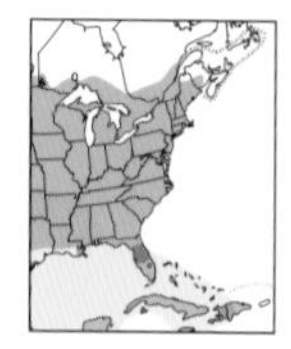

INDIGO BUNTING *Passerina cyanea* Common

5½ in. (14 cm). *Adult spring/summer male: Rich deep blue all over. First-spring/-summer male:* Blue is duller and variably mottled brown. *Fall/winter adult male:* Brown like female but usually with some blue in wings and tail. *Female and first-fall/-winter male:* Medium brown to olive-brown; breast slightly paler with faint *blurry* streaks; paler wing bars indistinct. **VOICE:** Song lively, high, and strident; measured phrases, usually paired: *sweet-sweet, chew-chew,* etc. Call a sharp, thin *spit* and a dry buzz (in flight). **SIMILAR SPECIES:** Blue Grosbeak much larger, has rusty wing bars. See Lazuli Bunting. **HABITAT:** Overgrown brushy fields, riparian thickets, bushy wood edges.

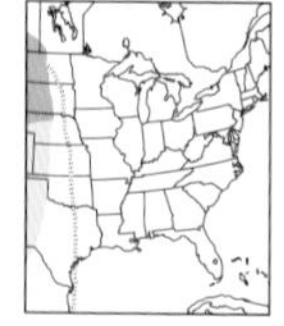

LAZULI BUNTING *Passerina amoena* Rare

5½ in. (14 cm). *Adult male:* Bright turquoise blue; orangey breast; white belly and *wing bars. Female and first-fall/-winter male:* Unstreaked with two distinct whitish wing bars; breast washed deep buff; juvenile may retain fine, sharp streaks into fall. **VOICE:** Song similar to Indigo Bunting but faster. Calls also similar. **SIMILAR SPECIES:** Female and juvenile told from Indigo Bunting by more distinct, whitish wing bars; unstreaked and warmer colored breast; juveniles with finer and sharper streaks than Indigo. Indigo and Lazuli Buntings occasionally hybridize. **RANGE:** Casual vagrant to East Coast, often in winter at feeders. **HABITAT:** Open brush, grassy hillsides with scattered oaks, riparian shrubs, chaparral, weedy fields and ditches.

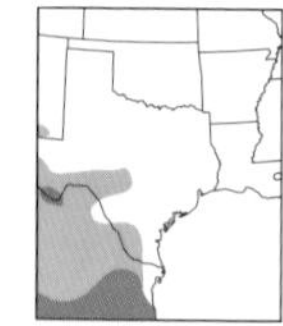

VARIED BUNTING *Passerina versicolor* Scarce, local

5½ in. (14 cm). *Adult male:* Plum-purple body (looks black at a distance). Crown, face, and rump blue, with *bright red patch on nape. Female and first-year male: Gray-brown* with lighter underparts. *No strong wing bars, breast streaks, or distinctive marks of any kind.* Bill smaller and ridge more curved than in other buntings. **VOICE:** Song thin, bright, more distinct than Painted Bunting. **SIMILAR SPECIES:** Female Indigo Bunting more olive-brown, with hint of blurry breast streaks. Female Lazuli Bunting has noticeable wing bars. **HABITAT:** Riparian thickets, mesquite and other scrub in washes.

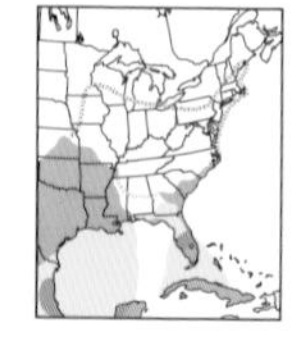

PAINTED BUNTING *Passerina ciris* Uncommon

5½ in. (14 cm). The most gaudily colored N. American songbird. *Adult male:* A patchwork of *blue-violet* on head, *green* on back, *red* on rump and underparts, red orbital ring. *Female and first-year male: Electric green above,* paling to lemon yellow below; *no other small finch is so green. Juvenile:* Grayer above with only tinge of green, duller gray to buff below. **VOICE:** Song a wiry warble; suggests Warbling Vireo. Call a sharp *chip.* **RANGE:** Widespread vagrant west and north of range. Sometimes at feeders, although beware cage-bird escapees. **HABITAT:** Riparian undergrowth, brushy hedgerows, woodland edges, weedy fields.

BUNTINGS
female and
first-year male
adult
male
BLUE BUNTING
INDIGO
BUNTING
adult male
first-spring/
summer male
female
and
first-fall
male
adult male
LAZULI
BUNTING
female and
first-fall
male
female and
first-year male
adult male
VARIED BUNTING
adult male
female
and
first-year
male
PAINTED BUNTING

NORTHERN CARDINAL *Cardinalis cardinalis* Common

8¾ in. (22 cm). *Male: All red* with pointed *crest* and black patch at base of heavy, *triangular reddish bill. Female:* Brown tinged pinkish buff, with some red on wings and tail. *Crest, dark face,* and *heavy reddish orange bill* distinctive. *Juvenile:* Similar to female but with blackish bill. **VOICE:** Song clear, slurred whistles, repeated. Several variations: *what-cheer cheer cheer,* etc.; *whoit whoit whoit* or *birdy birdy birdy,* etc.; usually two-parted. Call a short, sharp *tik.* **SIMILAR SPECIES:** Pyrrhuloxia. Male Summer and Hepatic Tanagers lack cardinal's crest and black face. **RANGE:** Wanders casually well north of range. **HABITAT:** Woodland edges, thickets, deserts, towns, gardens, feeders.

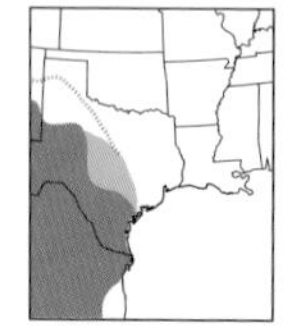

PYRRHULOXIA *Cardinalis sinuatus* Uncommon, local

8¾ in. (22 cm). A *slender, gray and red* bird, with spiky crest and *stubby yellow bill. Male:* Has *long, spiky crest. Female:* Has gray back, buff breast, and touch of red in wings. **VOICE:** Song a clear *quink quink quink quink quink* and a slurred *what-cheer, what-cheer,* etc. **SIMILAR SPECIES:** Best told from Northern Cardinal by bill color and shape, also by grayer color, spikier crest, lack of black mask in female, red face and throat in male. **RANGE:** Casual vagrant east of range. **HABITAT:** Mesquite, thorn scrub, deserts, feeders.

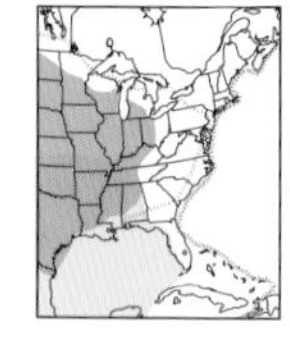

DICKCISSEL *Spiza americana* Fairly common

6¼ in. (16 cm). A grassland and farmland bird; migrants often travel in large flocks. Sings from fenceposts and wires. *Adult male:* Suggests a miniature meadowlark, with yellow underparts and black bib, but has chestnut shoulder patch. In fall, bib obscure. *Female and first-year male:* Duller and plainer; dullest females recall female House Sparrow but with bolder stripe over eye (often, but not always, tinged yellowish), brighter underparts, touch of yellow on breast, and blue-gray bill. **VOICE:** Song a staccato *dick-ciss-ciss-ciss* or *chup-chup-klip-klip-klip.* Call a short, hard buzz, often given in flight. **SIMILAR SPECIES:** Meadowlarks, female House Sparrow. **RANGE:** Scarce vagrant to East Coast. **HABITAT:** Alfalfa and other fields, meadows, weedy patches.

BLACKBIRDS and ORIOLES Family Icteridae

Varied color patterns; sharp bills. Some black and iridescent; orioles are highly colored. Sexes usually unlike, and in most species, males noticeably larger than females. **FOOD:** Insects, fruit, seeds, waste grain, small aquatic life. **RANGE:** New World; most species occur in Tropics.

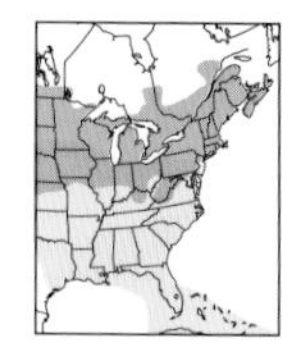

BOBOLINK *Dolichonyx oryzivorus* Fairly common

7 in. (18 cm). *Spring/summer male:* Our only songbird that is *solid black below and partially white above.* Has buff-yellow nape. Returning migrants in spring have extensive brownish tips to dark feathering. *Female and fall/winter male:* A bit larger than House Sparrow; rich buff-yellow, with dark striping on crown and back. Bill more like a sparrow's than a blackbird's. Note pointed tail feathers. **VOICE:** Song, in hovering flight and quivering descent, ecstatic and bubbling: starts with low, reedy notes and rollicks upward. Flight call a clear *ink,* often heard overhead in migration. **SIMILAR SPECIES:** Male Lark Bunting has white confined to wings. Female Red-winged Blackbird heavily striped below; has longer bill, less buff-yellow overall. Grasshopper Sparrow much smaller. **HABITAT:** Hayfields, moist meadows, marsh edges.

CARDINALS, DICKCISSEL, BOBOLINK
NORTHERN CARDINAL
juvenile
female
male
PYRRHULOXIA
female
male
DICKCISSEL
female
adult male
spring/ summer male
BOBOLINK
spring/ summer female
spring/summer male
fall/winter

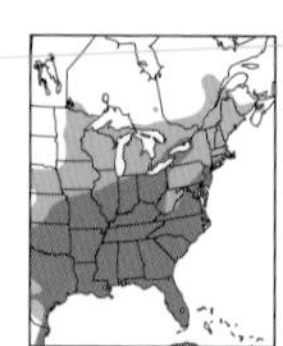

EASTERN MEADOWLARK
Uncommon to locally common

Sturnella magna

9½ in. (24 cm). A chunky, brown, starling-shaped bird of grassy habitats. Warm, reddish brown above, with blacker crown. Chest shows bright yellow crossed by black V; flanks buffier. Ages and sexes similar. When flushed, meadowlarks show conspicuous white sides on short tail. Several shallow, snappy wingbeats alternate with short glides—like a Spotted Sandpiper. Walking, it flicks tail open and shut. **VOICE:** Song comprises two clear, slurred whistles, musical and pulled out, *tee-yah, tee-yair* (last note slurred and descending). Call a rasping or buzzy *dzrrt;* also a guttural chatter. **SIMILAR SPECIES:** Western Meadowlark, Dickcissel. **HABITAT:** Open fields and pastures, meadows, marsh edges.

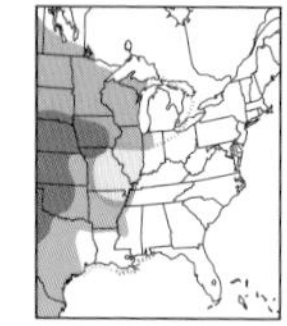

WESTERN MEADOWLARK
Uncommon to fairly common

Sturnella neglecta

9½ in. (24 cm). Nearly identical to Eastern Meadowlark but paler above and on flanks; yellow of throat invades farther into malar area behind bill. Crown stripes paler, more streaked with buff. Best identified by vocalizations. **VOICE:** Song variable; 7–10 flutelike notes, gurgling and double-noted, unlike clear whistles of Eastern Meadowlark. Calls *chupp* or *chuck* and a dry rattle. Occasionally gives *dzzrt* call like Eastern. **SIMILAR SPECIES:** Eastern Meadowlark. **RANGE:** Scarce winter visitor to Southeast coast; casual vagrant to Northeast. **HABITAT:** Similar to Eastern Meadowlark.

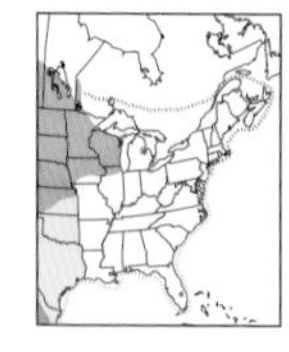

YELLOW-HEADED BLACKBIRD
Fairly common

Xanthocephalus xanthocephalus

9–9¾ in. (23–25 cm). Gregarious. *Adult male:* A robin-sized blackbird, with *rich yellow head and breast;* in flight, shows *white wing patch. Female and first-year male:* Smaller (female) and browner; yellow more confined to throat and chest; lower breast streaked with white; white wing patch restricted. **VOICE:** Song consists of low, hoarse rasping notes produced with much effort; suggests rusty hinges. Call a low *kruck* or *kack.* **RANGE:** Rare to casual vagrant to East Coast. **HABITAT:** Nests in freshwater marshes. Forages in farm fields, open country, feedlots. Often associates with other blackbirds in mixed fall/winter flocks.

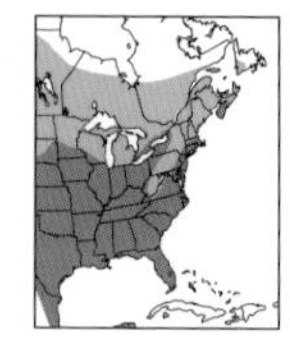

RED-WINGED BLACKBIRD *Agelaius phoeniceus*
Common

8½–8¾ in. (22 cm). *Adult male:* Black, with *bright red or orange-red, yellow-margined epaulets* (wing-covert patches), most conspicuous in breeding display. Much of the time red is concealed and only yellowish or off-whitish margin shows. *First-year male:* Sooty brown, mottled (like larger version of female), but with dull red epaulets. *Female:* Brownish, with sharply pointed bill and *well-defined dark streaking* below; adult females may have pinkish or dull red tinge on throat or shoulder. Gregarious, traveling and roosting in flocks in fall and winter. **VOICE:** Calls a loud *check* and a high, slurred *tee-err.* Song a liquid, gurgling *konk-la-ree* or *o-ka-lay.* **SIMILAR SPECIES:** Other blackbird species. **HABITAT:** Breeds in marshes, brushy swamps, fields, pastures; forages also in cultivated land, feedlots, towns, feeders, etc.

MEADOWLARKS, BLACKBIRDS
EASTERN MEADOWLARK
WESTERN MEADOWLARK
YELLOW-HEADED BLACKBIRD
adult male
female
red epaulets hidden
female
adult male
RED-WINGED BLACKBIRD
first-year male

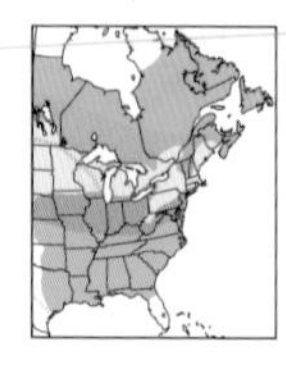

RUSTY BLACKBIRD *Euphagus carolinus* Uncommon

9 in. (23 cm). *Spring/summer male:* A medium-sized blackbird with pale yellow eye. Black head may show faint *greenish* gloss. *Spring/summer female:* Slate colored, with *light eye. Fall/winter:* Feathers variably *fringed rusty,* creating overall rusty appearance, *buffy eyebrow, narrow dark patch through eye.* **VOICE:** Call *chack.* "Song" a split creak, like a rusty hinge: *kush-a-lee,* alternating with *ksh-lay.* **SIMILAR SPECIES:** Brewer's Blackbird. Common Grackle larger, bill stronger. **HABITAT:** River groves, wooded swamps, muskeg, pond edges; in winter, also muddy fields, with other blackbirds.

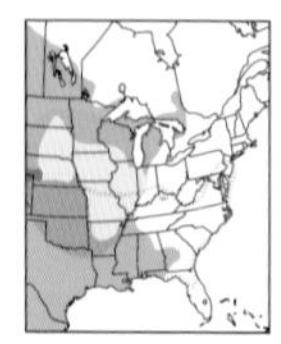

BREWER'S BLACKBIRD *Euphagus cyanocephalus* Uncommon

9 in. (23 cm). *Male:* All black with whitish eye; in good light, *purplish* reflections may be seen on head and neck, greenish on body. *Fresh first-fall male:* Can be fringed olive-brown. *Female:* Brownish gray, usually with *dark* eye. **VOICE:** Song a creaking *ksh-eee.* Call *chack.* **SIMILAR SPECIES:** Male Rusty Blackbird flatter black with dull *greenish* head reflections; bill slightly longer. Female Rusty has *light* eye. Beware first-fall male Brewer's, which can be fairly heavily fringed but is a muddier brown, not rusty. **RANGE:** Rare to casual winter visitor to East Coast. **HABITAT:** Fields, farms, feedlots, towns, parks, lawns, shopping malls.

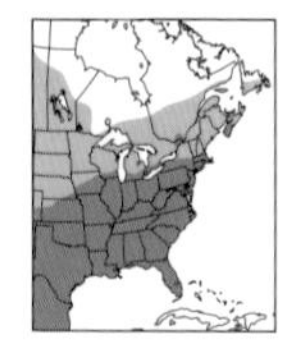

BROWN-HEADED COWBIRD *Molothrus ater* Common

7½ in. (19 cm). A small blackbird with sparrowlike bill. *Male:* Black with *brown head (appears all black in poor light). Female:* Gray-brown with lighter throat; note short *finchlike bill. Juvenile:* Buffy gray, with soft breast streaking and pale scaling above. *Molting first-fall male:* Splotched black. A nest parasite (never builds its own nest); juveniles are often seen being fed by smaller birds. Cowbirds feed on ground with tail lifted high. **VOICE:** Flight call a *weee-titi* (high whistle, two lower notes). Song a bubbly and creaky *glug-glug-gleeee.* Call *chuck.* **SIMILAR SPECIES:** Female told from female blackbirds by its *stubby bill* and smaller size. Juvenile cowbirds can resemble sparrows. **HABITAT:** In nesting season, deciduous forests and woodlands; also farms, fields, feedlots, roadsides, towns, parks, lawns, feeders.

BRONZED COWBIRD *Molothrus aeneus* Uncommon

8½–8¾ in. (21–22 cm). *Male:* Slightly larger and more *bull-headed* than Brown-headed Cowbird. Head *blackish;* bill longer; *eye red.* Conspicuous *ruff* on nape for courting. *Female:* Smaller nape ruff; dark brown to sooty overall, darker than female Brown-headed; eye reddish. *Juvenile:* Like a large-billed juvenile Brown-headed. **VOICE:** High-pitched mechanical creakings during animated display. **SIMILAR SPECIES:** Other cowbirds. **RANGE:** Rare vagrant along Gulf Coast. **HABITAT:** Cropland, brush, semi-open country, feedlots.

SHINY COWBIRD *Molothrus bonariensis* Scarce, local

7½ in. (19 cm). *Male:* Black with overall violet gloss, thin pointed bill. *Female:* Warm brown, slightly thinner, blacker bill compared with Brown-headed Cowbird. **VOICE:** Series of liquid burbles, ending in thin whistled note. **SIMILAR SPECIES:** Other cowbirds. **RANGE:** An invader to s. FL since 1985, with scattered records from as far north as NB and west to OK; has since declined in U.S. and Canada. **HABITAT:** Agricultural areas, disturbed habitats, suburban lawns.

ICTERIDS (BLACKBIRDS, ETC.)

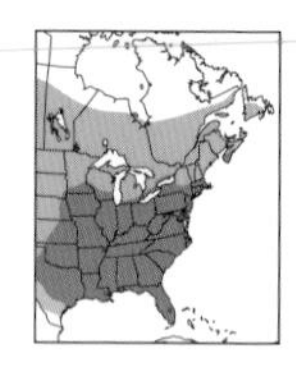

COMMON GRACKLE *Quiscalus quiscula* **Common**

12½ in. (32 cm). *Male:* A large, *iridescent,* yellow-eyed blackbird, larger than an American Robin, with long, *keel-shaped tail.* In good light, iridescent purple-blue on head. "Bronzed" Grackle (subspecies *Q. q. versicolor*) of New England and west of Appalachians deep bronze on back and belly; "Purple" Grackles (*Q. q. quiscula* and *Q. q. stonei*) of se. U.S. have glossy purple head and greener tinge to back. *Female:* Smaller and somewhat duller, with less wedge-shaped tail. Juveniles of both sexes dull sooty brown with dark eyes. **VOICE:** Call a *chuck* or *chack.* "Song" a split rasping note, *zhreep zhrap,* etc. **SIMILAR SPECIES:** Boat-tailed and Great-tailed Grackles, Rusty and Brewer's Blackbirds. **HABITAT:** Cropland, towns, parks, feeders, groves; swampy woods; often nests in conifers.

BOAT-TAILED GRACKLE *Quiscalus major* **Fairly common, local**

Male 16½ in. (42 cm); female 14½ in. (37 cm). *Male:* A very large blackbird; larger than Common Grackle, with longer, more ample tail. More rounded head than other grackles. Males of Atlantic Coast north of FL (subspecies *Q. m. torreyi*) have bright yellow eyes; those of Gulf region and FL (*Q. m. major* and *Q. m. westoni,* respectively) have brown to dull yellow eyes. *Female:* Smaller than male; much browner than female Common Grackle and with pale brownish breast. Juveniles of both sexes paler than female, with more distinct supercilium. **VOICE:** A harsh *check check check;* harsh whistles and clucks. **SIMILAR SPECIES:** Common Grackle shorter tailed, not often found in same coastal habitats. From LA westward, see Great-tailed Grackle. **RANGE:** Casual vagrant along Atlantic Coast north of range. **HABITAT:** Largely resident near salt water along coasts, marshes; more widespread habitats in FL.

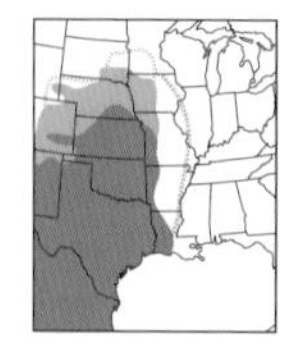

GREAT-TAILED GRACKLE *Quiscalus mexicanus* **Common**

Male 18 in. (46 cm); female 15 in. (38 cm). Like several other blackbirds, often found in large flocks, including while roosting, when they are quite noisy. *Male:* A very large, purple-glossed blackbird, distinctly larger than Common and Boat-tailed Grackles and with longer, more ample tail. *Female:* Smaller than male; dark gray-brown above, warm brown below. Adults of both sexes have yellow eyes. Juveniles of both sexes have dark eyes and are indistinctly streaked below. **VOICE:** A harsh *check check check;* also a high *kee-kee-kee-kee.* Shrill, discordant notes, whistles, and clucks. A rapid, upward-slurring *ma-ree.* **SIMILAR SPECIES:** Common Grackle much smaller, tail not nearly as keel-shaped; female blacker. Boat-tailed Grackle slightly smaller, with dark eyes (where ranges overlap), rounder crown (male), and slightly shorter, more rounded tail. **RANGE:** Casual vagrant well north and east of range. **HABITAT:** Groves, farms, feedlots, towns, city parks, parking lots.

male
"Purple"
COMMON
GRACKLE
male
female
male
Atlantic
Coast
"Bronzed"
female
FL and
Gulf Coast
male
BOAT-TAILED
GRACKLE
male
female
GREAT-TAILED
GRACKLE

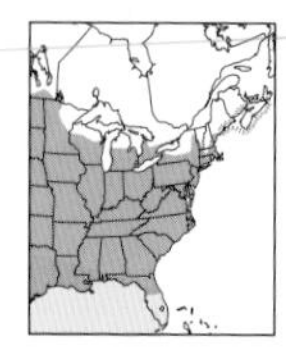

ORCHARD ORIOLE *Icterus spurius* **Fairly common**

7–7¼ in. (18 cm). A small, short- and straight-billed oriole. Often flicks tail sideways. *Adult male:* All dark; rump and underparts *deep chestnut. Female and juvenile:* Olive or greenish gray above, yellowish below; two white wing bars. *First-year male:* Develops black bib down to chest in fall and winter. **VOICE:** Song a fast-moving outburst interspersed with piping whistles and guttural notes. Suggests Purple or House Finch. A strident slurred *wheeer!* at or near end is distinctive. Call a soft *chuck.* **SIMILAR SPECIES:** See female and juvenile Hooded Oriole. Baltimore Oriole slightly larger, more orange. **HABITAT:** Wood edges, orchards, shade trees; more likely than other orioles to be seen in brushy areas.

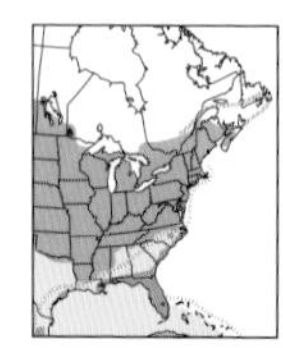

BALTIMORE ORIOLE *Icterus galbula* **Fairly common**

8¼–8½ in. (21–22 cm). *Adult male:* Flame orange and black, with solid black head, tail boldly patterned, black with orange sides. *Adult female:* Olive-brown above, burnt orange-yellow below; two white wing bars; variable amount of black on head, often suggesting hood of male; orange tail. *First-fall/-winter:* Duller, with grayer back and limited to some orange on underparts. Both sexes variably develop adultlike pattern over first winter and spring. **VOICE:** Song rich, piping whistles: *hew-hee-hee-hew-hee-hew-hew,* etc. Call a low, whistled *hewli.* Chatter call not as rough as Bullock's Oriole. **SIMILAR SPECIES:** Female Orchard Oriole smaller and yellow-green, less orange. See Bullock's Oriole. **HABITAT:** Open deciduous woods, elms, shade trees.

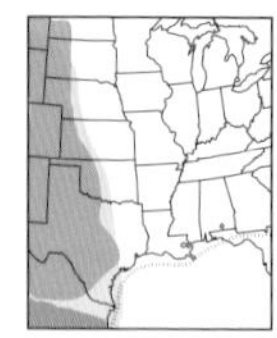

BULLOCK'S ORIOLE *Icterus bullockii* **Uncommon, local**

8¼–8½ in. (21–22 cm). *Adult male:* Note *orange cheeks* and *dark eye line, large white wing patches, and black-tipped tail. Female and juvenile:* Dark eye line, yellowish supercilium, plain gray back, *whitish belly. First-year male:* Similar to female, but orange feathering brighter and develops black throat during first fall and winter. **VOICE:** A phrase of doubled musical whistles and rattles: *jet-jet whichy-whichy ju-ju tthat-tthat,* etc. Calls include a rough chatter and low *churp.* **SIMILAR SPECIES:** Dull first-fall Baltimore Oriole much like female Bullock's but has less-distinct dark eye line and yellowish supercilium, darker brownish back with heavier dark mottling, yellower flanks not contrasting with undertail coverts; Bullock's and Baltimore can hybridize where ranges meet. See also Hooded and Orchard Orioles. **HABITAT:** Deciduous and riparian woods, oaks, shade trees, ranch yards.

SPOT-BREASTED ORIOLE *Icterus pectoralis* **Uncommon, local**

9¼–9½ in. (24 cm). A large, robin-sized oriole. *Adult male:* Note *orange crown,* black bib, and black spots on sides of breast. Much white in wing, including bases of primaries, but no wing bar. *Female and first-year male:* Similar but duller, often yellower. **VOICE:** Song a long, melodic series of whistles, slower than other orioles. **SIMILAR SPECIES:** Baltimore Oriole smaller with white wing-covert bar; most have black in crown. **HABITAT:** Flowering trees, residential areas.

ORIOLES
first-year male
adult male
ORCHARD ORIOLE
juvenile and female
adult female
adult male
BALTIMORE ORIOLE
juvenile and first-fall/winter
adult male
BULLOCK'S ORIOLE
juvenile and female
SPOT-BREASTED ORIOLE
adult male
female similar but duller

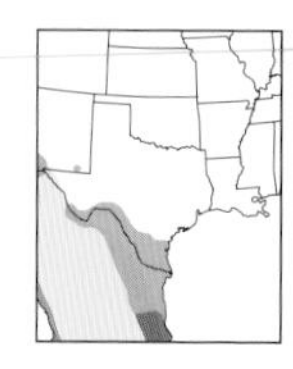

HOODED ORIOLE *Icterus cucullatus* — Uncommon, local

7½–8 in. (19–20 cm). *Adult male:* Orange and black, with black throat and *orange crown.* In winter, back scaled yellow or orange. *Female:* Similar to female Bullock's Oriole, but bill longer, slightly curved; more extensively greenish yellow below; back olive-gray; head and tail more yellowish. Call very different. *First-year male:* Like female, with slightly shorter bill (much like female Orchard Oriole); develops black throat during first winter. **VOICE:** Song consists of rambling, grating notes and piping whistles: *chut chut chut whew whew;* opening notes throaty. Call a distinctive, up-slurred, whistled *eek* or *wheenk.* **SIMILAR SPECIES:** Scott's Oriole. Female and first-year male Orchard and Hooded Orioles can be difficult to separate, but note Hooded's thinner based and more curved bill, longer tail, weaker wing bars, and *quite different calls.* Beware especially juvenile Hooded Orioles with developing bill and tail lengths; these only rarely overlap in range with Orchard Oriole in summer. **HABITAT:** Open woods, shade trees, towns, gardens, palms.

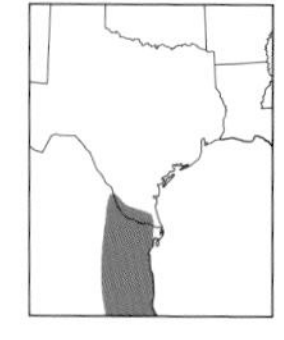

ALTAMIRA ORIOLE *Icterus gularis* — Uncommon, local

10 in. (25 cm). *Adult:* Similar to male Hooded Oriole but larger, with thicker bill. Upperwing bar yellow or orange, not white. Sexes similar. *First-year:* Yellower, with less pure black. **VOICE:** Song disjointed whistled notes. A harsh "fuss" note. **SIMILAR SPECIES:** Other orange orioles. **HABITAT:** Scrubby woodlands, often near water. "Altamira," in Spanish, means "look high," and species is often found in treetops.

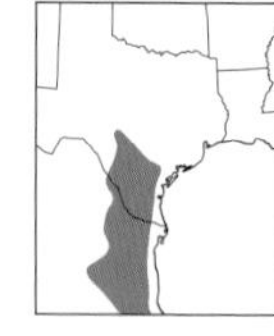

AUDUBON'S ORIOLE *Icterus graduacauda* — Uncommon, local

9½ in. (24 cm). *Adult:* A yellow oriole with black wings, head, and tail. Yellowish back distinctive. Other male orioles have black back. Sexes similar. *First-year:* Duller, head mixed yellowish and black, tail green or mixed green and black. *Juvenile:* Lacks black in head. **VOICE:** Disjointed notes suggesting a child learning to whistle. **SIMILAR SPECIES:** Scott's Oriole. Green Jay at a distance can appear yellow with a black head. **HABITAT:** Riparian woods.

SCOTT'S ORIOLE *Icterus parisorum* — Uncommon, local

8¾–9 in. (22–23 cm). *Adult male:* Solid black head and back and *lemon yellow* underparts distinguish this oriole. *Female:* More greenish yellow below and more olive-gray and streaked above than other female orioles. *Juvenile and first-year:* Both sexes lack black at first but variably develop blackish in throat (female) and/or head (male) by spring. **VOICE:** Song rich fluty whistles; suggests Western Meadowlark. Call a harsh *chuck.* **SIMILAR SPECIES:** Female Hooded and Bullock's Orioles. **HABITAT:** Dry woods and scrub, pinyon-juniper, sugar-water feeders. Also eucalyptus and date palms in winter.

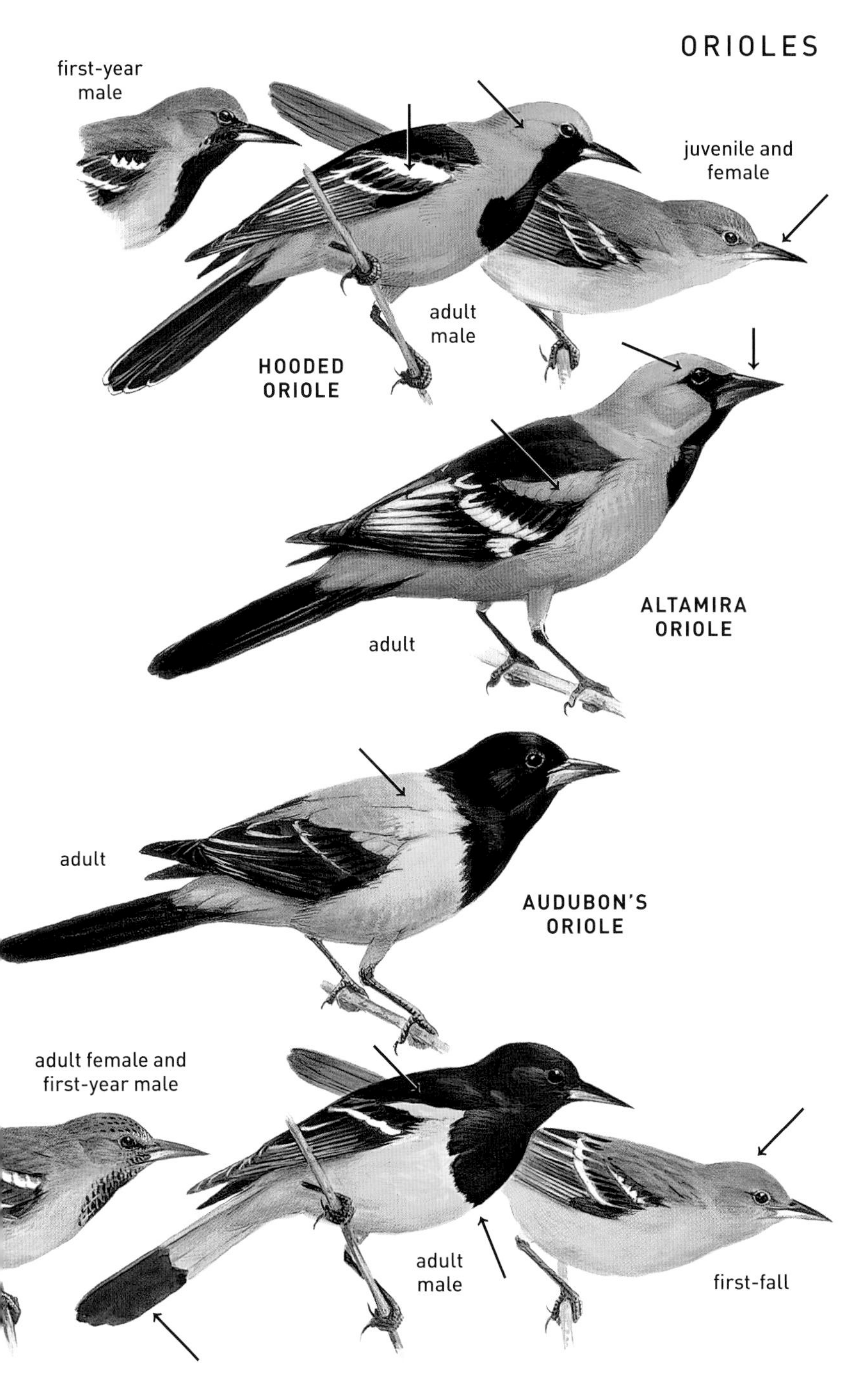
first-year
male
juvenile and
female
adult
male
HOODED
ORIOLE
ALTAMIRA
ORIOLE
adult
adult
AUDUBON'S
ORIOLE
adult female and
first-year male
adult
male
first-fall
SCOTT'S ORIOLE

STARLINGS Family Sturnidae

A varied family; some blackbirdlike. Sharp-billed, usually short-tailed. Gregarious. **FOOD:** Insects, seeds, berries. **RANGE:** Widespread in Old World. Introduced in New World.

EUROPEAN STARLING *Sturnus vulgaris* **Common, introduced**

8½ in. (22 cm). Introduced from Europe in 1890. A gregarious, garrulous species; shape of a meadowlark with *short tail* and *sharply pointed bill.* In flight, shows *triangular wings;* flies swiftly and directly. *Spring/summer:* Plumage iridescent, bill *yellow,* blue-based in male, pink-based in female. *Fall/winter: Heavily speckled with white,* bill dark. *Juvenile:* Dusky gray-brown, a bit like a female cowbird but stockier, tail shorter, bill longer. **VOICE:** A harsh, wheezy *tseeeer;* a whistled *whooee.* Also clear whistles, clicks, chuckles; often mimics other birds. **SIMILAR SPECIES:** Cedar Waxwing, in flight. **HABITAT:** Cities, suburbs, parks, feeders, farms, livestock pens, open groves, fields. Has had negative impact on several native cavity-nesting species.

COMMON HILL MYNA *Gracula religiosa* **Unestablished exotic**

10½ in. (27 cm). *Glossy black* body, orange bill, yellow face wattles and legs. *White wing patches* stand out in flight. Sexes alike. **VOICE:** Squawks, buzzes, whistles; excellent mimic. **SIMILAR SPECIES:** European Starling. **RANGE:** Exotic from Asia, formerly established in s. FL; escapees occasionally still encountered there and elsewhere. **HABITAT:** Lush suburban neighborhoods and parks.

COMMON MYNA *Acridotheres tristis* **Common, local, exotic**

10 in. (25 cm). A *brown-bodied* relative of European Starling, with black head and *white undertail.* Bill, face, and legs bright yellow. **VOICE:** Starlinglike gurgles, squeaks, and cackles. **SIMILAR SPECIES:** European Starling. **RANGE:** Introduced from s. Asia. Widespread and increasing in s. and cen. FL. **HABITAT:** Urban and suburban habitats. Often occurs in noisy flocks.

BULBULS Family Pycnonotidae

Native to Old World. One species introduced in FL. **FOOD:** Insects, fruit.

RED-WHISKERED BULBUL **Uncommon, local, exotic**
Pycnonotus jocosus

7 in. (18 cm). Note black crest, red cheek patch, black half-collar, and red undertail coverts; juvenile has duller head pattern. **VOICE:** Noisy chattering. **SIMILAR SPECIES:** Waxwings. **RANGE:** This native of se. Asia was established locally in s. Miami, FL, in early 1960s, where it still forms a small breeding population. **HABITAT:** Heavy vegetation in suburban neighborhoods.

STARLING, MYNAS,
AND BULBUL
spring/
summer
fall/
winter
juvenile
EUROPEAN
STARLING
COMMON
HILL MYNA
COMMON
MYNA
adult
RED-WHISKERED
BULBUL

LIFE LIST INDEX

The following pages contain the American Birding Association's checklist as of December 2018 (also available online at http://listing.aba.org/aba-checklist/). The ABA Checklist "includes ABA-area breeding species, regular visitors, and casual and accidental species from other regions that are believed to have strayed here without direct human aid, and well-established introduced species that are now part of our avifauna."

Scientific names are not given below but can be found in the species accounts throughout the book. Note that the sequence here does not follow that of the plates in this book, which have been arranged as much for ease of identification as in accordance with our understanding of current (and frequently changing) phylogenetic sequence.

DUCKS, GEESE, AND SWANS (ANATIDAE)

________ Black-bellied Whistling-Duck

________ Fulvous Whistling-Duck

________ Emperor Goose

________ Snow Goose

________ Ross's Goose

________ Graylag Goose

________ Greater White-fronted Goose

________ Lesser White-fronted Goose

________ Taiga Bean-Goose

________ Tundra Bean-Goose

________ Pink-footed Goose

________ Brant

________ Barnacle Goose

________ Cackling Goose

________ Canada Goose

________ Hawaiian Goose

________ Mute Swan

________ Trumpeter Swan

________ Tundra Swan

________ Whooper Swan

________ Egyptian Goose

________ Common Shelduck

________ Muscovy Duck

________ Wood Duck

________ Baikal Teal

________ Garganey

________ Blue-winged Teal

________ Cinnamon Teal

________ Northern Shoveler

________ Gadwall

________ Falcated Duck

________ Eurasian Wigeon

________ American Wigeon

________ Laysan Duck

________ Hawaiian Duck

________ Eastern Spot-billed Duck

________ Mallard

________ American Black Duck

________ Mottled Duck

________ White-cheeked Pintail

________ Northern Pintail

________ Green-winged Teal

________ Canvasback

________ Redhead

________ Common Pochard

________ Ring-necked Duck

________ Tufted Duck

________ Greater Scaup

________ Lesser Scaup

________ Steller's Eider

________ Spectacled Eider

________ King Eider

________ Common Eider

________ Harlequin Duck

________ Labrador Duck

________ Surf Scoter

________ White-winged Scoter

________ Common Scoter

________ Black Scoter

________ Long-tailed Duck

________ Bufflehead

________ Common Goldeneye

________ Barrow's Goldeneye

________ Smew

________ Hooded Merganser

________ Common Merganser

________ Red-breasted Merganser

________ Masked Duck

________ Ruddy Duck

CURASSOWS AND GUANS (CRACIDAE)

________ Plain Chachalaca

NEW WORLD QUAIL (ODONTOPHORIDAE)

________ Mountain Quail

________ Northern Bobwhite

________ Scaled Quail

________ California Quail

________ Gambel's Quail

________ Montezuma Quail

PARTRIDGES, GROUSE, TURKEYS, AND OLD WORLD QUAIL (PHASIANIDAE)

________ Chukar

________ Gray Francolin

________ Black Francolin

________ Erckel's Francolin

________ Himalayan Snowcock

________ Gray Partridge

________ Red Junglefowl

________ Kalij Pheasant

________ Ring-necked Pheasant

________ Indian Peafowl

________ Ruffed Grouse

________ Greater Sage-Grouse

________ Gunnison Sage-Grouse

________ Spruce Grouse

________ Willow Ptarmigan

________ Rock Ptarmigan

________ White-tailed Ptarmigan

________ Dusky Grouse

________ Sooty Grouse

________ Sharp-tailed Grouse

________ Greater Prairie-Chicken

________ Lesser Prairie-Chicken

________ Wild Turkey

FLAMINGOS (PHOENICOPTERIDAE)

________ American Flamingo

GREBES (PODICIPEDIDAE)

________ Least Grebe
________ Pied-billed Grebe
________ Horned Grebe
________ Red-necked Grebe
________ Eared Grebe
________ Western Grebe
________ Clark's Grebe

SANDGROUSES (PTEROCLIDAE)

________ Chestnut-bellied Sandgrouse

PIGEONS AND DOVES (COLUMBIDAE)

________ Rock Pigeon
________ Scaly-naped Pigeon
________ White-crowned Pigeon
________ Red-billed Pigeon
________ Band-tailed Pigeon
________ Oriental Turtle-Dove
________ European Turtle-Dove
________ Eurasian Collared-Dove
________ Spotted Dove
________ Zebra Dove
________ Passenger Pigeon
________ Inca Dove
________ Common Ground-Dove
________ Ruddy Ground-Dove
________ Ruddy Quail-Dove
________ Key West Quail-Dove
________ White-tipped Dove
________ White-winged Dove
________ Zenaida Dove
________ Mourning Dove

CUCKOOS, ROADRUNNERS, AND ANIS (CUCULIDAE)

________ Common Cuckoo
________ Oriental Cuckoo
________ Yellow-billed Cuckoo
________ Mangrove Cuckoo
________ Black-billed Cuckoo
________ Greater Roadrunner
________ Smooth-billed Ani
________ Groove-billed Ani

GOATSUCKERS (CAPRIMULGIDAE)

________ Lesser Nighthawk
________ Common Nighthawk
________ Antillean Nighthawk
________ Common Pauraque
________ Common Poorwill
________ Chuck-will's-widow
________ Buff-collared Nightjar
________ Eastern Whip-poor-will
________ Mexican Whip-poor-will
________ Gray Nightjar

SWIFTS (APODIDAE)

________ Black Swift
________ White-collared Swift
________ Chimney Swift
________ Vaux's Swift
________ White-throated Needletail
________ Mariana Swiftlet
________ Common Swift
________ Fork-tailed Swift
________ White-throated Swift
________ Antillean Palm-Swift

HUMMINGBIRDS (TROCHILIDAE)

________ Mexican Violetear
________ Green-breasted Mango
________ Rivoli's Hummingbird

_________ Plain-capped Starthroat
_________ Amethyst-throated Hummingbird
_________ Blue-throated Hummingbird
_________ Bahama Woodstar
_________ Lucifer Hummingbird
_________ Ruby-throated Hummingbird
_________ Black-chinned Hummingbird
_________ Anna's Hummingbird
_________ Costa's Hummingbird
_________ Bumblebee Hummingbird
_________ Broad-tailed Hummingbird
_________ Rufous Hummingbird
_________ Allen's Hummingbird
_________ Calliope Hummingbird
_________ Broad-billed Hummingbird
_________ Berylline Hummingbird
_________ Buff-bellied Hummingbird
_________ Cinnamon Hummingbird
_________ Violet-crowned Hummingbird
_________ White-eared Hummingbird
_________ Xantus's Hummingbird

RAILS, GALLINULES, AND COOTS (RALLIDAE)

_________ Yellow Rail
_________ Black Rail
_________ Corn Crake
_________ Ridgway's Rail
_________ Clapper Rail
_________ King Rail
_________ Virginia Rail
_________ Rufous-necked Wood-Rail
_________ Sora
_________ Laysan Rail
_________ Hawaiian Rail
_________ Paint-billed Crake
_________ Spotted Rail
_________ Purple Gallinule
_________ Purple Swamphen
_________ Common Gallinule
_________ Common Moorhen
_________ Eurasian Coot
_________ Hawaiian Coot
_________ American Coot

SUNGREBES (HELIORNITHIDAE)

_________ Sungrebe

LIMPKINS (ARAMIDAE)

_________ Limpkin

CRANES (GRUIDAE)

_________ Sandhill Crane
_________ Common Crane
_________ Whooping Crane

THICK-KNEES (BURHINIDAE)

_________ Double-striped Thick-knee

STILTS AND AVOCETS (RECURVIROSTRIDAE)

_________ Black-winged Stilt
_________ Black-necked Stilt
_________ American Avocet

OYSTERCATCHERS (HAEMATOPODIDAE)

_________ Eurasian Oystercatcher
_________ American Oystercatcher
_________ Black Oystercatcher

LAPWINGS AND PLOVERS (CHARADRIIDAE)

_________ Northern Lapwing
_________ Black-bellied Plover
_________ European Golden-Plover
_________ American Golden-Plover
_________ Pacific Golden-Plover

_________ Lesser Sand-Plover
_________ Greater Sand-Plover
_________ Collared Plover
_________ Snowy Plover
_________ Wilson's Plover
_________ Common Ringed Plover
_________ Semipalmated Plover
_________ Piping Plover
_________ Little Ringed Plover
_________ Killdeer
_________ Mountain Plover
_________ Eurasian Dotterel

JACANAS (JACANIDAE)

_________ Northern Jacana

SANDPIPERS, PHALAROPES, AND ALLIES (SCOLOPACIDAE)

_________ Upland Sandpiper
_________ Bristle-thighed Curlew
_________ Whimbrel
_________ Little Curlew
_________ Eskimo Curlew
_________ Long-billed Curlew
_________ Far Eastern Curlew
_________ Slender-billed Curlew
_________ Eurasian Curlew
_________ Bar-tailed Godwit
_________ Black-tailed Godwit
_________ Hudsonian Godwit
_________ Marbled Godwit
_________ Ruddy Turnstone
_________ Black Turnstone
_________ Great Knot
_________ Red Knot
_________ Surfbird
_________ Ruff
_________ Broad-billed Sandpiper
_________ Sharp-tailed Sandpiper
_________ Stilt Sandpiper
_________ Curlew Sandpiper
_________ Temminck's Stint
_________ Long-toed Stint
_________ Spoon-billed Sandpiper
_________ Red-necked Stint
_________ Sanderling
_________ Dunlin
_________ Rock Sandpiper
_________ Purple Sandpiper
_________ Baird's Sandpiper
_________ Little Stint
_________ Least Sandpiper
_________ White-rumped Sandpiper
_________ Buff-breasted Sandpiper
_________ Pectoral Sandpiper
_________ Semipalmated Sandpiper
_________ Western Sandpiper
_________ Short-billed Dowitcher
_________ Long-billed Dowitcher
_________ Jack Snipe
_________ Eurasian Woodcock
_________ American Woodcock
_________ Solitary Snipe
_________ Pin-tailed Snipe
_________ Common Snipe
_________ Wilson's Snipe
_________ Terek Sandpiper
_________ Common Sandpiper
_________ Spotted Sandpiper
_________ Green Sandpiper
_________ Solitary Sandpiper
_________ Gray-tailed Tattler
_________ Wandering Tattler
_________ Lesser Yellowlegs
_________ Willet
_________ Spotted Redshank
_________ Common Greenshank

_________ Greater Yellowlegs
_________ Common Redshank
_________ Wood Sandpiper
_________ Marsh Sandpiper
_________ Wilson's Phalarope
_________ Red-necked Phalarope
_________ Red Phalarope

PRATINCOLES (GLAREOLIDAE)

_________ Oriental Pratincole

SKUAS AND JAEGERS (STERCORARIIDAE)

_________ Great Skua
_________ South Polar Skua
_________ Pomarine Jaeger
_________ Parasitic Jaeger
_________ Long-tailed Jaeger

AUKS, MURRES, AND PUFFINS (ALCIDAE)

_________ Dovekie
_________ Common Murre
_________ Thick-billed Murre
_________ Razorbill
_________ Great Auk
_________ Black Guillemot
_________ Pigeon Guillemot
_________ Long-billed Murrelet
_________ Marbled Murrelet
_________ Kittlitz's Murrelet
_________ Scripps's Murrelet
_________ Guadalupe Murrelet
_________ Craveri's Murrelet
_________ Ancient Murrelet
_________ Cassin's Auklet
_________ Parakeet Auklet
_________ Least Auklet
_________ Whiskered Auklet
_________ Crested Auklet
_________ Rhinoceros Auklet
_________ Atlantic Puffin
_________ Horned Puffin
_________ Tufted Puffin

GULLS, TERNS, AND SKIMMERS (LARIDAE)

_________ Swallow-tailed Gull
_________ Black-legged Kittiwake
_________ Red-legged Kittiwake
_________ Ivory Gull
_________ Sabine's Gull
_________ Bonaparte's Gull
_________ Gray-hooded Gull
_________ Black-headed Gull
_________ Little Gull
_________ Ross's Gull
_________ Laughing Gull
_________ Franklin's Gull
_________ Belcher's Gull
_________ Black-tailed Gull
_________ Heermann's Gull
_________ Mew Gull
_________ Ring-billed Gull
_________ Western Gull
_________ Yellow-footed Gull
_________ California Gull
_________ Herring Gull
_________ Yellow-legged Gull
_________ Iceland Gull
_________ Lesser Black-backed Gull
_________ Slaty-backed Gull
_________ Glaucous-winged Gull
_________ Glaucous Gull
_________ Great Black-backed Gull
_________ Kelp Gull
_________ Brown Noddy
_________ Black Noddy

_________ Blue-gray Noddy
_________ White Tern
_________ Sooty Tern
_________ Gray-backed Tern
_________ Bridled Tern
_________ Aleutian Tern
_________ Little Tern
_________ Least Tern
_________ Large-billed Tern
_________ Gull-billed Tern
_________ Caspian Tern
_________ Black Tern
_________ White-winged Tern
_________ Whiskered Tern
_________ Roseate Tern
_________ Common Tern
_________ Arctic Tern
_________ Forster's Tern
_________ Royal Tern
_________ Great Crested Tern
_________ Sandwich Tern
_________ Elegant Tern
_________ Black Skimmer

TROPICBIRDS (PHAETHONTIDAE)

_________ White-tailed Tropicbird
_________ Red-billed Tropicbird
_________ Red-tailed Tropicbird

LOONS (GAVIIDAE)

_________ Red-throated Loon
_________ Arctic Loon
_________ Pacific Loon
_________ Common Loon
_________ Yellow-billed Loon

ALBATROSSES (DIOMEDEIDAE)

_________ Yellow-nosed Albatross
_________ White-capped Albatross
_________ Chatham Albatross
_________ Salvin's Albatross
_________ Black-browed Albatross
_________ Light-mantled Albatross
_________ Wandering Albatross
_________ Laysan Albatross
_________ Black-footed Albatross
_________ Short-tailed Albatross

SHEARWATERS AND PETRELS (PROCELLARIIDAE)

_________ Northern Fulmar
_________ Great-winged Petrel
_________ Providence Petrel
_________ Kermadec Petrel
_________ Trindade Petrel
_________ Herald Petrel
_________ Murphy's Petrel
_________ Mottled Petrel
_________ Bermuda Petrel
_________ Black-capped Petrel
_________ Juan Fernandez Petrel
_________ Hawaiian Petrel
_________ White-necked Petrel
_________ Bonin Petrel
_________ Black-winged Petrel
_________ Fea's Petrel
_________ Zino's Petrel
_________ Cook's Petrel
_________ Stejneger's Petrel
_________ Tahiti Petrel
_________ Bulwer's Petrel
_________ Jouanin's Petrel
_________ White-chinned Petrel
_________ Parkinson's Petrel
_________ Streaked Shearwater
_________ Cory's Shearwater

_________ Cape Verde Shearwater

_________ Wedge-tailed Shearwater

_________ Buller's Shearwater

_________ Short-tailed Shearwater

_________ Sooty Shearwater

_________ Great Shearwater

_________ Pink-footed Shearwater

_________ Flesh-footed Shearwater

_________ Christmas Shearwater

_________ Manx Shearwater

_________ Newell's Shearwater

_________ Bryan's Shearwater

_________ Black-vented Shearwater

_________ Audubon's Shearwater

_________ Barolo Shearwater

SOUTHERN STORM-PETRELS (OCEANITIDAE)

_________ Wilson's Storm-Petrel

_________ White-faced Storm-Petrel

_________ Black-bellied Storm-Petrel

NORTHERN STORM-PETRELS (HYDROBATIDAE)

_________ European Storm-Petrel

_________ Fork-tailed Storm-Petrel

_________ Ringed Storm-Petrel

_________ Swinhoe's Storm-Petrel

_________ Leach's Storm-Petrel

_________ Townsend's Storm-Petrel

_________ Ashy Storm-Petrel

_________ Band-rumped Storm-Petrel

_________ Wedge-rumped Storm-Petrel

_________ Black Storm-Petrel

_________ Tristram's Storm-Petrel

_________ Least Storm-Petrel

STORKS (CICONIIDAE)

_________ Jabiru

_________ Wood Stork

FRIGATEBIRDS (FREGATIDAE)

_________ Magnificent Frigatebird

_________ Great Frigatebird

_________ Lesser Frigatebird

BOOBIES AND GANNETS (SULIDAE)

_________ Masked Booby

_________ Nazca Booby

_________ Blue-footed Booby

_________ Brown Booby

_________ Red-footed Booby

_________ Northern Gannet

CORMORANTS (PHALACROCORACIDAE)

_________ Brandt's Cormorant

_________ Neotropic Cormorant

_________ Double-crested Cormorant

_________ Great Cormorant

_________ Red-faced Cormorant

_________ Pelagic Cormorant

DARTERS (ANHINGIDAE)

_________ Anhinga

PELICANS (PELECANIDAE)

_________ American White Pelican

_________ Brown Pelican

BITTERNS, HERONS, AND ALLIES (ARDEIDAE)

_________ American Bittern

_________ Yellow Bittern

_________ Least Bittern

_________ Bare-throated Tiger-Heron

_________ Great Blue Heron

_________ Gray Heron

_________ Great Egret

_________ Intermediate Egret

_________ Chinese Egret
_________ Little Egret
_________ Western Reef-Heron
_________ Snowy Egret
_________ Little Blue Heron
_________ Tricolored Heron
_________ Reddish Egret
_________ Cattle Egret
_________ Chinese Pond-Heron
_________ Green Heron
_________ Black-crowned Night-Heron
_________ Yellow-crowned Night-Heron

IBISES AND SPOONBILLS (THRESKIORNITHIDAE)

_________ White Ibis
_________ Scarlet Ibis
_________ Glossy Ibis
_________ White-faced Ibis
_________ Roseate Spoonbill

NEW WORLD VULTURES (CATHARTIDAE)

_________ Black Vulture
_________ Turkey Vulture
_________ California Condor

OSPREYS (PANDIONIDAE)

_________ Osprey

HAWKS, KITES, EAGLES, AND ALLIES (ACCIPITRIDAE)

_________ White-tailed Kite
_________ Hook-billed Kite
_________ Swallow-tailed Kite
_________ Golden Eagle
_________ Double-toothed Kite
_________ Northern Harrier
_________ Chinese Sparrowhawk
_________ Sharp-shinned Hawk
_________ Cooper's Hawk
_________ Northern Goshawk
_________ Black Kite
_________ Bald Eagle
_________ White-tailed Eagle
_________ Steller's Sea-Eagle
_________ Mississippi Kite
_________ Crane Hawk
_________ Snail Kite
_________ Common Black Hawk
_________ Great Black Hawk
_________ Roadside Hawk
_________ Harris's Hawk
_________ White-tailed Hawk
_________ Gray Hawk
_________ Red-shouldered Hawk
_________ Broad-winged Hawk
_________ Hawaiian Hawk
_________ Short-tailed Hawk
_________ Swainson's Hawk
_________ Zone-tailed Hawk
_________ Red-tailed Hawk
_________ Rough-legged Hawk
_________ Ferruginous Hawk

BARN OWLS (TYTONIDAE)

_________ Barn Owl

TYPICAL OWLS (STRIGIDAE)

_________ Oriental Scops-Owl
_________ Flammulated Owl
_________ Western Screech-Owl
_________ Eastern Screech-Owl
_________ Whiskered Screech-Owl
_________ Great Horned Owl
_________ Snowy Owl
_________ Northern Hawk Owl
_________ Northern Pygmy-Owl

_________ Ferruginous Pygmy-Owl
_________ Elf Owl
_________ Burrowing Owl
_________ Mottled Owl
_________ Spotted Owl
_________ Barred Owl
_________ Great Gray Owl
_________ Long-eared Owl
_________ Stygian Owl
_________ Short-eared Owl
_________ Boreal Owl
_________ Northern Saw-whet Owl
_________ Northern Boobook

TROGONS (TROGONIDAE)

_________ Elegant Trogon
_________ Eared Quetzal

HOOPOES (UPUPIDAE)

_________ Eurasian Hoopoe

KINGFISHERS (ALCEDINIDAE)

_________ Ringed Kingfisher
_________ Belted Kingfisher
_________ Amazon Kingfisher
_________ Green Kingfisher

WOODPECKERS AND ALLIES (PICIDAE)

_________ Eurasian Wryneck
_________ Lewis's Woodpecker
_________ Red-headed Woodpecker
_________ Acorn Woodpecker
_________ Gila Woodpecker
_________ Golden-fronted Woodpecker
_________ Red-bellied Woodpecker
_________ Williamson's Sapsucker
_________ Yellow-bellied Sapsucker
_________ Red-naped Sapsucker
_________ Red-breasted Sapsucker
_________ American Three-toed Woodpecker
_________ Black-backed Woodpecker
_________ Great Spotted Woodpecker
_________ Downy Woodpecker
_________ Nuttall's Woodpecker
_________ Ladder-backed Woodpecker
_________ Red-cockaded Woodpecker
_________ Hairy Woodpecker
_________ White-headed Woodpecker
_________ Arizona Woodpecker
_________ Northern Flicker
_________ Gilded Flicker
_________ Pileated Woodpecker
_________ Ivory-billed Woodpecker

CARACARAS AND FALCONS (FALCONIDAE)

_________ Collared Forest-Falcon
_________ Crested Caracara
_________ Eurasian Kestrel
_________ American Kestrel
_________ Red-footed Falcon
_________ Merlin
_________ Eurasian Hobby
_________ Aplomado Falcon
_________ Gyrfalcon
_________ Peregrine Falcon
_________ Prairie Falcon

PARAKEETS, MACAWS, AND PARROTS (PSITTACIDAE)

_________ Monk Parakeet
_________ Carolina Parakeet
_________ Nanday Parakeet
_________ Green Parakeet
_________ Thick-billed Parrot
_________ White-winged Parakeet
_________ Red-crowned Parrot

LORIES, LOVEBIRDS, AND AUSTRALASIAN PARROTS (PSITTACULIDAE)

_________ Rose-ringed Parakeet

_________ Rosy-faced Lovebird

BECARDS, TITYRAS, AND ALLIES (TITYRIDAE)

_________ Masked Tityra

_________ Gray-collared Becard

_________ Rose-throated Becard

TYRANT FLYCATCHERS (TYRANNIDAE)

_________ Northern Beardless-Tyrannulet

_________ Greenish Elaenia

_________ White-crested Elaenia

_________ Dusky-capped Flycatcher

_________ Ash-throated Flycatcher

_________ Nutting's Flycatcher

_________ Great Crested Flycatcher

_________ Brown-crested Flycatcher

_________ La Sagra's Flycatcher

_________ Great Kiskadee

_________ Social Flycatcher

_________ Sulphur-bellied Flycatcher

_________ Piratic Flycatcher

_________ Variegated Flycatcher

_________ Crowned Slaty Flycatcher

_________ Tropical Kingbird

_________ Couch's Kingbird

_________ Cassin's Kingbird

_________ Thick-billed Kingbird

_________ Western Kingbird

_________ Eastern Kingbird

_________ Gray Kingbird

_________ Loggerhead Kingbird

_________ Scissor-tailed Flycatcher

_________ Fork-tailed Flycatcher

_________ Tufted Flycatcher

_________ Olive-sided Flycatcher

_________ Greater Pewee

_________ Western Wood-Pewee

_________ Eastern Wood-Pewee

_________ Cuban Pewee

_________ Yellow-bellied Flycatcher

_________ Acadian Flycatcher

_________ Alder Flycatcher

_________ Willow Flycatcher

_________ Least Flycatcher

_________ Hammond's Flycatcher

_________ Gray Flycatcher

_________ Dusky Flycatcher

_________ Pine Flycatcher

_________ Pacific-slope Flycatcher

_________ Cordilleran Flycatcher

_________ Buff-breasted Flycatcher

_________ Black Phoebe

_________ Eastern Phoebe

_________ Say's Phoebe

_________ Vermilion Flycatcher

SHRIKES (LANIIDAE)

_________ Red-backed Shrike

_________ Brown Shrike

_________ Loggerhead Shrike

_________ Northern Shrike

VIREOS (VIREONIDAE)

_________ Black-capped Vireo

_________ White-eyed Vireo

_________ Thick-billed Vireo

_________ Cuban Vireo

_________ Bell's Vireo

_________ Gray Vireo

_________ Hutton's Vireo

_________ Yellow-throated Vireo

_________ Cassin's Vireo

________ Blue-headed Vireo
________ Plumbeous Vireo
________ Philadelphia Vireo
________ Warbling Vireo
________ Red-eyed Vireo
________ Yellow-green Vireo
________ Black-whiskered Vireo
________ Yucatan Vireo

JAYS AND CROWS (CORVIDAE)

________ Canada Jay
________ Brown Jay
________ Green Jay
________ Pinyon Jay
________ Steller's Jay
________ Blue Jay
________ Florida Scrub-Jay
________ Island Scrub-Jay
________ California Scrub-Jay
________ Woodhouse's Scrub-Jay
________ Mexican Jay
________ Clark's Nutcracker
________ Black-billed Magpie
________ Yellow-billed Magpie
________ Eurasian Jackdaw
________ American Crow
________ Northwestern Crow
________ Tamaulipas Crow
________ Fish Crow
________ Hawaiian Crow
________ Chihuahuan Raven
________ Common Raven

MONARCH FLYCATCHERS (MONARCHIDAE)

________ Kauai Elepaio
________ Oahu Elepaio
________ Hawaii Elepaio

LARKS (ALAUDIDAE)

________ Eurasian Skylark
________ Horned Lark

SWALLOWS (HIRUNDINIDAE)

________ Purple Martin
________ Cuban Martin
________ Gray-breasted Martin
________ Southern Martin
________ Brown-chested Martin
________ Tree Swallow
________ Mangrove Swallow
________ Violet-green Swallow
________ Bahama Swallow
________ Northern Rough-winged Swallow
________ Bank Swallow
________ Cliff Swallow
________ Cave Swallow
________ Barn Swallow
________ Common House-Martin

CHICKADEES AND TITMICE (PARIDAE)

________ Carolina Chickadee
________ Black-capped Chickadee
________ Mountain Chickadee
________ Mexican Chickadee
________ Chestnut-backed Chickadee
________ Boreal Chickadee
________ Gray-headed Chickadee
________ Bridled Titmouse
________ Oak Titmouse
________ Juniper Titmouse
________ Tufted Titmouse
________ Black-crested Titmouse

VERDIN (REMIZIDAE)

________ Verdin

BUSHTITS (AEGITHALIDAE)

________ Bushtit

NUTHATCHES (SITTIDAE)

________ Red-breasted Nuthatch
________ White-breasted Nuthatch
________ Pygmy Nuthatch
________ Brown-headed Nuthatch

CREEPERS (CERTHIIDAE)

________ Brown Creeper

WRENS (TROGLODYTIDAE)

________ Rock Wren
________ Canyon Wren
________ House Wren
________ Pacific Wren
________ Winter Wren
________ Sedge Wren
________ Marsh Wren
________ Carolina Wren
________ Bewick's Wren
________ Cactus Wren
________ Sinaloa Wren

GNATCATCHERS AND GNATWRENS (POLIOPTILIDAE)

________ Blue-gray Gnatcatcher
________ California Gnatcatcher
________ Black-tailed Gnatcatcher
________ Black-capped Gnatcatcher

DIPPERS (CINCLIDAE)

________ American Dipper

BULBULS (PYCNONOTIDAE)

________ Red-vented Bulbul
________ Red-whiskered Bulbul

KINGLETS (REGULIDAE)

________ Golden-crowned Kinglet
________ Ruby-crowned Kinglet

BUSH-WARBLERS (CETTIIDAE)

________ Japanese Bush-Warbler

LEAF WARBLERS (PHYLLOSCOPIDAE)

________ Willow Warbler
________ Common Chiffchaff
________ Wood Warbler
________ Dusky Warbler
________ Pallas's Leaf Warbler
________ Yellow-browed Warbler
________ Arctic Warbler
________ Kamchatka Leaf Warbler

SYLVIID WARBLERS (SYLVIIDAE)

________ Lesser Whitethroat
________ Wrentit

WHITE-EYES (ZOSTEROPIDAE)

________ Japanese White-eye

LAUGHINGTHRUSHES (TIMALIIDAE)

________ Greater Necklaced Laughingthrush
________ Hwamei
________ Red-billed Leiothrix

REED WARBLERS (ACROCEPHALIDAE)

________ Thick-billed Warbler
________ Millerbird
________ Sedge Warbler
________ Blyth's Reed Warbler

GRASSBIRDS (LOCUSTELLIDAE)

________ Middendorff's Grasshopper-Warbler
________ River Warbler

_____ Lanceolated Warbler

OLD WORLD FLYCATCHERS (MUSCICAPIDAE)

_____ Gray-streaked Flycatcher

_____ Asian Brown Flycatcher

_____ Spotted Flycatcher

_____ Dark-sided Flycatcher

_____ White-rumped Shama

_____ European Robin

_____ Siberian Rubythroat

_____ Bluethroat

_____ Siberian Blue Robin

_____ Rufous-tailed Robin

_____ Red-flanked Bluetail

_____ Narcissus Flycatcher

_____ Mugimaki Flycatcher

_____ Taiga Flycatcher

_____ Common Redstart

_____ Stonechat

_____ Northern Wheatear

_____ Pied Wheatear

THRUSHES (TURDIDAE)

_____ Eastern Bluebird

_____ Western Bluebird

_____ Mountain Bluebird

_____ Townsend's Solitaire

_____ Brown-backed Solitaire

_____ Kamao

_____ Amaui

_____ Olomao

_____ Omao

_____ Puaiohi

_____ Orange-billed Nightingale-Thrush

_____ Black-headed Nightingale-Thrush

_____ Veery

_____ Gray-cheeked Thrush

_____ Bicknell's Thrush

_____ Swainson's Thrush

_____ Hermit Thrush

_____ Wood Thrush

_____ Eurasian Blackbird

_____ Eyebrowed Thrush

_____ Dusky Thrush

_____ Fieldfare

_____ Redwing

_____ Mistle Thrush

_____ Song Thrush

_____ Clay-colored Thrush

_____ White-throated Thrush

_____ Rufous-backed Robin

_____ American Robin

_____ Red-legged Thrush

_____ Varied Thrush

_____ Aztec Thrush

MOCKINGBIRDS AND THRASHERS (MIMIDAE)

_____ Blue Mockingbird

_____ Gray Catbird

_____ Curve-billed Thrasher

_____ Brown Thrasher

_____ Long-billed Thrasher

_____ Bendire's Thrasher

_____ California Thrasher

_____ LeConte's Thrasher

_____ Crissal Thrasher

_____ Sage Thrasher

_____ Bahama Mockingbird

_____ Northern Mockingbird

STARLINGS (STURNIDAE)

_____ European Starling

_____ Common Myna

WAXWINGS (BOMBYCILLIDAE)

________ Bohemian Waxwing
________ Cedar Waxwing

HAWAIIAN MOHOS (MOHOIDAE)

________ Kauai Oo
________ Oahu Oo
________ Bishop's Oo
________ Hawaii Oo
________ Kioea

SILKY-FLYCATCHERS (PTILIOGONATIDAE)

________ Gray Silky-flycatcher
________ Phainopepla

OLIVE WARBLERS (PEUCEDRAMIDAE)

________ Olive Warbler

ACCENTORS (PRUNELLIDAE)

________ Siberian Accentor

WAXBILLS (ESTRILDIDAE)

________ Common Waxbill
________ Red Avadavat
________ African Silverbill
________ Java Sparrow
________ Scaly-breasted Munia
________ Chestnut Munia

OLD WORLD SPARROWS (PASSERIDAE)

________ House Sparrow
________ Eurasian Tree Sparrow

WAGTAILS AND PIPITS (MOTACILLIDAE)

________ Eastern Yellow Wagtail
________ Citrine Wagtail
________ Gray Wagtail
________ White Wagtail
________ Tree Pipit
________ Olive-backed Pipit
________ Pechora Pipit
________ Red-throated Pipit
________ American Pipit
________ Sprague's Pipit

FRINGILLINE AND CARDUELINE FINCHES AND ALLIES (FRINGILLIDAE)

________ Common Chaffinch
________ Brambling
________ Evening Grosbeak
________ Hawfinch
________ Common Rosefinch
________ Pallas's Rosefinch
________ Poo-uli
________ Akikiki
________ Oahu Alauahio
________ Kakawahie
________ Maui Alauahio
________ Palila
________ Laysan Finch
________ Nihoa Finch
________ Kona Grosbeak
________ Lesser Koa-Finch
________ Greater Koa-Finch
________ Ula-ai-hawane
________ Akohekohe
________ Laysan Honeycreeper
________ Apapane
________ Iiwi
________ Hawaii Mamo
________ Black Mamo
________ Ou
________ Lanai Hookbill
________ Maui Parrotbill
________ Kauai Nukupuu
________ Oahu Nukupuu
________ Maui Nukupuu
________ Akiapolaau

_________ Lesser Akialoa
_________ Kauai Akialoa
_________ Oahu Akialoa
_________ Maui-nui Akialoa
_________ Anianiau
_________ Hawaii Amakihi
_________ Oahu Amakihi
_________ Kauai Amakihi
_________ Greater Amakihi
_________ Hawaii Creeper
_________ Akekee
_________ Oahu Akepa
_________ Maui Akepa
_________ Hawaii Akepa
_________ Pine Grosbeak
_________ Eurasian Bullfinch
_________ Asian Rosy-Finch
_________ Gray-crowned Rosy-Finch
_________ Black Rosy-Finch
_________ Brown-capped Rosy-Finch
_________ House Finch
_________ Purple Finch
_________ Cassin's Finch
_________ Oriental Greenfinch
_________ Yellow-fronted Canary
_________ Common Redpoll
_________ Hoary Redpoll
_________ Red Crossbill
_________ Cassia Crossbill
_________ White-winged Crossbill
_________ Eurasian Siskin
_________ Pine Siskin
_________ Lesser Goldfinch
_________ Lawrence's Goldfinch
_________ American Goldfinch
_________ Island Canary

LONGSPURS AND SNOW BUNTINGS (CALCARIIDAE)

_________ Lapland Longspur
_________ Chestnut-collared Longspur
_________ Smith's Longspur
_________ McCown's Longspur
_________ Snow Bunting
_________ McKay's Bunting

EMBERIZIDS (EMBERIZIDAE)

_________ Pine Bunting
_________ Yellow-browed Bunting
_________ Little Bunting
_________ Rustic Bunting
_________ Yellow-throated Bunting
_________ Yellow-breasted Bunting
_________ Gray Bunting
_________ Pallas's Bunting
_________ Reed Bunting

TOWHEES AND SPARROWS (PASSERELLIDAE)

_________ Olive Sparrow
_________ Green-tailed Towhee
_________ Spotted Towhee
_________ Eastern Towhee
_________ Rufous-crowned Sparrow
_________ Canyon Towhee
_________ California Towhee
_________ Abert's Towhee
_________ Rufous-winged Sparrow
_________ Botteri's Sparrow
_________ Cassin's Sparrow
_________ Bachman's Sparrow
_________ American Tree Sparrow
_________ Chipping Sparrow
_________ Clay-colored Sparrow
_________ Brewer's Sparrow

_________ Field Sparrow
_________ Worthen's Sparrow
_________ Black-chinned Sparrow
_________ Vesper Sparrow
_________ Lark Sparrow
_________ Five-striped Sparrow
_________ Black-throated Sparrow
_________ Sagebrush Sparrow
_________ Bell's Sparrow
_________ Lark Bunting
_________ Savannah Sparrow
_________ Grasshopper Sparrow
_________ Baird's Sparrow
_________ Henslow's Sparrow
_________ LeConte's Sparrow
_________ Seaside Sparrow
_________ Nelson's Sparrow
_________ Saltmarsh Sparrow
_________ Fox Sparrow
_________ Song Sparrow
_________ Lincoln's Sparrow
_________ Swamp Sparrow
_________ White-throated Sparrow
_________ Harris's Sparrow
_________ White-crowned Sparrow
_________ Golden-crowned Sparrow
_________ Dark-eyed Junco
_________ Yellow-eyed Junco

SPINDALISES (SPINDALIDAE)

_________ Western Spindalis

YELLOW-BREASTED CHATS (ICTERIIDAE)

_________ Yellow-breasted Chat

BLACKBIRDS (ICTERIDAE)

_________ Yellow-headed Blackbird
_________ Bobolink
_________ Eastern Meadowlark
_________ Western Meadowlark
_________ Black-vented Oriole
_________ Orchard Oriole
_________ Hooded Oriole
_________ Streak-backed Oriole
_________ Bullock's Oriole
_________ Spot-breasted Oriole
_________ Altamira Oriole
_________ Audubon's Oriole
_________ Baltimore Oriole
_________ Black-backed Oriole
_________ Scott's Oriole
_________ Red-winged Blackbird
_________ Tricolored Blackbird
_________ Tawny-shouldered Blackbird
_________ Shiny Cowbird
_________ Bronzed Cowbird
_________ Brown-headed Cowbird
_________ Rusty Blackbird
_________ Brewer's Blackbird
_________ Common Grackle
_________ Boat-tailed Grackle
_________ Great-tailed Grackle

WOOD-WARBLERS (PARULIDAE)

_________ Ovenbird
_________ Worm-eating Warbler
_________ Louisiana Waterthrush
_________ Northern Waterthrush
_________ Bachman's Warbler
_________ Golden-winged Warbler
_________ Blue-winged Warbler
_________ Black-and-white Warbler
_________ Prothonotary Warbler
_________ Swainson's Warbler
_________ Crescent-chested Warbler
_________ Tennessee Warbler

_________ Orange-crowned Warbler
_________ Colima Warbler
_________ Lucy's Warbler
_________ Nashville Warbler
_________ Virginia's Warbler
_________ Connecticut Warbler
_________ Gray-crowned Yellowthroat
_________ MacGillivray's Warbler
_________ Mourning Warbler
_________ Kentucky Warbler
_________ Common Yellowthroat
_________ Hooded Warbler
_________ American Redstart
_________ Kirtland's Warbler
_________ Cape May Warbler
_________ Cerulean Warbler
_________ Northern Parula
_________ Tropical Parula
_________ Magnolia Warbler
_________ Bay-breasted Warbler
_________ Blackburnian Warbler
_________ Yellow Warbler
_________ Chestnut-sided Warbler
_________ Blackpoll Warbler
_________ Black-throated Blue Warbler
_________ Palm Warbler
_________ Pine Warbler
_________ Yellow-rumped Warbler
_________ Yellow-throated Warbler
_________ Prairie Warbler
_________ Grace's Warbler
_________ Black-throated Gray Warbler
_________ Townsend's Warbler
_________ Hermit Warbler
_________ Golden-cheeked Warbler
_________ Black-throated Green Warbler
_________ Fan-tailed Warbler
_________ Rufous-capped Warbler
_________ Golden-crowned Warbler
_________ Canada Warbler
_________ Wilson's Warbler
_________ Red-faced Warbler
_________ Painted Redstart
_________ Slate-throated Redstart

CARDINALS, PIRANGA TANAGERS AND ALLIES (CARDINALIDAE)

_________ Hepatic Tanager
_________ Summer Tanager
_________ Scarlet Tanager
_________ Western Tanager
_________ Flame-colored Tanager
_________ Crimson-collared Grosbeak
_________ Northern Cardinal
_________ Pyrrhuloxia
_________ Yellow Grosbeak
_________ Rose-breasted Grosbeak
_________ Black-headed Grosbeak
_________ Blue Bunting
_________ Blue Grosbeak
_________ Lazuli Bunting
_________ Indigo Bunting
_________ Varied Bunting
_________ Painted Bunting
_________ Dickcissel

TANAGERS AND ALLIES (THRAUPIDAE)

_________ Red-crested Cardinal
_________ Yellow-billed Cardinal
_________ Saffron Finch
_________ Red-legged Honeycreeper
_________ Bananaquit
_________ Yellow-faced Grassquit
_________ Black-faced Grassquit
_________ Morelet's Seedeater

INDEX

1
2
3
4
5
6
7
8
9
10
11
11
12
13
14
15

SHORE SILHOUETTES
1 Forster's Tern
2 Black Tern
3 Herring Gull
4 Cormorant
5 Loon
6 Great Blue Heron
7 Mallard
8 Pied-billed Grebe
9 Marbled Godwit
10 Greater Yellowlegs
11 Dowitcher
12 Clapper Rail
13 Whimbrel
14 Black-bellied Plover
15 Turnstone
16 Night-Heron
17 Phalarope
18 Least Sandpiper
19 Semipalmated Plover
20 Sanderling
21 Spotted Sandpiper
22 Killdeer
23 Coot
24 Green-backed Heron
16
3
4
1
23
16
17
24
19
21
22
18
20

1
2
3
4
5
6
7
8
9
10
11
12
13
14
15
16
17
18
19

20
21
22
23
24
25
26
FLIGHT SILHOUETTES
1 Barn Swallow
2 Cliff Swallow
3 Purple Martin
4 Chimney Swift
5 Starling
6 Common Grackle
7 Blackbird
8 Bluebird
9 Robin
10 Goldfinch
11 House Sparrow
12 Belted Kingfisher
13 Blue Jay
14 Flicker
15 Mourning Dove
16 Meadowlark
17 Bobwhite
18 Ruffed Grouse
19 Pheasant
20 Nighthawk
21 Crow
22 Sharp-shinned Hawk
23 Kestrel
24 Killdeer
25 Wilson's Snipe
26 Woodcock

1
2
3
4
5
6
7
12
13
14
15
16
19
20
21
22
27
28
29

ROADSIDE SILHOUETTES
1 Mourning Dove
2 House Sparrow
3 Common Grackle
4 Starling
5 Cowbird
6 Blackbird
7 Belted Kingfisher
8 Scissor-tailed Flycatcher
9 Purple Martin
10 Barn Swallow
11 Cliff Swallow
12 Blue Jay
13 Mockingbird
14 Song Sparrow
15 Shrike
16 Flicker
17 Kestrel
18 Cardinal
19 Bluebird
20 Magpie
21 Nighthawk
22 Robin
23 Meadowlark
24 Kingbird
25 Horned Lark
26 Phoebe
27 Killdeer
28 Pheasant
29 Burrowing Owl
30 Bobwhite
31 California Quail
32 Crow
8
9
10
11
17
18
23
24
25
26
30
31
32

ONE-PAGE INDEX

Pages listed are first occurrences.